Methods in Molecular Biology

Series Editor
John M. Walker
School of Life and Medical Sciences
University of Hertfordshire
Hatfield, Hertfordshire AL10 9AB, UK

For further volumes:
http://www.springer.com/series/7651

Microbial Metabolomics

Methods and Protocols

Edited by

Edward E.K. Baidoo

Joint BioEnergy Institute, Lawrence Berkeley National Laboratory, Emeryville, CA, USA

Humana Press

Editor
Edward E.K. Baidoo
Joint BioEnergy Institute
Lawrence Berkeley National Laboratory
Emeryville, CA, USA

ISSN 1064-3745 ISSN 1940-6029 (electronic)
Methods in Molecular Biology
ISBN 978-1-4939-8756-6 ISBN 978-1-4939-8757-3 (eBook)
https://doi.org/10.1007/978-1-4939-8757-3

Library of Congress Control Number: 2018957074

This Humana Press imprint is published by the registered company Springer Science+Business Media, LLC, part of Springer Nature.
The registered company address is: 233 Spring Street, New York, NY 10013, U.S.A.

Preface

Since the term metabolomics was first coined, it was thought that comprehensive metabolomic data sets would be the shape of things to come. However, the structural diversity that exists within the metabolome would make it extremely difficult to achieve this with current analytical technologies. This left scientists to ponder the questions, how much of the metabolome was necessary to characterize metabolism and what aspects of metabolomics data were truly meaningful to a research study? This book explores current approaches to answering these two very fundamental questions in microbiological research.

The protocols in this volume represent the breadth of microbial metabolomics approaches to both large-scale and small-scale experiments. The intention of this volume is to cover protocols that can be used for applications ranging from environmental microbiology to human disease. All protocols described in this volume utilize mass spectrometry as their primary measurement tool, which is reflective of the popularity of this technique. This volume covers four major areas: (1) microbial metabolomics, metabolism, and microbial physiology, (2) metabolite sample preparation, (3) current analytical techniques used to profile primary and secondary metabolites and lipids, and (4) establishing data analysis workflows for targeted metabolomics, untargeted metabolomics, analysis of metabolic fluxes, and genome-scale models.

Part II of this volume provides background information on metabolomics in the context of microbial metabolism and physiology. The protocols described in this volume provide guidance for both novice and advanced users. Extra information on the steps of each protocol can be found in the notes section at the end of each chapter. The information covered in these protocols can serve as reference material and can be adapted to similar analytical platforms or customized to suit the needs of the researcher. I would like to thank all the authors for their outstanding efforts and for kindly agreeing to contribute to this issue on *Microbial Metabolomics*. I would also like to thank and acknowledge John M. Walker for his invitation to edit this volume of *Methods in Molecular Biology* and for his invaluable advice and encouragement during the preparation of the issue.

Emeryville, CA, USA **Edward E.K. Baidoo**

Contents

Contributors

PATRÍCIA VERARDI ABDELNUR • *Brazilian Agricultural Research Corporation, Embrapa Agroenergy, Brasília, DF, Brazil; Institute of Chemistry, Federal University of Goiás, Goiânia, GO, Brazil*

MARY ABERNATHY • *Department of Environmental, Energy and Chemical Engineering, Washington University in St. Louis, St. Louis, MO, USA*

JOÃO RICARDO MOREIRA DE ALMEIDA • *Brazilian Agricultural Research Corporation, Embrapa Agroenergy, Brasília, DF, Brazil; Graduate Program on Chemical and Biological Technologies, Institute of Chemistry, University of Brasília, Brasília, DF, Brazil*

DAVID ANDO • *Biological Systems and Engineering Division, Lawrence Berkeley National Laboratory, Berkeley, CA, USA; Joint BioEnergy Institute, Emeryville, CA, USA*

JOSÉ ANTÔNIO DE AQUINO RIBEIRO • *Brazilian Agricultural Research Corporation, Embrapa Agroenergy, Brasília, DF, Brazil*

EDWARD E.K. BAIDOO • *Biological Systems and Engineering Division, Lawrence Berkeley National Laboratory, Berkeley, CA, USA; Joint BioEnergy Institute, Emeryville, CA, USA*

CORAL BARBAS • *Centro de Metabolómica y Bioanállisis (CEMBIO), Facultad de Farmacia, Universidad CEU San Pablo, Campus Monteprincipe, Madrid, Spain*

DERYLIZE BEUKES • *Human Metabolomics, North-West University, Potchefstroom, South Africa*

BARBARA BOJKO • *Department of Chemistry, University of Waterloo, Waterloo, ON, Canada*

YUPING CAI • *Interdisciplinary Research Center on Biology and Chemistry, Shanghai Institute of Organic Chemistry, Chinese Academy of Sciences, Shanghai, P. R. China; University of Chinese Academy of Sciences, Shanghai, P. R. China*

CHRISTIANE GONÇALVES CAMPOS • *Brazilian Agricultural Research Corporation, Embrapa Agroenergy, Brasília, DF, Brazil; Institute of Chemistry, Federal University of Goiás, Goiânia, GO, Brazil*

SIMON A. COBBOLD • *Department of Biochemistry and Molecular Biology, Bio21 Institute of Molecular Science and Biotechnology, The University of Melbourne, Parkville, Victoria, Australia*

HÉCTOR GARCÍA MARTÍN • *Biological Systems and Engineering Division, Lawrence Berkeley National Laboratory, Berkeley, CA, USA; Joint BioEnergy Institute, Emeryville, CA, USA*

JOSHUA HEINEMANN • *Environmental Genomics and Systems Biology, Lawrence Berkeley National Laboratory, Berkeley, CA, USA; Joint BioEnergy Institute, Emeryville, CA, USA*

CHIJIOKE J. JOSHUA • *Biological Systems and Engineering Division, Lawrence Berkeley National Laboratory, Berkeley, CA, USA; Joint BioEnergy Institute, Emeryville, CA, USA*

JAY D. KEASLING • *Biological Systems and Engineering Division, Lawrence Berkeley National Laboratory, Berkeley, CA, USA; Joint BioEnergy Institute, Emeryville, CA, USA; Department of Chemical Engineering, University of California, Berkeley, CA, USA; Department of Bioengineering, University of California, Berkeley, CA, USA; Novo Nordisk Foundation Center for Biosustainability, Technical University of Denmark, Kgs Lyngby, Denmark; Center for Synthetic Biochemistry, Institute of Synthetic Biology Research, Shenzhen Institutes of Advanced Technologies, Shenzhen, Guangdong, China*

YUJING LI • *College of Life Sciences, Nankai University, Tianjin, China*

Zhucui Li • *iHuman Institute, ShanghaiTech University, Shanghai, China; Tianjin Institute of Industrial Biotechnology, Chinese Academy of Sciences, Tianjin, China; University of Chinese Academy of Sciences, Beijing, China*

Du Toit Loots • *Human Metabolomics, North-West University, Potchefstroom, South Africa*

Ángeles López-Gonzálvez • *Centro de Metabolómica y Bioanállisis (CEMBIO), Facultad de Farmacia, Universidad CEU San Pablo, Campus Montepríncipe, Madrid, Spain*

Malcolm J. McConville • *Department of Biochemistry and Molecular Biology, Bio21 Institute of Molecular Science and Biotechnology, The University of Melbourne, Parkville, Victoria, Australia*

Fatemeh Mousavi • *Donnelly Centre for Cellular and Biomolecular Research, University of Toronto, Toronto, ON, Canada; Department of Chemistry, University of Waterloo, Waterloo, ON, Canada*

Trent R. Northen • *DOE Joint Genome Institute, Walnut Creek, CA, USA; Environmental Genomics and Systems Biology Division, Lawrence Berkeley National Laboratory, Berkeley, CA, USA*

Janusz Pawliszyn • *Department of Chemistry, University of Waterloo, Waterloo, ON, Canada*

Ilse du Preez • *Human Metabolomics, North-West University, Potchefstroom, South Africa*

Betania Ferraz Quirino • *Brazilian Agricultural Research Corporation, Embrapa Agroenergy, Brasília, DF, Brazil; Graduate Program on Chemical and Biological Technologies, Institute of Chemistry, University of Brasília, Brasília, DF, Brazil; Genomic Sciences and Biotechnology Program, Universidade Católica de Brasilia, Brasília, DF, Brazil*

David Rojo • *Centro de Metabolómica y Bioanállisis (CEMBIO), Facultad de Farmacia, Universidad CEU San Pablo, Campus Montepríncipe, Madrid, Spain*

Wenqing Shui • *iHuman Institute, ShanghaiTech University, Shanghai, China*

Tami L. Swenson • *Environmental Genomics and Systems Biology Division, Lawrence Berkeley National Laboratory, Berkeley, CA, USA*

Ralf Takors • *Institute of Biochemical Engineering, University of Stuttgart, Stuttgart, Germany*

Yinjie J. Tang • *Department of Environmental, Energy and Chemical Engineering, Washington University in St. Louis, St. Louis, MO, USA*

Veronica Teixeira Benites • *Biological Systems and Engineering Division, Lawrence Berkeley National Laboratory, Berkeley, CA, USA; Joint BioEnergy Institute, Emeryville, CA, USA*

Attila Teleki • *Institute of Biochemical Engineering, University of Stuttgart, Stuttgart, Germany*

Ni Wan • *Department of Environmental, Energy and Chemical Engineering, Washington University in St. Louis, St. Louis, MO, USA*

George Wang • *Biological Systems and Engineering Division, Lawrence Berkeley National Laboratory, Berkeley, CA, USA; Joint BioEnergy Institute, Emeryville, CA, USA*

Yuxin Yin • *Institute of Systems Biomedicine, School of Basic Medical Sciences, Peking University Health Science Center, Beijing, China*

Juntuo Zhou • *Institute of Systems Biomedicine, School of Basic Medical Sciences, Peking University Health Science Center, Beijing, China*

Zheng-Jiang Zhu • *Interdisciplinary Research Center on Biology and Chemistry, Shanghai Institute of Organic Chemistry, Chinese Academy of Sciences, Shanghai, P. R. China; University of Chinese Academy of Sciences, Shanghai, P. R. China*

Chapter 1

Microbial Metabolomics: A General Overview

Edward E.K. Baidoo

Abstract

In the biosciences, there has been growing interest in the elucidation of gene function. Consequently, metabolomics has garnered a lot of attention of late due to its provision of metabolic information pertaining to both function and phenotype. Furthermore, when metabolomics data is integrated with other "omics" data, precise characterization of metabolic activity can be achieved. This chapter briefly introduces a few important aspects of the metabolome, the challenges faced when acquiring metabolomic information and the steps that are necessary to overcoming them. This chapter also briefly covers current analytical technologies and some microbial metabolomic applications.

Key words Mass spectrometry, Metabolomics, Microbial, LC-MS, GC-MS, CE-MS, Metabolic quenching, Metabolite extraction, Data analysis, Microbial communities, Human disease

Progression in biology related scientific research has always depended on technological advances, which has enabled the study of living organisms and the components the that are necessary for life. Over past 70 years, concerted efforts have been made to understand biological systems through the elucidation of chemical structure, function, development, adaptation, and evolution. In the early years, biochemical information was thought to flow directionally from the transcription of genomic DNA to mRNA, the translation of the latter to proteins, which subsequently produce metabolites [1]. From the 1980s onward, molecular biology "omics" technologies explored the geneses, interactions, associations, and roles of various types of molecules that make up an organism, to provide a deeper understanding of the principles of life. This led to the discovery that the flow of biochemical information is not only unidirectional but rather a complex system of interactions between DNA, RNA, protein, and metabolite [1], with the metabolome (the complete set of metabolites formed by the cell in association with metabolism [2]) being able to influence the preceding "omes" by many forms of regulatory control

Edward E.K. Baidoo (ed.), *Microbial Metabolomics: Methods and Protocols*, Methods in Molecular Biology, vol. 1859,
https://doi.org/10.1007/978-1-4939-8757-3_1,

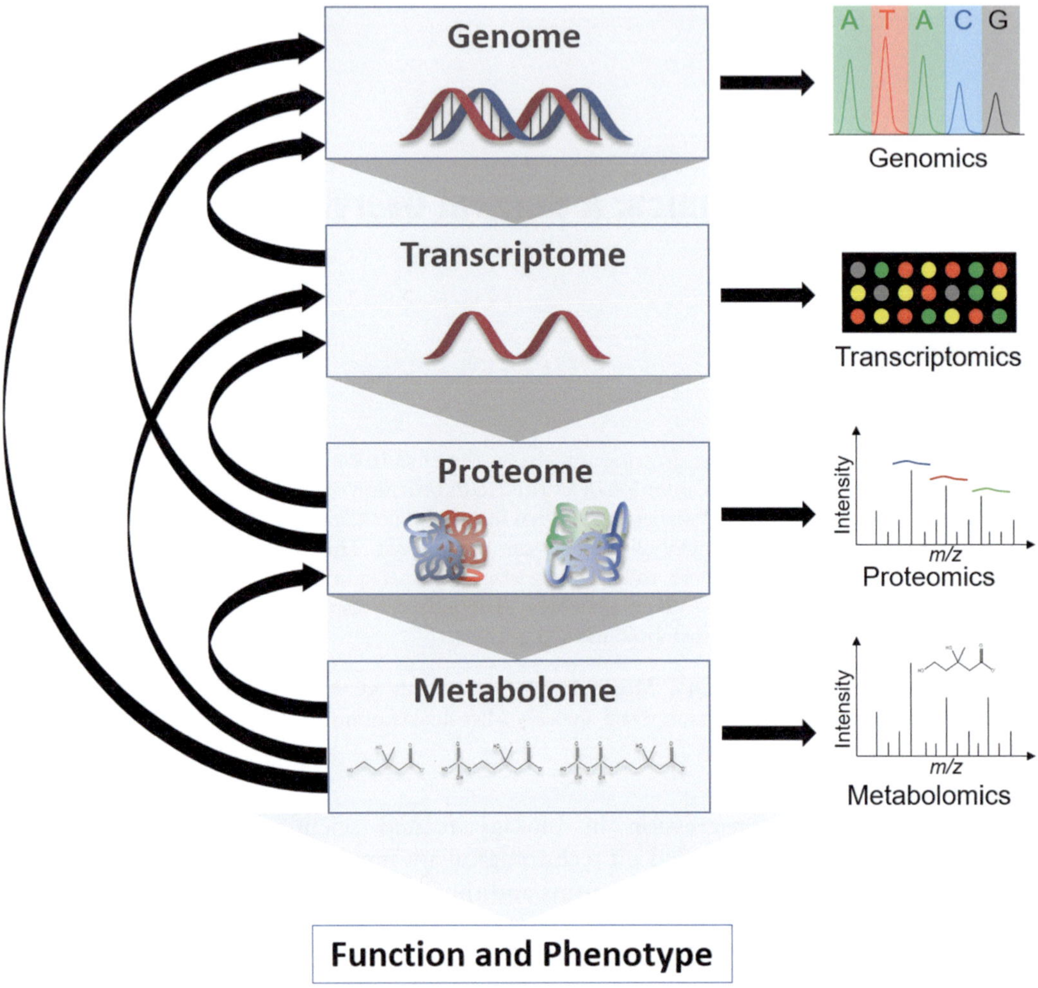

Fig. 1 "Omics" overview. This is an illustration of the flow of molecular information from genes to metabolites to function and phenotype, the interactions between the "omes" and the "omics" techniques used to measure them. This diagram was adapted from Hollywood et al. [10]

mechanisms (feedback loops, chromatin regulation, global biological regulation, etc.) (Fig. 1) [3–7].

Metabolites mediate regulation of microbial metabolism in response to continual environmental changes [8]. These regulatory mechanisms allow microorganisms to adapt to various ecological situations (e.g., changes to nutrient supply, etc.) but must also be reversible as the environment can revert to its original state. Since metabolites show the fastest response times to changes in the environment, it is clear to see why they are utilized in numerous regulatory mechanisms. Aside from regulation, metabolites are mainly known for participating in biochemical reactions (with environmental nutrients) that are involved in maintaining the living state of cells (metabolism) to generate energy (catabolism) for synthesis of the building blocks of cellular materials such as nucleic acids, proteins, polysaccharides, and lipids (anabolism) [9]. These biochemical reactions are also responsible for producing secondary metabolites such as alkaloids and terpenoids, which are involved in cell function and survival [9]. Thus, metabolites play a major role in the response of a biological system to abiotic or biotic perturbations, thereby linking genotype to phenotype (which is generated by the flux of metabolites within a biological system) [8, 10]. It is for these reasons that molecular information from the metabolome is crucial to understanding metabolic networks and determining genetic function.

Metabolomics, the newest member of the "omics" family of techniques, seeks to measure the low molecular weight chemical products of enzyme-catalyzed reactions (metabolites) in biological systems (e.g., cells, compartments, tissues, organisms) at specific points in time [2, 9]. The importance of metabolomics to biological research is highlighted by the fact that biochemical information from genes, transcripts, and proteins do not always correlate, due to not all genes being under transcriptional control and the incomplete prediction of the proteome from the transcriptome owing to post-translational modifications [9]. However, the biopolymeric nature of genes, transcripts, and proteins makes their analyses easier than that of the metabolome because they encode information based on sequences of well-defined monomeric nucleotides and amino acids [1, 10]. By contrast, metabolites are distinct chemical entities from biochemical transformations, which complicates both the measurement and data analysis processes. At present, there is no single analytical method that can measure the entire metabolome of an organism and to achieve something close to that, multiple methods would have to be used. Sample preparation, which is arguably the most critical step in metabolomics experiments, is still challenging. Both the chemical diversity within the metabolome and the complexity of cell wall structure across the microbial diaspora complicate this process. Therefore, it is currently not possible to extract all metabolites from a given biological system with the same level of efficiency. This is further compounded by the rapid change in metabolite levels due to their constant formation and transformation in biochemical reactions within a cell. Furthermore, some metabolites are required in small amounts by their prospective reactions (e.g., signaling molecules such as cyclic adenosine monophosphate, cAMP, which plays a role in regulating gene expression in catabolite repression [11, 12]), while others are needed in large amounts (e.g., adenosine triphosphate, ATP, which provides energy for physiological processes upon hydrolysis) [2]. Thus, stopping (i.e., quenching) these biochemical reactions quickly is critical to capturing accurate, quantitative metabolic information. Additionally, efficient extraction of metabolites from biological sample matrices with appropriate

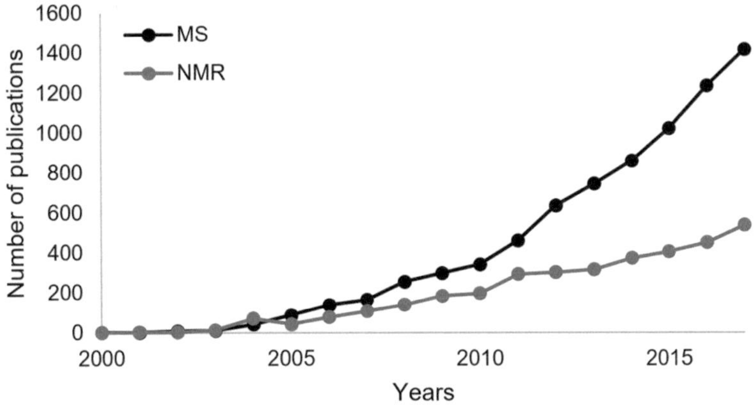

Fig. 2 Popularity of MS and NMR from 2000 to 2017. The popularity was based on the number of publications obtained from Web of Science. The search criteria are as follows: (1) metabolomics and mass spectrometry and years, and (2) metabolomics and NMR and years. NMR was used instead of nuclear magnetic resonance because it generated more publications

extraction agents (e.g., methanol, chloroform, water, etc.) is necessary for obtaining quantitative metabolite measurements.

The two major measurement techniques used in microbial metabolomic analyses are nuclear magnetic resonance (NMR) and mass spectrometry (MS). Although the former can provide high-throughput metabolic fingerprinting (i.e., analysis providing a fingerprint of metabolites produced by a cell [2]), unambiguous identification of metabolites through comprehensive structural elucidation, and nondestructive metabolite measurements, it is rather costly, lacks sensitivity, and is semiquantitative. On the other hand, MS-based technologies can provide high-throughput accurate mass determination and structural elucidation, while operating at high sensitivities, enabling the use of quantitative metabolite measurements. Further, MS instrumentation can also be available at much lower costs, depending on the level of analytical resolution required. It is for these reasons that MS remains the most popular metabolomic measurement technique to date (Fig. 2). MS systems are normally coupled to separation devices via electron ionization (EI) or electrospray ionization (ESI). Separation technologies such as liquid chromatography (LC separates hydrophobic and hydrophilic compounds), gas chromatography (GC separates volatile hydrophobic compounds), and capillary electrophoresis (CE efficiently separates hydrophilic compounds) improve metabolite identification and quantitation by reducing the complexity of sample matrices. These separation technologies provide complementary (orthogonal) methods, and because their devices are relatively inexpensive, are often purchased together with MS instrumentation.

Over the past 14 years, there have been significant advances in separation and MS system integration (i.e., better MS ion source interfacing), chromatographic separation efficiency (e.g., UHPLC and 2D-GC), MS resolution (e.g., time-of-flight and orbitrap), detector sensitivity, data acquisition, system robustness, and computational power. Furthermore, the multiplexing of LC techniques has enabled scientists to achieve target-based metabolite profiling across many different classes of organic compounds [13]. These advancements have led to better metabolite identification, quantitation, throughput, and reproducibility, and therefore greater capture of metabolite information. But to achieve the desired goal of global metabolite profiling, these boundaries need to be pushed further in addition to developments in efficient automated microbial metabolite sample preparation (i.e., by improving sampling and providing better, universally applied, metabolic quenching and metabolite extraction methodologies).

Data analysis is still challenging though. While there are numerous computational tools to automate the process, issues regarding widely varying metabolite abundances, reliable detection, and mass spectral identification [14], make it difficult to achieve precise metabolic feature extraction and, therefore, rapid automated data analysis. As more data is being generated, data analysis workstations are constantly requiring upgrades in software, processing power, and RAM management to meet these demands.

Untargeted and target metabolomics are the two major strategies used to obtain metabolic information. In untargeted metabolomics experiments, the available metabolite information is obtained from a biological sample via comprehensive feature extractions of measurable analytes [1] and treated with the relevant statistical analyses to identify metabolic features that are unique among sample types. This approach requires excellent chromatographic separation and high-resolution accurate mass spectrometry (i.e., HRAMS such as from the quadrupole time-of-flight and orbitrap). The resulting accurate mass metabolic features can then be identified by searching against known MS databases and/or MS/MS spectral libraries. Even though untargeted metabolomics can take a considerable amount of time and effort, it can be used very effectively in discovery experiments and to identify diagnostic biological markers. Targeted metabolomics, however, measures specific groups of known metabolites (e.g., classes of organic compounds, biosynthetic pathways, etc.), highlighting only those portions of metabolomics data that are meaningful to a research study, thus, providing more immediate answers to biological questions. Other advantages of targeted metabolomics experiments are that they can provide quantitative information (i.e., via external calibration curves and/or internal standardization), do not require expensive HRAMS or elaborate chromatographic separations, can be used with high-throughput separation-MS methodologies, and achieve faster data analysis times. For these reasons, targeted metabolomics has emerged as the favored

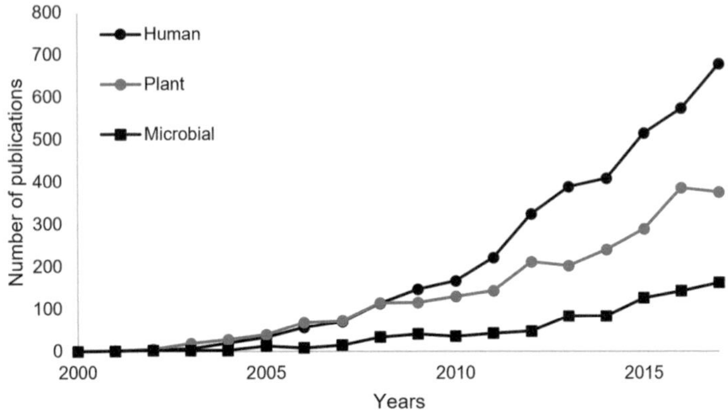

Fig. 3 Popularity of metabolomics in the fields of human biology, plant biology, and microbiology from 2000 to 2017. The popularity was based on the number of publications obtained from Web of Science. The search criteria are as follows: (1) metabolomics and human and year, (2) metabolomics and plant and year, and (3) metabolomics and microbial and year

approach of most laboratories. Algorithms are now able to overlay metabolomics data on top of metabolic pathways. Furthermore, the data from these approaches can be modeled to reveal fluxes in metabolic pathways and to predict the outcomes of future biological experiments.

At present, metabolomics data is more widely applied to human related studies, followed by plant studies, and then microbiological studies (Fig. 3). In systems microbiology research, multi-omics technologies (including metabolomics) are being used to characterize metabolic networks and microbial interactions. Metabolite data from these experiments can also being leveraged to provide signaling pathway information to strengthen predictive mathematical models. Advancements in systems biology "omics" technologies are leading to developments in synthetic biology research. In synthetic biology studies, metabolomics is enabling researchers to identify bottlenecks in engineered pathways (through pathway flux assessments), identify components of cellular toxicity (e.g., accumulating toxic intermediates or products), and assess strain variation (e.g., carbon utilization and energy production). In addition to this, when metabolomics is utilized with the other "omics" technologies and microbial physiology is incorporated in the design of an experiment, the information gained can be used to identify the microbial regulatory mechanisms that circumvent engineered biosynthetic pathways [15]. This approach, together with machine learning, can improve the rational design, development, and optimization of engineered strains.

In the study of infectious human diseases, metabolomics is helping researchers to identify biomarkers of infection and new antimicrobial compounds. By understanding the mechanisms involved in microbial infection, pathogenicity, and immunity, researchers can develop more effective strategies to reduce the occurrence of infectious disease. For the noninfectious category, metabolomics is being used to find biomarkers of microbial induced disease (diabetes, cancer, atherosclerosis, etc.) from microorganisms and human related tissues and bodily fluids. A deeper learning of host–microbe interaction will provide a greater understanding of the factors influencing microbial growth and population maintenance within host organisms. It will also allow us to understand how the actions of these factors are countered by the host immune system [16] and will be necessary to establish reliable cultured isolates from patient specimens, which is required to develop and hence implement effective treatments.

Metabolomics studies of the human gut microbiome are enabling researchers to identify mechanisms that predispose individuals to disease and the roles the microbiome plays in drug metabolism, aging, and cancer. In the study of environmental microbial communities, metabolomics is being used in combination with other "omics" technologies to characterize community structure and functional diversity. Knowledge from these studies can be used to unravel microbial interactions within communities, the environment, and host organisms to address environmental challenges and human disease.

Metabolomics has come a long way since its inception and is now routinely applied to many biological experiments. As researchers continue to push the envelope in metabolomic technological advancements, measurements are likely to become more reliable, readily available, and comprehensive, and could be used to obtain global metabolite information on a regular basis. Further, the integration of multi-omic data (that includes high-quality metabolomic measurements) can be used to generate accurate predictive models, with the potential to revolutionize the biological sciences.

Acknowledgments

The authors would also like to acknowledge that this work was part of the DOE Joint BioEnergy Institute (http://www.jbei.org) supported by the US Department of Energy, Office of Science, Office of Biological and Environmental Research, through contract DE-AC02-05CH11231 between Lawrence Berkeley National Laboratory and the US Department of Energy.

References

1. Roberts LD, Souza AL, Gerszten RE et al (2012) Targeted metabolomics. Curr Protoc Mol Biol 1:1–24
2. Villas-Bôas SG, Roessner U, Hansen MAE et al (2007) Metabolome analysis: an introduction. John Wiley & Sons, Inc, Hoboken

3. van der KJA, Verrijzer CP (2016) Undercover: gene control by metabolites and metabolic enzymes. Genes Dev 30:2345–2369
4. Lee SJ, Trostel A, Adhya S (2014) Metabolite changes signal genetic regulatory mechanisms for robust cell behavior. MBio 5:1–8
5. Kochanowski K, Gerosa L, Brunner SF et al (2017) Few regulatory metabolites coordinate expression of central metabolic genes in *Escherichia coli*. Mol Syst Biol 13:903
6. Penn K, Wang J, Fernando SC et al (2014) Secondary metabolite gene expression and interplay of bacterial functions in a tropical freshwater cyanobacterial bloom. ISME J 8:1866–1878
7. Berthoumieux S, De JH, Baptist G et al (2013) Shared control of gene expression in bacteria by transcription factors and global physiology of the cell. Mol Syst Biol 9:1–11
8. Aretz I, Meierhofer D (2016) Advantages and pitfalls of mass spectrometry based metabolome profiling in systems biology. Int J Mol Sci 17:E632
9. Baidoo EEK, Benke PI, Keasling JD (2012) Mass spectrometry-based microbial metabolomics. In: Navid A (ed) Microbial systems biology: methods and protocols. Springer, New York, pp 215–278
10. Hollywood K, Brison DR, Goodacre R (2006) Metabolomics: current technologies and future trends. Proteomics 6:4716–4723
11. Botsford JL, Harman JG (1992) Cyclic AMP in prokaryotes. Microbiol Rev 56:100–122
12. Buettner MJ, Spitz E, Rickenberg HV (1973) Cyclic adenosine 3′, 5′-monophosphate in Escherichia coli. J Bacteriol 14:1068–1073
13. Wei R, Li G, Seymour AB (2010) High-throughput and multiplexed LC/MS/MRM method for targeted metabolomics. Anal Chem 82:5527–5533
14. Baran R (2017) Untargeted metabolomics suffers from incomplete raw data processing. Metabolomics 13:107
15. Baidoo EEK, Keasling JD (2013) Microbial metabolomics: welcome to the real world! Metabolomics 9:755–756
16. Lorenz MC (2007) Host. Microbe 449:7164–7164

Part I

Microbial Metabolomics, Metabolism, and Physiology

Chapter 2

Mass Spectrometry-Based Microbial Metabolomics: Techniques, Analysis, and Applications

Edward E.K. Baidoo and Veronica Teixeira Benites

Abstract

The demand for understanding the roles genes play in biological systems has steered the biosciences into the direction the metabolome, as it closely reflects the metabolic activities within a cell. The importance of the metabolome is further highlighted by its ability to influence the genome, transcriptome, and proteome. Consequently, metabolomic information is being used to understand microbial metabolic networks. At the forefront of this work is mass spectrometry, the most popular metabolomics measurement technique. Mass spectrometry-based metabolomic analyses have made significant contributions to microbiological research in the environment and human disease. In this chapter, we break down the technical aspects of mass spectrometry-based metabolomics and discuss its application to microbiological research.

Key words Mass spectrometry, Metabolomics, Microbial, LC-MS, GC-MS, CE-MS, Metabolic quenching, Metabolite extraction, Data analysis, Microbial communities, Human disease

1 Introduction

In recent years, there has been a push in the biosciences to characterize the functions and phenotypes associated with genes. In this regard, the metabolome (i.e., the complete set of metabolites from a cell, compartment or tissue [1]) provides both functional and phenotypic information since metabolites are the small molecule products of enzyme-catalyzed metabolic reactions. The measurement of metabolite profiles (metabolomics) can be achieved by spectra obtained from mass spectrometric analyses. Mass spectrometers either induce or facilitate ionization of metabolites and separate them based on the mass-to-charge ratios of their charged forms. To date, mass spectrometry (MS) is the most popular measurement technique for metabolomics studies due to its structural elucidation capability, high sensitivity, and quantitative ability. When MS is coupled to chromatographic or electrophoretic separation, these features are further enhanced. However, since each metabolite is structurally unique, at present, there is no single

Edward E.K. Baidoo (ed.), *Microbial Metabolomics: Methods and Protocols*, Methods in Molecular Biology, vol. 1859, https://doi.org/10.1007/978-1-4939-8757-3_2,

analytical method that can resolve all metabolites. Additionally, metabolomics sample preparation is further complicated by the varying rates at which metabolites are formed and transformed and the chemical diversity within the metabolome. Consequently, there are many metabolomics and sample preparation methods that are designed to capture either a broad range of metabolites (e.g., for untargeted metabolomics experiments) or focus on a specific subset (e.g., for targeted metabolomics experiments). Metabolomic data mining and integration can also be challenging due to varying metabolite abundances resulting from the data acquisition process [2]; and while recent developments in data analysis tools are designed to correct issues concerning variation and identification, obtaining precise metabolic feature extraction can still be difficult. Thus, the future of mass spectrometry-based metabolomics will most likely center around resolving these issues and automating much of the data acquisition and analysis processes.

2 Metabolic Quenching and Metabolite Extraction

Preparation of biological samples is of the upmost importance to analytical measurements as it can preserve the metabolic integrity of sampled biomass. Strategies involving efficient metabolic quenching and metabolite extraction are necessary to capture accurate metabolite information and should be determined during the design of the experiment (simple approaches with the least number of steps will often yield the best measurement precision). Metabolic quenching can be described as the cessation of all cellular activity that is governed by proteins and nucleic acids. It is achieved by dramatically changing the cell environment through the process of denaturation, in which proteins and nucleic acids lose their native structure (i.e., quaternary, tertiary, and secondary structures) [3, 4], by the application of a quenching agent (i.e., an external stress such as temperature, pH, or a chemical such as an organic solvent). To understand the immediate impact metabolic quenching has on the metabolome, it is important to consider the roles proteins play in the cell.

Proteins are required to maintain the structure, function, biochemical interactions, regulation, and transport function of a cell and can be generally categorized as structural proteins, regulatory proteins, transport proteins, and enzymes. Structural proteins are the largest class of proteins and, as suggested by their name, confer rigidity to cell components and are frequently involved in motility [3]. Transport proteins aid the passage of nutrients, metabolites, and other proteins in and out of the cell via the cell membrane. These proteins have a high degree of specificity for the substance being transported [5]. They are often used to excrete metabolites that are underutilized or toxic to the cell to prevent the intracellular accumulation of these compounds. Regulatory proteins can bind to determined sequences on a DNA strand (e.g., TATA binding protein) to turn genes on and off, thus, playing an important role in gene expression [6]. Lastly, enzymes are directly responsible for the biochemical interactions and molecular transformations in a cell. Enzyme catalysis directly results in the generation of a new metabolite or an intermediate to a final compound from a metabolic pathway in a cell. Enzymatic activity depends on the availability of substrates and cofactors and can also be affected by their own products and the products of other enzyme catalyzed reactions via feedback and feedforward cellular control mechanisms. Enzymes can recompose themselves after catalyzing substrates into products, which not only maintains their activity but cell metabolism as well. While most researchers often think of enzymes when designing sample preparation procedures, it is clear to see the importance of all classes of proteins to maintaining metabolism and hence the metabolome and why quenching their activities negatively impacts the metabolic processes in a cell.

High metabolic quenching efficiency is crucial to obtaining instantaneous metabolic information. However, this is often difficult to achieve due to the rapid turnover of certain metabolites, some of which have half-lives of ≤ 1 s and others that are intermediates of several metabolic reactions [1]. Furthermore, the quenching agent needs to rapidly transit the cell membrane barrier and affect the intracellular environment, which provides another major obstacle. Since microorganisms have varying morphologies, it also is difficult to devise a single effective quenching and metabolite extraction strategy for them all. Some of the fastest quenching methods directly transfer the microbial culture to a quenching agent, followed by the separation of biomass from the culture medium by either centrifugation or filtration. The disadvantage of this is that intracellular metabolites from microorganisms with weak cell walls, such as gram-negative bacteria, can be leaked from the cell [1]. On the contrary, Koning and Dam found evidence to suggest that *S. cerevisiae* cells do not leak metabolites when the culture is quenched with −40 °C methanol [7]. Methods that separate biomass from the culture medium first, on average, do not quench metabolism as efficiently but are not as prone to metabolite leakage. In the past, this has been especially true of centrifugation, whereas fast filtration can provide good quenching efficiency [8, 9]. In any case, the goal of the research study (and, therefore, the type of metabolites targeted) will dictate the type of quenching strategy to use. Metabolic quenching efficiency is often determined via the adenylate energy charge (i.e., [ATP] + 0.5 [ADP]/[ATP] + [ADP] + [AMP]) as ATP is rapidly utilized by the cell as soon as it is formed [4]. Since the energy charge of most cells falls within 0.8–0.95, obtaining values within this range normally indicates that the quenching process is efficient [4].

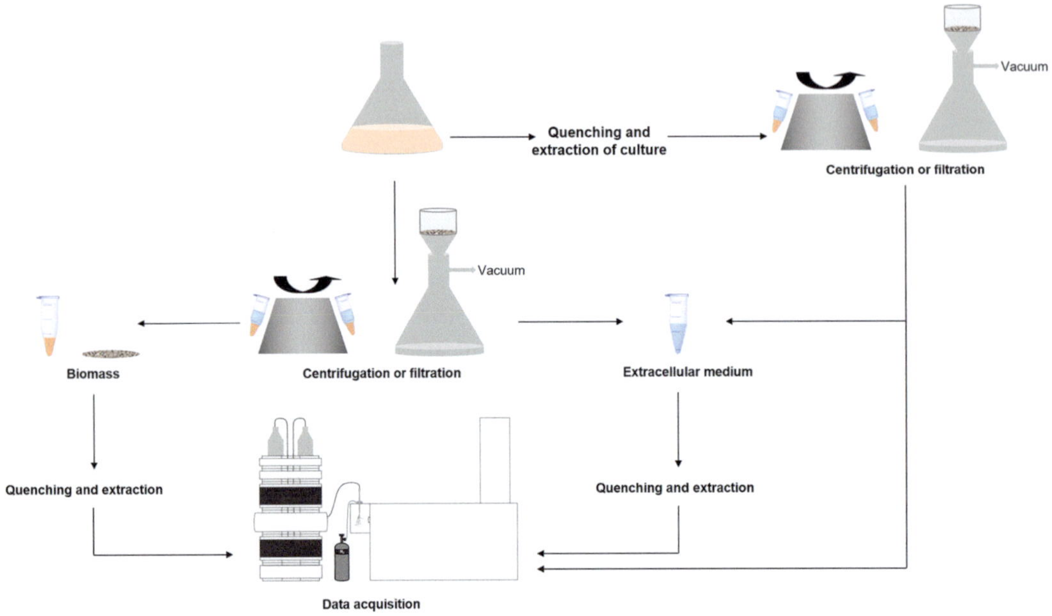

Fig. 1 Sample preparation workflow. Biomass can be separated from the culture medium before or after quenching and extraction. Prepared samples are then run on the data acquisition system

Extracellular metabolites can simply be obtained by separating the microbial biomass from the culture medium (by centrifugation or filtration) and collecting the latter (followed by the addition of quenching agents if extracellular enzymes are suspected) (Fig. 1). Since most proteins are found intracellularly or bound to cell membranes, metabolic quenching of the broth is not as significant, but it is still recommended by many researchers. The exometabolome can provide metabolic footprints of excreted metabolites [10], whereas the endometabolome (intracellular metabolites) yields metabolic fingerprints of metabolites produced by a cell [1]. While these approaches are normally qualitative, high quality extra- and intracellular extracts can be used for quantitative experiments. Of the two, however, it is far more difficult to extract intracellular metabolites.

To extract metabolites from a cell, factors influencing cell lysis such as permeability, disruption, and degradation of the cell membrane and wall should be considered. It is, therefore, important for researchers to understand the physical characteristics and chemical composition of the microbial cell membrane and wall under study, as they will confer different barrier strengths to each microorganism. For example, gram-positive bacteria are more difficult to lyse than gram-negative as they contain a thick peptidoglycan layer outside of the cell membrane. Further, Yeast cells are also difficult to lyse because they contain a thick cell wall comprised of mannoproteins, glucans, and chitin. In these cases, harsher extraction agents may be required. Organic solvents are the most widely used extraction agents due to their affinity for most cell membrane structures, as they enable cell permeabilization and the efficient release of cell contents. Additionally, knowing the chemical and physical properties of the analytes of interest will help to improve extraction efficiency (i.e., the extent to which metabolites are recovered from a cell) as methods can be tailored to those metabolites. For example, in methanol–water–chloroform extraction of microbial metabolites, methanol and water can interact with polar metabolites while chloroform interacts with the nonpolar fraction [11–15]. The multipurpose nature of methanol and chloroform allows them to be used to disrupt cell membranes and quench metabolism. As well as being used as quenching agents, most organic solvents tend not to degrade metabolites because of their pH neutrality and weak oxidizing capabilities. However, when they are used in conjunction with extreme pH variation (e.g., via perchloric acid) and high temperatures (e.g., boiling ethanol), metabolites could be degraded or deformed, leading to reduced yields. Of the numerous extraction methods, methanol-based extraction procedures appear to be the most popular for intracellular metabolites. This is due to methanol's quenching and extraction (causing cell lysis and interacting with polar to mid-polar metabolites) capabilities, and lack of artifact production [2]. Cell lysis can also be brought about via enzymatic (e.g., lysozyme) and mechanical (e.g., sonication and bead-beating) cell wall/membrane disruption techniques [1, 2]. Enzymatic techniques should not be used with quenching agents as enzyme activities can be arrested. Care should also be taken with mechanical disruption techniques, as they can often raise the temperature of sample lysates, leading to the possible degradation of thermally labile compounds (in such cases the sample disruption technique can be performed at low temperatures). Thus, efficient metabolite extraction from microbial cells is achieved by complete cell lysis, strong analyte-extracting agent interaction, and minimal loss and degradation of analytes.

Clean-up strategies can be employed to enhance the overall quality of the metabolite extract by reducing its complexity and ensuring reproducibility by minimizing interferences from highly abundant sample matrix ions [16]. There are numerous examples of metabolite sample cleanup strategies. Molecular weight cut-off filtration (MW/CO) is a separation process using a membrane filter under centrifugation to remove macromolecules from a biological extract. Liquid-liquid extraction uses solvent phase separation to purify metabolites and remove interfering compounds. The best example of this is the previously described methanol–water–chloroform extraction procedure [11–15]. Solid phase extraction (SPE) allows the retention of metabolites on a solid stationary phase support (via chromatography), while unretained substances are washed off. Metabolites can then be eluted from the SPE column/cartridge via a solvent or solution it has a strong interaction with

[16, 17]. Solid phase microextraction (SPME) purifies metabolites by their adsorption on a fiber coated with an extraction phase. Retained metabolites are transferred to the injection port of the separating device, where they are subsequently desorbed and transferred to the mass spectrometer [18]. Cleanup strategies are becoming more desirable to analytical laboratories as they become part of automated sample preparation platforms that are helping to streamline metabolomic workflows and improve the consistency of analytical measurements [18].

3 Analytical Separation Technologies

3.1 Introduction to Separation

It is without doubt that separation, whether chromatographic or electrophoretic, is one of the most important components of metabolomic measurements. By separating metabolites in in real-time, they will elute from the separating device at different times, thereby reducing the complexity of biological extracts before MS detection. Without separation, all sampled metabolites would enter the MS simultaneously, making it difficult to distinguish between those of similar chemical structure (e.g., isomeric and isobaric compounds). Further, when multiple unseparated compounds enter the MS ion source, they compete for available charges [19]. Consequently, the most ionizable volatile compounds yield the best detector responses and dominate the acquired mass spectra. This is especially true of sample matrix components involved in the now infamous ion suppression phenomenon. Such components affect the detection capability, precision, and accuracy of analytical methods. Separation reduces these effects and, therefore, improves upon quantitation.

Chromatography is the most popular separation technique in metabolomic analyses and relies on the differential distribution and equilibration of sample components (according to their distribution coefficients) between two phases [20]. The phases employed are typically a moving mobile (liquid or gas) phase and a stationary phase adsorbent (i.e., microscopic particles on which analytes are retained by physicochemical interactions). The mobile phase carries components from the sample through the absorptive stationary phase material. The extent of analyte partitioning between the phases determines the length of time spent in the chromatography column. Individual components are detected as they leave the column and the eluted compound is observed as a Gaussian distribution plot of the detector response as a function of elution time and resides within a chromatogram (a graphical representation of detector response versus time for the duration of the analytical run). The area under the Gaussian distribution plot is used to calculate the peak area for quantitative purposes. The amount of time a compound is retained in chromatography column is known as the retention time (RT). Acquisition and analysis software typically measure the RT from sample injection to the elution of the compound from the column and is normally measured at the apex of a chromatographic peak. The RT helps to differentiate between compounds with similar structures (e.g., isomers) and, when coupled to MS detection, the two can be used to reliably determine the identity of a compound.

Liquid chromatography (LC) separation is dependent on interactions between the analyte, stationary phase, and the liquid mobile phase. A change in the composition of the liquid mobile phase can alter analyte chemistry, interaction, and separation. Gas chromatography (GC) separation is dependent on analyte volatility and interaction with the stationary phase. An unfortunate consequence of the GC temperature gradient (which is employed to encourage volatilization), is its adverse effects (i.e., degradation and/or decomposition) on thermally labile compounds. Low GC temperatures are needed to analyze these compounds. Where LC excels is in its versatility. LC is generally used to separate nonpolar, polar, and charged analytes, which span the entire spectrum of functionalities within the metabolome, whereas GC is ideal for nonpolar, volatile analytes. This is, in part, due to the broad stationary phase portfolio of LC, which spans numerous chromatography methods to suit any metabolomic application. LC can cover a wider range of metabolite classes without the need for the time-consuming analyte derivatization steps (which are required to render nonvolatile compounds volatile) found in GC methods. Conventional GC methods, however, provide better separation efficiency and reproducibility than their LC counterparts. Further, the simple construction of a GC system (i.e., a column housed in an oven) creates a very cost-effective, robust, easy-to-use separation device. GC is also easier to couple to MS, since it utilizes a temperature-controlled carrier gas mobile phase as opposed to the liquid mobile phase of LC. Hence, volatiles are transferred directly in the gas phase from GC to MS, thereby simplifying the coupling of these two systems. On the other hand, LC requires elaborate evaporation/desolvation procedures to produce gaseous ions before they enter the MS.

Capillary electrophoresis (CE) utilizes an applied electric field to separate charged analytes, by the process electrophoresis, through a fused silica capillary (FSC) filled with a conductive liquid medium (e.g., a buffer or electrolyte) [2]. Migration and elution order largely depend on the charge-to-size ratios of analytes, which is complementary to MS detection. Consequently, the electrophoretic mobility of an analyte relies on its charge state (q), its radius (r), and the viscosity of the liquid separation medium (η), which can be described as follows: $\mu_{EP} = q/6\pi\eta r$ [21]. Like other electrophoretic techniques, size and charge play key roles in analyte elution order, with smaller highly charged ions migrating faster than larger ions with lower charge states [2]. The time it takes for the analyte to migrate through the FSC and be detected is known as the migration time (MT). The eluted compound is observed as a Gaussian distribution plot of the detector response versus migration time and resides within an electropherogram (a graphical representation of detector response versus time for the duration of the analytical run). As mentioned previously, the Gaussian distribution plot is used to calculate the peak area.

CE provides higher separation efficiencies than GC and LC but is incapable of separating uncharged neutral compounds without the aid of surfactants that are generally incompatible with MS. CE separation is also less stable than that of GC and LC. The small diameter of the FSC makes CE susceptible to changes in environmental temperature, which can cause internal FSC temperature fluctuations, leading to varying buffer/electrolyte viscosities and, hence, MTs. Sample and reagent costs are minimized with CE, however, as it utilizes smaller sample injection volumes (nL range) and lower flow rates (nL/min), but this can also reduce sensitivity. Of the three techniques, LC has the highest reagent and instrument costs. LC is also more complex than GC and CE as its separation can be based on distribution, adsorption, affinity, size, or ion interaction [22]. The ability to vary liquid mobile phase composition(s) at high pressure further complicates LC instrumentation. While all three separation techniques can be used independently, they should also be viewed as being complementary to each other. This is an important point, especially because there is currently no single method that can simultaneously separate all the components of the metabolome. Therefore, the take home message should be that LC is the most versatile and popular separation technique (Fig. 2), GC provides the most robust analytical separations and is the easiest technique to interface with MS, and CE provides the highest separation efficiencies.

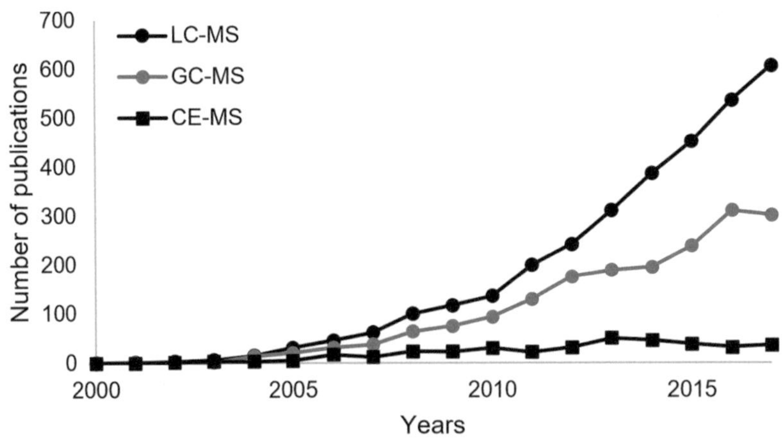

Fig. 2 Popularity of separation-MS techniques from 2000 to 2017. The popularity was based on the number of publications obtained from Web of Science. The search criteria are as follows: (1) metabolomics and liquid chromatography and mass spectrometry and year, (2) metabolomics and gas chromatography and mass spectrometry and year, and (3) metabolomics and capillary electrophoresis and mass spectrometry and year

3.2 LC Separation

LC separates hydrophobic, hydrophilic, and thermally labile compounds (which cover the vast majority of metabolites) by partitioning them between a liquid mobile phase and a solid particulate stationary phase. Metabolites are normally retained by having an affinity for the LC column's stationary phase via hydrophobic interactions (i.e., Van der Waals forces of attraction), hydrogen bonding, and electrostatic interactions [23]. They are carried through the LC column by a liquid mobile phase and eluted from it when the level of interaction with the mobile phase is greater than the stationary phase. The mobile phase also interacts with the columns stationary phase; and as the level of the eluting reagent (i.e., the portion of the liquid mobile phase that interacts with both the analyte and stationary phase) increases, the analyte interacts less with the stationary phase and more with the mobile phase. This eventually leads to the analyte being displaced from the stationary phase and carried off to the detector. This mode of separation is often referred to as gradient elution. Isocratic elution occurs when the mobile phase composition remains constant during the entire length of a chromatographic run. Gradient separation can produce more selective chromatography whereas isocratic separation is more robust. With the latter, there needs to be a balance between the eluting reagent and retention reagent (i.e., to provide both retention and elution). An understanding of LC column chemistry and analyte chemical properties are necessary for achieving the best separations.

At present, many LC metabolomics methods employ reversed phase (RP), RP-ion pairing (RP-IP), and hydrophilic interaction liquid chromatography (HILIC). Most chromatographic applications are based on RP-LC because of its ability to separate numerous classes of compounds. Compounds that are unionized (i.e., neutral) and hydrophobic (nonpolar) are well retained and separated by RP chromatography. Polar compounds can be made nonpolar by the adjustment of mobile phase pH. At low pH, most acidic compounds will be neutralized and well retained by the RP column, while basic compounds are unionized and retained at higher pH. Mobile phases can also be buffered to prevent drastic changes in mobile phase pH and ionic strength (which can improve retention time reproducibility and chromatographic peak shape). RP stationary phases tend to be hydrophobic, and the most commonly used is octadecyldimethyl-silane (C18). In this mode of chromatography, analytes are partitioned between a polar mobile phase and a nonpolar stationary phase; and the extent to which they partition is dependent on their polarity and hydrophobicity. Thus, in RP separation, analyte retention is greatest when the mobile phase is aqueous, causing the analyte to interact more strongly with the hydrophobic stationary phase. Conversely, the analyte is eluted from the LC column as the organic solvent content of the mobile phase is increased (e.g., from 20% to 90% organic) (Fig. 3). Analytes are

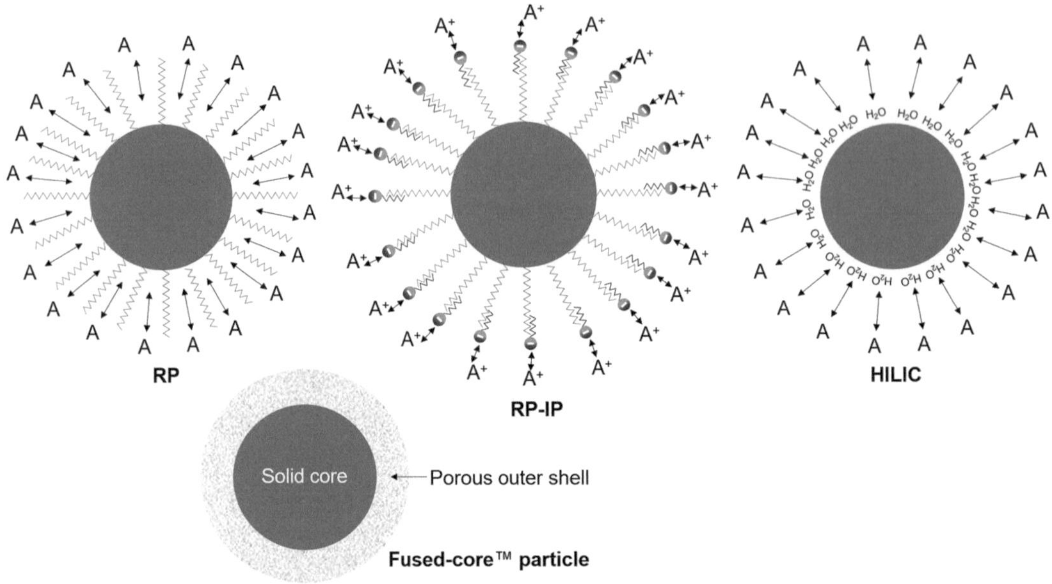

Fig. 3 LC retention mechanisms and Fused-core™ particle. Retention is mediated as follows: (1) in RP chromatography the analyte (i.e., A) interacts with hydrophobic functional groups via Van der Waals forces of attraction, (2) in RP-IP chromatography, the analyte interacts electrostatically with the IP agent, and (3) in HILIC the analyte interacts with the water layer surrounding the stationary phase by hydrogen bonding

eluted in order of decreasing hydrophobicity [2]. The separation of hydrophilic compounds requires the use of a highly aqueous eluent to provide some retention. Most manufactured C18 columns have a limit of 95% aqueous. Above this level, the dewetting phenomenon is observed [24]. Here, the oil-like stationary phases tend to minimize their surface energy through self-aggregation. Consequently, the C18 functional groups on the silica stationary phase support collapse on themselves, leading to a loss of retention [24]. RP columns with wider pore sizes and shorter alkyl chain stationary phases are less susceptible to dewetting; but due to their lack of retention of very polar compounds, they are not commonly used for metabolite analysis [23]. However, column chemistries utilizing polar embedded groups or polar end-capping are able to overcome this phenomenon and can be used at up to a 100% aqueous mobile phase composition. To improve the retention of polar metabolites, reversed phase columns have incorporated additional functionalities (e.g., -OH, -NH, and $-CONR_2$ groups). Further, alternative RP column chemistries offer extra chromatographic selectivity (e.g., the use of phenyl-hexyl and biphenyl columns for the separation of aromatic compounds).

The use of ion-pairing (IP) reagents with RP-LC improves separations for hydrophilic metabolites. One end of the IP reagent is ionic and the other nonpolar. When the IP reagent is added to the mobile phase and the LC column is allowed to equilibrate, its nonpolar hydrophobic end interacts strongly with the RP columns stationary phase, leaving the ionic end exposed to the mobile phase. Metabolites of opposing charge are electrostatically attracted to the immobilized ionic end, causing chromatographic retention [23]. Negatively charged analytes are retained by positively charged reagents (e.g., tetrabutyl ammonium chloride) and vice versa (Fig. 3). Typical IP reagents for positively charged analytes are alkyl sulphonic acids [23]. A major disadvantage of using IP reagents is the suppression of MS analyte detection (IP reagents yield very strong ions that outcompete analytes for MS detection) and contamination of MS instrumentation after prolonged use [2]. Therefore, ion pairing should be used with dedicated LC-MS instrumentation.

Arguably the most popular LC technique for separating hydrophilic metabolites (such as phosphorylated compounds, nucleotides, organic and amino acids) is HILIC [20, 25]. It employs traditional polar stationary phases such as silica, amino or cyano and utilizes mobile phases similar to that of RP-LC. Here, retention is obtained via hydrogen bonding between the analyte and an immobilized water-enriched layer that is adsorbed onto the hydrophilic stationary phase (Fig. 3) [20]. The more hydrophilic the analyte, the greater the level of interaction it will have with the water layer on the stationary phase. This shifts the partitioning equilibrium toward the immobilized water layer and, as a result, the analyte is retained [20]. Thus, analytes are eluted in the order of decreasing polarity and hydrophilicity. During HILIC separation, the mobile phase is highly organic (i.e., mainly composed of an organic solvent) causing analytes to interact more with the water layer on the stationary phase. On the other hand, elution occurs when the mobile phase is aqueous as analytes interact more with this phase than with the water layer on the stationary phase. Since HILIC separation relies on the water-enriched layer, maintenance of this layer is crucial to reproducible chromatography. Consequently, HILIC gradient elution methods require long LC column equilibration times. Typically, up to 20 column volumes is required to reequilibrate a HILIC column after gradient elution [20]. This is much longer than what is required for RP-LC. HILIC isocratic elution methods, however, require shorter equilibration times due their constant mobile phase compositions. Factors that influence HILIC separation and equilibration are ionic strength, buffer type, pH, and the type of mobile phase organic solvent used, all of which can disrupt the water-enriched layer on the surface of the HILIC stationary phase [20, 23]. In addition to this, temperature can also affect separation and equilibration. The two main advantages of HILIC separation are its applicability to hydrophilic metabolites and its use of highly organic volatile mobile phases, which improve MS detection by providing more efficient electrospray ionization (ESI) [2, 20, 23].

Now that we have covered the three most widely used LC separation techniques for metabolomic studies, we can turn our attention to the stationary phases upon which they are built. At present, there is no single stationary phase that can provide the complete separation of all the components of the metabolome [2]. The most commonly used stationary phase materials are manufactured from high purity silica to produce spherical porous particles [23]. The silica used can be completely bare (e.g., in HILIC) or its surface chemically modified via covalent bonding (e.g., in RP-LC) [2]. For RP-LC, not all available silanol (Si-OH) groups on the stationary phase particle can be covalently modified due to steric hindrances from bulky alkyl groups. When the exposed silanols interact with basic compounds, significant chromatographic peak tailing can result. It is for this reason that manufacturers have adopted the practice of masking the effects of these groups by chemically modifying them with short alkyl-chain groups such as trimethylsilyl [2, 22]. This process of end-capping serves to improve chromatographic peak shape. The spherical porous stationary phase particles can tolerate a pH range of 2–8. The stationary phase siloxane Si–O–Si linkages hydrolyze at low pH (<2) and silica dissolves at high pH (>8). For a wider working pH range, manufacturers have modified (e.g., via ethylene bridges) the silica-based stationary phases to protect the siloxane bonds and the silica core [2, 22].

Stationary phase particle size and LC column dimension can have a dramatic effect on chromatographic resolution (R) and separation efficiency (N). The following equations describe the influence of plate height (H) on N and how the latter is related to R, retention time (t), and peak width at base (w) [22, 23]:

$$H \approx A + \frac{B}{u_x} + Cu_x \tag{1}$$

$$H = \frac{L}{N} \tag{2}$$

$$N \approx \frac{3500\ \mathrm{L(cm)}}{\mathrm{dp}\ (\mu\mathrm{m})} \tag{3}$$

$$R = \frac{\sqrt{N}}{4} \frac{(\alpha - 1)}{\alpha} \frac{k}{(k+1)} \tag{4}$$

$$N = 16\left(\frac{t}{w}\right)^2 \tag{5}$$

$$R = 2\frac{(t_2 - t_1)}{(w_1 - w_2)} \tag{6}$$

where term A is eddy diffusion, B is longitudinal diffusion, C is mass transfer, u_x is linear velocity, L is column length, dp is particle diameter, α is selectivity or relative retention, and k is the capacity factor. α is the ability of the chromatographic system to "chemically" distinguish between sample components [26], and k is a measure of analyte retention. Thus, R or the separation of analytes depends on the narrowness of a chromatographic peak (i.e., N), the distance between adjacent peaks (i.e., α), and the retention of a peak (i.e., k) [22, 23, 26].

The Van Deemter (Eq. 1) describes the relationship between height equivalent to a theoretical plate H (i.e., the height of a theoretical plate), diffusion (terms A and B), analyte movement between the mobile and stationary phases (C), and (u_x) [22, 23]. A theoretical plate can be described as a hypothetical zone in the LC column in which the sample reaches equilibrium with the mobile and stationary phases [22, 23]. Moreover, it is hypothesized that the LC column contains many theoretical plates and that the analyte is transferred from one plate to the next via an equilibrated mobile phase until elution. The more theoretical plates there are, the better the separation efficiency (N). Thus, a reduction in H increases the number of theoretical plates and, hence, N (Eq. 2). In reality, a theoretical plate is just an imaginary term that is used to describe the efficiency with which a column separates an analyte(s) [22, 23]. Term A is caused by turbulence in the analyte molecule's flow path through an LC column packed with stationary phase and term B is the movement of the analyte from the center to the edge of its injected band. Term A leads to broadening of the analyte band as the different paths it can take through the stationary phase (via Eddy diffusion currents) are at random and of varying lengths. Whereas band broadening resulting from term B is due to longitudinal diffusion (along the axis of the column), since the analyte spreads outward from a higher concentration at the center of the band to a lower concentration at its edge. Diffusion, therefore, leads to broader chromatographic peaks. Higher flow rates reduce the effect of term B (as the analyte spends less time in the column) but have little effect on term A; while wider analyte bands resulting from the effect of term C can occur with increasing flow rates.

Other factors affecting chromatographic separation are LC column diameter and length, and stationary phase particle size. A reduction in LC column diameter (which reduces diffusion and H, and increases N) and an increase in column length (which increases the number of stationary phase particles, thus, increasing the functional group surface area for greater analyte interaction and retention) (Eq. 3) can provide better chromatographic separation. Smaller stationary phase particles yield shorter diffusion path lengths, allowing analytes to transit the particle much more quickly. Consequently, analytes experience reduced diffusion as they spend less time inside the stationary phase particle (where diffusion occurs) and have a wider optimum velocity/flow rate range. Thus, a reduction in particle size can decrease H, increase N (Eq. 3) and reduce analyte dispersion across a wider range of flow rates [22]. However, these changes increase the pressure within the LC system (known as the back pressure) and could exceed the pressure limits of conventional HPLC systems (which are normally limited to 400 bar).

The development of ultrahigh-performance liquid chromatography (UHPLC) by Waters (who coined the name ultra performance liquid chromatography, UPLC™) has enabled analytical chemists to capitalize on the advantages of small stationary phase particle separation while operating under very high back pressures (≥1000 bar) [27–29]. In this configuration, the stationary phase particle, LC column, and LC system are built to withstand these excessive pressures. Improvements have been made in stationary phase particle mechanical strength, toughness, ruggedness, and stability by bridging methyl groups in the silica matrix [23, 27, 28]. LC column ruggedness was achieved by engineering a smoother interior surface, establishing precise stationary phase particle packing uniformity, and improving the robustness of the LC end frits matrix [23, 27, 28]. UHPLC utilization of sub-2 μm stationary phase particles at high mobile phase flow rates significantly reduces H and increases N (Eq. 2). The use of smaller stationary phase particles minimizes the space between them, thereby decreasing the column void volume (which is the volume of mobile phase in a column) and reducing diffusion; whereas the increased number of stationary phase particles provide improved retention and separation. UHPLC also relies on the minimization of LC system dwell volume (i.e., the volume of liquid in the LC system from the point at which the gradient mobile phases mix and the inlet of the column) for optimal separation by using smaller diameter connecting tubing around the injector. UHPLC dramatically reduces LC run cycle time (<5 min with peak widths of <10 s) and reduces analysis costs by decreasing analysis times and solvent consumption [2]. Further, UHPLC improves peak shape, signal-to-noise ratio of chromatographic peaks, and LC-MS method sensitivity. However, UHPLC systems are fairly costly compared to their conventional HPLC counterparts and are, therefore, not as commonly employed. In relation to this, the recent development of UHPLC has meant that it still lags behind HPLC in popularity. But as time moves on and HPLC systems become obsolete, UHPLC systems will gradually replace them.

Superficially porous stationary phases or Fused-core™ particles were developed to provide UHPLC-type separations at reduced costs and to prolong the use of HPLC systems. These stationary phase particles generally consist of a homogenous porous outer shell of various functionalities (RP-C18, HILIC, etc.) that is grown on a solid silica core (Fig. 3) [23]. The uniformity of the densely populated packed particles creates an optimal bed structure that promotes high separation efficiencies by reducing the band broadening effects caused by diffusion. Since the outer shell of the

Fused-core™ particle is porous, the back pressure within the LC column and the LC system can be significantly reduced. As a result, superficially porous stationary phases are generally compatible with standard HPLC systems. The sizes of these particles can range from 1.3 to 5 μm, of which the most commonly used are the 2.6 and 2.7 μm. While the latter offer slightly reduced separation efficiencies (i.e., ~20% reduction) compared to their sub-2 μm UHPLC counterparts, the development of sub-2 μm (e.g., 1.3 and 1.7 μm) superficially porous stationary phase particles can provide equivalent separation efficiencies [23, 30, 31]. Further, reduced back pressures should prolong the lifespan of the LC column as well as the LC system. The popularity and use of superficially porous stationary phase-based columns is increasing exponentially in metabolomics applications.

LC sample injection tends to be less problematic than GC, since there is no need to evaporate the sample before it reaches the column. The injector mechanism is generally comprised of an injection loop with a fixed volume (e.g., 1–100 μL) and operates as follows: (1) The appropriate volume of sample is loaded on to the loop, (2) a rotating valve diverts the mobile phase to the loop, and (3) the sample is transferred from the loop and delivered to the column by the mobile phase (Fig. 4) [23]. The injection loop is subsequently washed by the continuous flow of the mobile phase

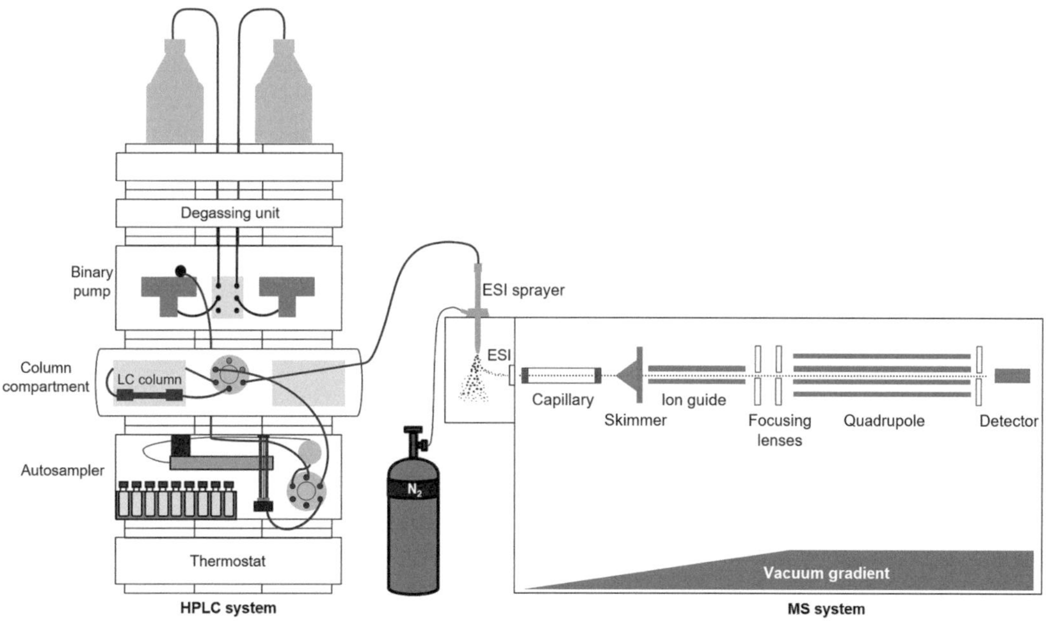

Fig. 4 The LC-MS system. Mobile phase delivered from the binary pump flows through the LC column en route to the MS system following sample injection. Electrospray ionization (ESI) facilitates the transfer of ionized gas phase analytes from the LC column effluent to the MS. Ions are transmitted to the mass analyzer where they are separated and then detected

until the next injection is ready to proceed. The sample medium is normally composed of a weaker solvent than the eluent to ensure minimal sample broadening [2].

3.3 GC Separation

GC separates volatile analytes by partitioning them between the gas phase and the internal coating in the column. Most metabolomics analyses are generally carried out by capillary GC columns, as they provide increased sensitivity and very high separation efficiencies. The internal coating (which can also be described as the stationary phase) is chemically bonded to the inner surface of the GC column, and has a general thickness of 0.1–5 μm [32]. Of the limited number of stationary phases available for GC, the most commonly used are based on methyl-silicon (e.g., dimethyl-polysiloxanes) and polyethylene glycol polymer ("Carbowax") [2, 32]. Analytes are typically carried through the GC column by helium gas as it is inert and compatible with most detectors. The inability of helium to react with sample components yields reliable results and a robust analytical technique. The carrier gas is delivered at very high purity (99.99995%) which is maintained by precolumn purifiers [32]. Impurities from hydrocarbons can lead to high background, whereas oxygen and water can lead to column degradation. Water appears to affect columns containing polyethylene glycol (PEG) chemically bonded phases but not polysiloxane stationary phases. Copper and stainless-steel carrier gas connection tubing and fittings mitigate the effects of oxygen by significantly reducing its diffusion [32].

Since GC is limited to hydrophobic, nonpolar, volatile analytes that are <600 Da, polar, highly charged, hydrophilic metabolites require chemical derivatization to induce volatility and thermal stability [2]. This can be brought about by alkylation, acylation, and silylation of -COOH, -OH, -NH, -$CONR_2$, RCHO, enolizable ROR' and -SH functional groups. Compounds containing these groups can easily form hydrogen bonds, which reduces their volatility and thermal stability. These functional groups can also lead to the adsorption of metabolites to a column's stationary phase, which could negatively impact its performance and longevity. Derivatizing agents such as trimethylsilyl (TMS), replace active hydrogens on apolar groups (e.g., amines, amides, alcohols, thiols, enolizable ketones and aldehydes, and acids) or through chemical interactions with polar functional groups [2]. However, the derivatization process can be rather laborious and time consuming [22, 32]. In addition to this, the chemically diverse nature of metabolites causes them to have different affinities for derivatization reagents [2]. This could result in inaccurate quantitation of metabolites if normalization procedures (with chemical standards) are not used. There is a lot of literature available to assist the researcher in choosing the most appropriate reagent for the biological experiment in question [32–35].

The split–splitless interface is the most common way of introducing a sample to GC column. It consists of a programed temperature vaporizer injection system that introduces the liquid sample through a gas tight neoprene or silicone rubber septum via a hypodermic needle and syringe and volatilizes it in a heated chamber consisting of a glass liner before it enters the column [32]. The injection port is maintained at a high temperature (e.g., 250 °C) to immediately vaporize the sample. In the splitless mode, all of the sample in the liner is transferred to the column by the carrier gas [32], whereas only a small fraction of the sample is transferred to the column in the split mode. The splitless and split modes are used for trace analysis experiments and to avoid sample overloading, respectively. Once injected, the sample is rapidly heated and gaseous analytes emerge, forming a sample plug. As analytes are carried through the GC column, they interact with the internal coating in the column, which provides retention. Further, as the GC oven temperature is increased (often referred to as being ramped), volatile analytes partition in favor of the carrier gas phase and eventually elute from the column; the more volatile the analyte, the earlier it will elute. Thus, the steepness and shallowness of the GC temperature gradient play a major role in determining the overall resolution and throughput of a GC method. Other factors affecting GC performance are carrier gas flow rate, column internal coating material type and thickness, column length, column internal diameter, and sample injection volume. By increasing the column length or film thickness of the stationary phase on the inner surface of the GC column, better resolution can be achieved at the expense of lower throughput. While reduced internal column diameters improve analyte separation, as with most separation techniques, it limits the amount of sample that can be loaded. In general, a column internal diameter of 0.25 μm is used for the vast majority of conventional metabolite analyses, whereas 0.1 μm is used for the second-dimension column in 2D GCxGC [2]. A typical column length used in most GC analyses is around 30 m for conventional one-dimensional (1D) GC experiments and 1–2 m for the second-dimension column in 2D GCxGC experiments. Regarding the latter, after analytes are separated and subsequently eluted from the 1D column, they are focused via liquid N_2 or a CO_2-cooled cryojet and confined to timed segments before their introduction to the 2D column as narrow bands [32]. Elution from the 2D column yields very narrow and sharp analyte peaks with extremely short separation times. Therefore, separation occurs in the first dimension, whereas the second dimension sharpens and focuses the peak to provide highly efficient chromatography. At present, 2D GCxGC is one of the most highly efficient separation techniques used for metabolomics experiments. It should be noted, however, that most analytical laboratories still use 1D GC systems for their simplicity and ease of use. Figure 5 shows an illustration of a typical 1D GC-MS system.

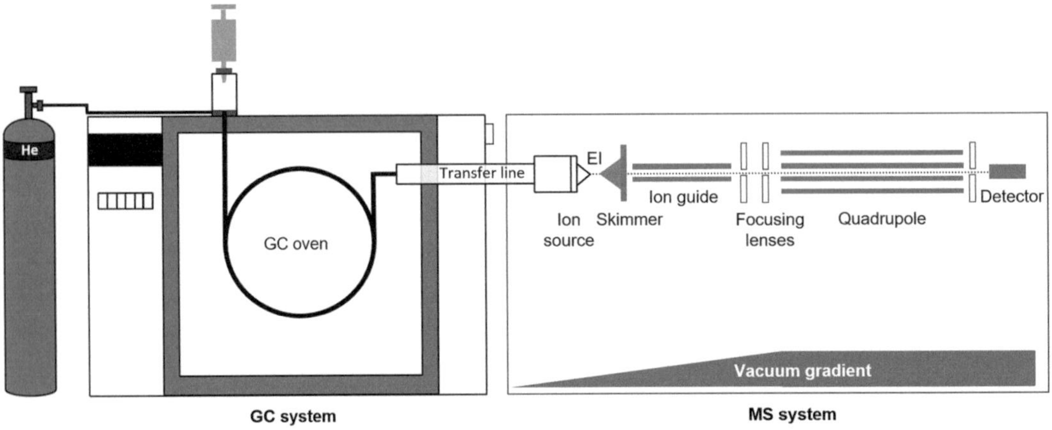

Fig. 5 The GC-MS system. After samples are injected, the carrier gas and temperature gradient (controlled by the column oven) cause volatile analytes to transit the GC column en route to the MS system. Once gaseous analytes leave the GC column, they are ionized via electron ionization and transmitted from the ion source to the mass analyzer, where they are separated and subsequently detected

3.4 CE Separation

CE separation is heavily influenced by the chemistry of the inner surface of the FSC and the manipulation of this surface can drastically alter the electrophoresis process. The easiest way of achieving the latter is by varying the pH of the liquid medium. When the pH is above 3, silanol (Si-OH) groups on the inner surface of the FSC are transformed to negatively charged silanoate (Si-O^-) groups [2, 21]. Cations from the liquid medium adhere to the Si-O^- groups, forming a fixed layer. However, the number of cations is not sufficient to neutralize all available Si-O^- groups and so a second outer layer of cations is formed. Since this layer of cations is further away from the Si-O^- groups and, therefore, not as strongly bound to the capillary wall, it is fairly fluid and often referred to as the mobile layer [21]. Cations from this layer are generally solvated and, upon the application of an electric field, migrate toward the cathode, carrying with them the bulk of the liquid medium, including slow migrating analytes. This movement of the mobile layer of cations is called the electroosmotic flow (EOF) (Fig. 6). Thus, the observed electrophoretic mobility (μOBS) is the combination of electrophoretic mobility of that ion (μEP) and the electroosmotic mobility (μEOF). That is, $\mu OBS = \mu EP + \mu EOF$ [21].

The EOF is slower at the FSC wall due to fictional drag and faster through its center. As a result, the EOF has a rather flat flow profile when compared to the laminar flow of HPLC. During the EOF, analytes migrate at the same velocity irrespective of their cross-sectional position and elute as thin bands that give rise to narrow electrophoretic peaks and highly efficient separations [21]. Whereas the pressure driven laminar flow of HPLC can produce rather broad peaks (Fig. 7). The EOF allows both anions

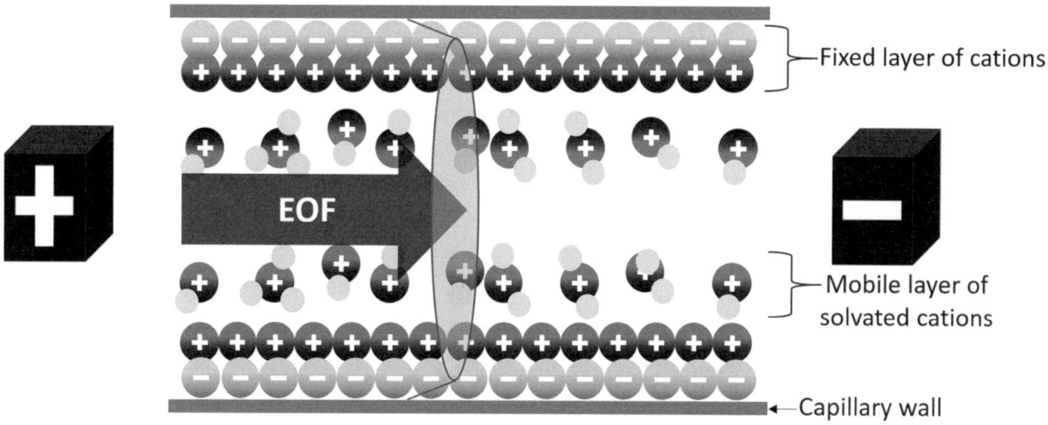

Fig. 6 Electroosmotic flow (EOF). Upon the application of an electric field, the solvated mobile layer of cations moves toward the cathode causing the bulk of the liquid medium to move in that direction

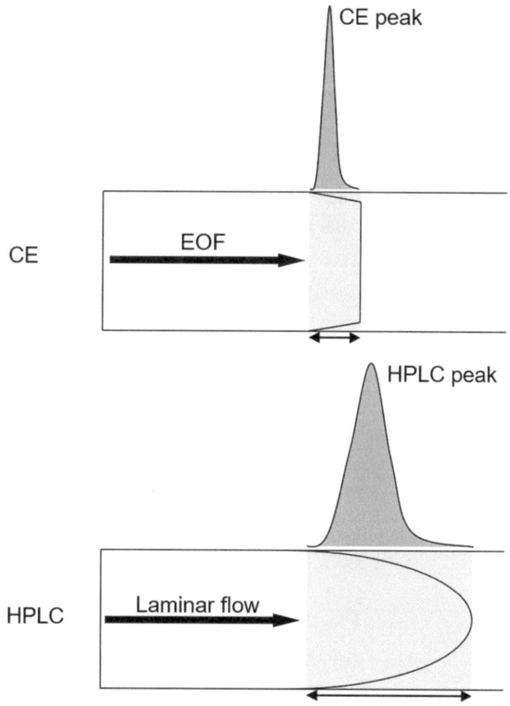

Fig. 7 EOF versus HPLC flow profile. The EOF produces a narrow flat flow profile, while HPLC produces a broad laminar flow profile

and cations can be separated in a single run cycle, but can easily be affected by changes to the liquid medium's composition, pH, ionic strength, temperature, and voltage [21]. When CE migration is in the direction of the cathode, cations move faster than the EOF as they are attracted to the cathode, neutrals move at the same rate, and anions move slower as they are attracted to the anode and are eventually carried to the cathode by the EOF [21]. The opposite is true when CE migration is in the direction of the anode.

The most popular mode of CE is capillary zone electrophoresis (CZE). During the CZE process, analytes migrate in tightly packed zones in the order of electrophoretic mobility toward electrodes of opposing potential. Ergo, cations migrate toward the negatively charged electrode (the cathode) and, conversely, anions migrate toward the positively charged electrode (the anode) [2, 21]. While this may appear to be a very simplistic view of CE migration, as we delve deeper we find that it can be a much more complex process. The CE techniques routinely employed for metabolomics studies are sample stacking and pressure-assisted capillary electrophoresis (PACE).

Sample stacking of cation analytes can be used to separate a wide range of cationic metabolites including amino acids, polyamines and nucleotide bases, and is arguably the most robust CE method used in metabolomics studies. This technique requires that the sample be dissolved in a dilute medium (i.e., in water or in one-tenth of the run electrolyte) [2, 21]. This results in the sample plug having a significant reduction in ionic strength and conductivity, leading to a higher electric field strength generated within this region than the rest of the liquid medium. The proportionality of electric field strength to electrophoretic velocity ensures that cations migrate rapidly through the dilute sample plug and slow down at the boundary where the plug meets the rest of the liquid medium due to a reduced electric field strength [21]. The resulting narrow, stacked zones of cations then migrate toward the outlet end of the FSC under the influence of their electrophoretic mobilities. This leads to the formation of very narrow, sharp electrophoretic peaks, and improved signal-to-noise ratio and sensitivity (Fig. 8) [21, 36]. Best results are achieved when the EOF is effectively removed or brought close to zero as it has little influence on the stacking process [37]. The easiest way of achieving this is to employ $\geq$1.0 M formic acid as the run electrolyte as it provides a pH of $\leq$1.8, which is enough to neutralize the inner surface of the FSC via protonation of Si-O$^-$ groups [36]. However, efficient stacking is achieved with low sample injection volumes (<30 nL), which limits the techniques ability to preconcentrate cationic analytes.

Sample stacking is also among the most popular anionic separation techniques. It provides excellent separation for many types of anions and operates under the same guiding principles as that of cations. The only differences being the polarities of the electrodes are reversed and the FSC is treated with a cationic polymer such as polybrene, resulting in a net positive charge on the inner surface of the FSC that attracts solvated anions from the run buffer. Upon the application of an electric field, the solvated layer of anions moves toward the anode, pulling the bulk of the buffer along with it. This

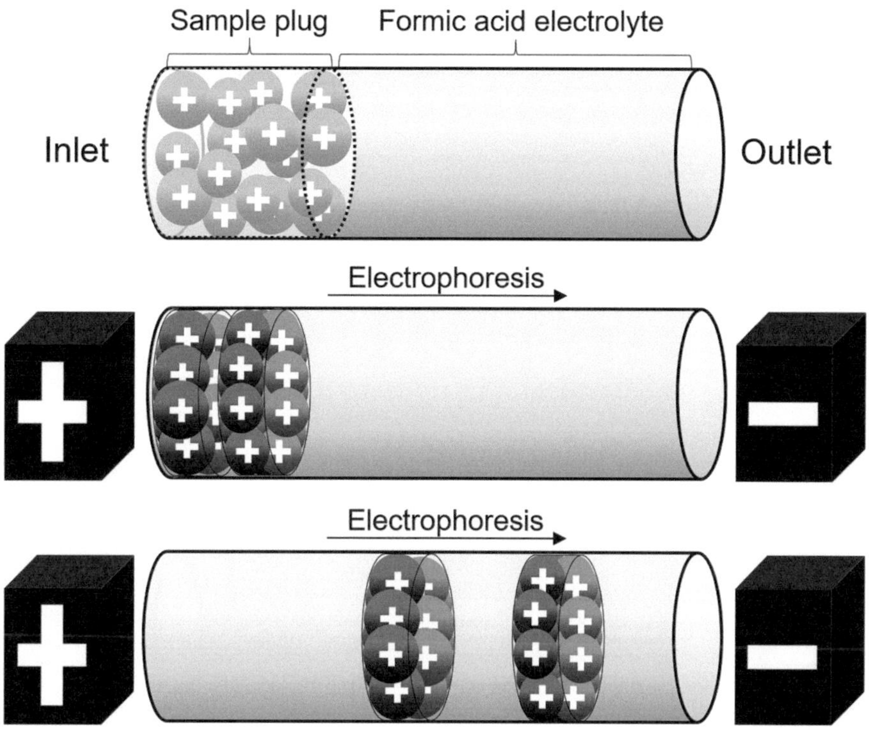

Fig. 8 Sample stacking of cations. An electric field is applied after the sample (which is in a dilute medium) is introduced hydrodynamically (under pressure), to a fused silica capillary that has been filled with the run electrolyte. The higher electric field strength generated within the sample plug than in the run buffer causes analyte ions to migrate rapidly through the dilute sample plug (due to electrophoretic velocity being proportional to electric field strength) and slow down when they reach the concentration boundary between the sample plug and the run buffer, leading to the formation of narrow, stacked zones that eventually migrate to the detector. There is little-to-no EOF due to neutralization of the silanol groups on the inner surface of the FSC by formic acid

generates a reversed EOF (Fig. 9) [21]. A drawback of this method is a reduction in reproducibility and robustness probably due to the instability of the cationic polymer compared to bare FSC. Also, there are concerns over peak tailing of several anionic metabolites (e.g., citrate) and loss of resolution of CoA compounds [38]. It is possible that these anionic compounds interact with the parts of the inner FSC wall that are not bound to the cationic polymer.

Pressure-assisted capillary electrophoresis (PACE) allows for the comprehensive analysis of a wide range of anionic metabolites (e.g., organic acids, sugar phosphates, nucleotides, and CoAs) [38, 39]. It can provide reproducible measurements as it employs unmodified bare FSC [39]. PACE utilizes a combination of EOF and an applied supplementary pressure to drive anions to the cathode in the positive mode of CE. Separation is, therefore, brought about via the electrostatic attraction each analyte has for the anode

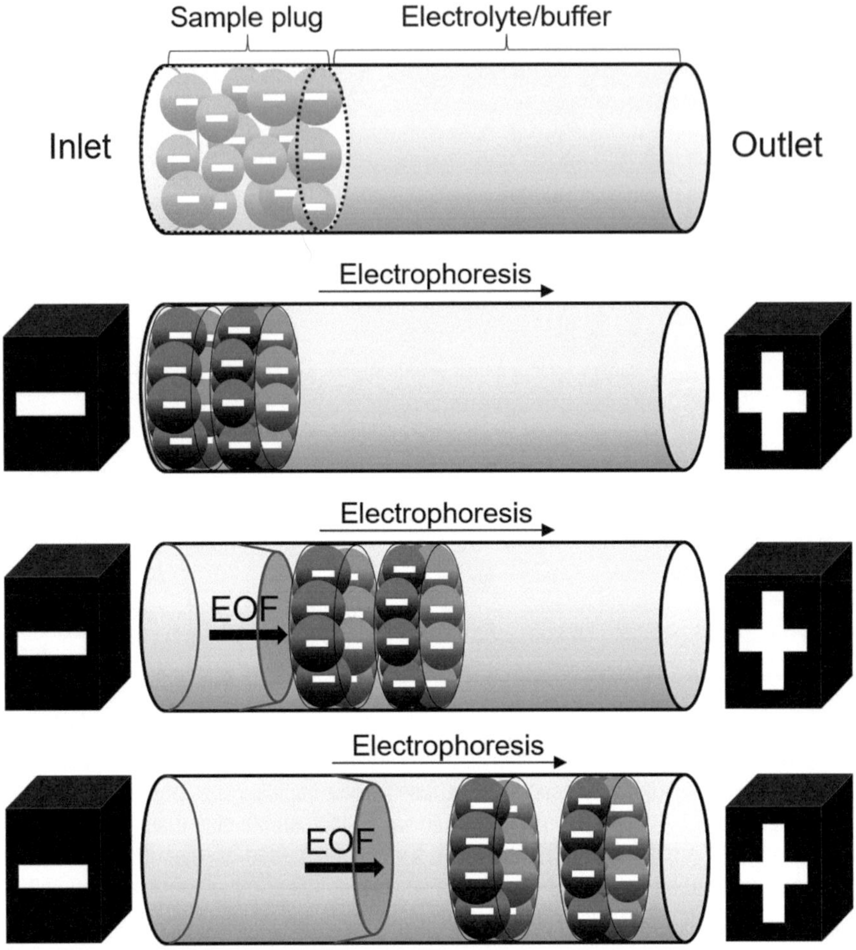

Fig. 9 Sample stacking of anions. The principle of separation is the same as that for cations. The only differences being the polarities of the electrodes are reversed and the FSC is treated with a cationic polymer. This causes a net positive charge on the inner surface of the FSC that attracts anions from the run buffer that are solvated. Therefore, after an electric field has been applied, the solvated layer of anions moves toward the anode, generating a reversed EOF in the direction of the anode. It should be noted that the EOF is constantly moving through the FSC

as well as charge-to-size ratio. Analytes that exhibit a strong negative charge will migrate slowly to the cathode and those with a weak negative charge will migrate quickly (Fig. 10). However, the disadvantage of this method is that it relies on the rather slow EOF and a supplementary pressure for analyte elution, which can result in rather long run times. Further, the supplementary pressure can change the flow profile within the FSC from flat plane to laminar, leading to broad analyte electrophoretic peaks.

Samples can either be introduced to the FSC by hydrodynamic injection (under pressure) or electrokinetic injection (under voltage) [21]. The former provides more reproducible measurements

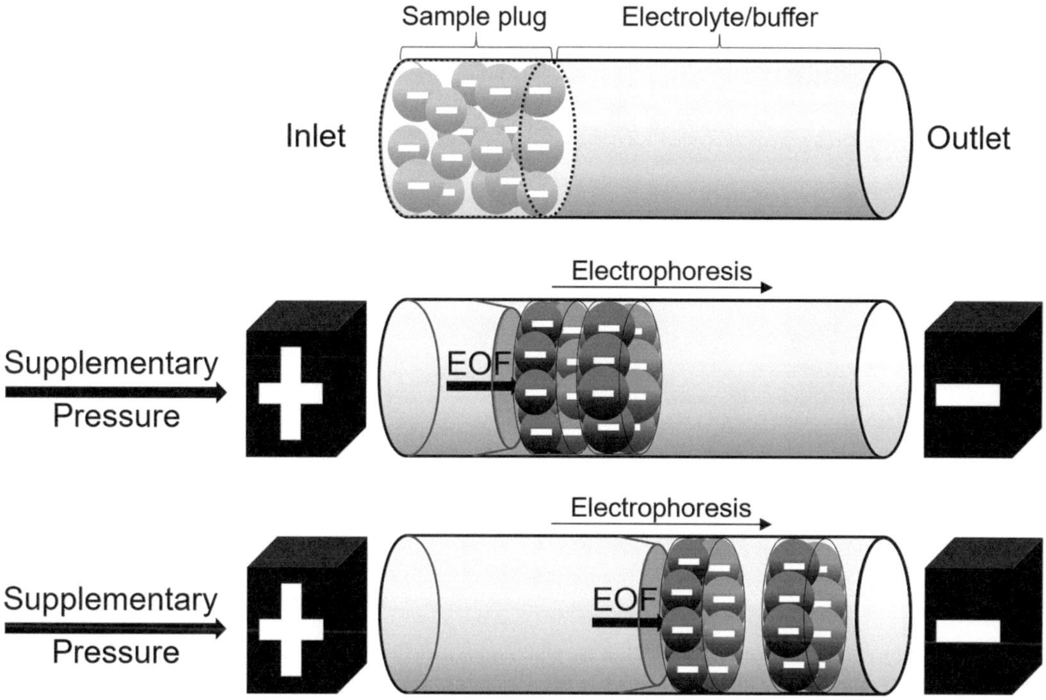

Fig. 10 Pressure-assisted capillary electrophoresis. The EOF and an applied supplementary pressure to drive anions to the cathode in the positive mode of CE, following sample introduction via hydrodynamic injection. Separation is based on electrostatic attraction each analyte has for the anode and charge-to-size ratio. It should be noted that this technique uses bare FSC and that the EOF is constantly moving through the FSC

and is the desired method of choice for most CE experiments. In contrast, electrokinetic injection, is less reproducible as it creates a sampling bias when analytes with higher electrophoretic mobilities are more readily introduced to the FSC than those with lower electrophoretic mobilities. Electrokinetic injection also causes analytes to be depleted from the sample, thereby reducing the number of times a sample can be rerun. Electrokinetic injection does, however, increase sensitivity by preconcentrating analytes in the FSC before separation.

While CE has a lot of upside, it is still relatively unpopular compared to the more established chromatography techniques (Fig. 2). When coupled to MS, the highest resolution analytical measurements can be achieved, but also the most unstable MS-based separations as well. The latter is due to the coupling of two voltage-driven techniques as opposed to a pressure–voltage approach (e.g., LC-MS), which is a more stable arrangement. In addition to this, CE requires a lot of practical experience and theoretical knowledge to achieve the best results. Fig. 11 shows an illustration of a typical CE-MS system.

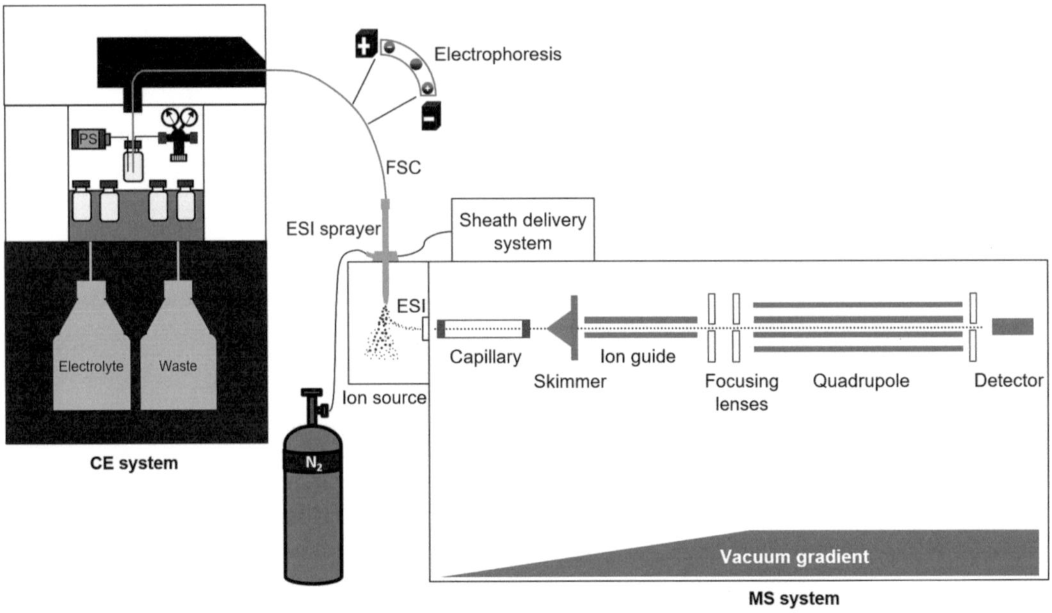

Fig. 11 CE-MS system. Analytes are separated via electrophoresis following hydrodynamic introduction to the FSC. The CE effluent and sheath liquid enter a coaxial sprayer. From there, ESI facilitates the transfer of ionized gas phase analytes to the MS. Ions are transmitted to the mass analyzer where they are separated and then detected. PS is an abbreviation for power supply

4 Mass Spectrometry

4.1 Compound Ionization

Once the separated metabolites are eluted from the chromatographic or electrophoretic separation system and delivered to the ion source of the mass spectrometer. Ionization takes place in the ion source, which is an interface between the separation device and MS system. The ion source converts analytes into gas phase ions and initiates their transfer to the MS system. Ions can be formed in a variety of ways including electron ejection, protonation, deprotonation, adduct formation, and transfer of charged species from the condensed phase [40]. Ionization methods include electron ionization (EI), electrospray ionization (ESI), chemical ionization, matrix-assisted laser desorption/ionization (MALDI), and photoionization. In this chapter we will discuss EI and ESI, the two most popular ionization techniques used for metabolomics measurements.

When choosing an ionization method, it is important to consider the physicochemical property of the analyte and internal energy (the energy contained within a system) transferred during the ionization process [40]. Fragmentation of the precursor molecule occurs when the internal energy received in the ion source is high. This is especially true for EI, which produces gas phase ions from volatile compounds as they pass through a focused beam of electrons emitted from a heated filament (i.e., tungsten or rhenium). The electron beam causes the removal of a molecular electron via electrostatic repulsion, resulting in the formation of a radical cation ion. The high kinetic energy supplied by the ionizing electrons (i.e., at 70 eV) cannot be contained within the bonds of organic molecules and, therefore, leads to extensive fragmentation of precursor ions [41]. The standardized 70 eV provides consistent fragmentation patterns across all EI-based MS instruments, resulting in standardized mass spectral libraries such as NIST. An important advantage GC has over LC and CE is that its eluted analytes are already in gaseous form and, because of this, the GC capillary outlet can be directly connected via a heated transfer line to the MS ion source [2].

The internal energy received during the ESI process, is considerably lower than EI, allowing for the measurement of intact molecular ions. Thus, ESI is generally considered to be a soft ionization technique and, as such, is suitable for labile (thermal, oxidative, etc.), nonvolatile compounds. The ESI source facilitates the transfer of gas phase ions, from a nebulized eluent to the MS, under atmospheric pressure. Nitrogen is generally used as the nebulizer gas. It is the process of nebulization that helps to create a fine spray of eluent droplets within which analyte molecules are contained. Analytes emanating from the eluent are already charged due to LC mobile phase and CE electrolyte conditions (buffer, pH, etc.). During the ESI process, an electric field is applied to the eluent passing through a narrow stainless-steel capillary inside the ESI sprayer. Analytes accumulate at the tip of the capillary and form of a Taylor cone [40], where the surface of the liquid forms a conical shape resulting from charge accumulation (due to electrostatic and hydrodynamic forces). Once the surface tension of the cone is broken, highly charged droplets are released, which become progressively smaller the farther away they are from the tip of the ESI sprayer. Upon subsequent evaporation, the charge density on the surface of these droplets increases, causing them to become unstable and release many smaller droplets. These smaller droplets accumulate similar charges, leading to a cascade of coulombic repulsions that eventually yield a single droplet containing a single ion (via the addition of adduct ions or deprotonations) and/or the emission of solvated ions from the charged droplets. These processes lead to the formation of gas phase ions after further evaporation [22, 40, 42–47]. Ions are either positively (e.g., via H^+, Na^+, K^+, and NH_4^+ cation adducts) or negatively charged (e.g., via -H deprotonations or $HCOO^-$, $CH3COO^-$, $CF3COO^-$, Cl^-, Br^- anion adducts). In the case of CE-MS, the CE system is usually connected to the MS via an ESI source by inserting the capillary outlet into the center of a coaxial sprayer. In this arrangement, the fused silica capillary (FSC) is surrounded by a narrow stainless-steel capillary, which is responsible for the delivery of a sheath liquid (e.g., methanol–water, 1:1, v/v, by an isocratic or syringe pump) to the outlet end of the FSC (Fig. 11). The sheath flow provides enough liquid at the tip of the FSC to ensure a stable electrical connection (acting as the outlet electrode) and is responsible for delivering the analytes to the MS. The stainless-steel sheath capillary is itself housed in a concentric tube, which delivers nitrogen gas to assist spray formation [48, 49]. The coaxial sheath flow interface generally exhibits poor detection limits when compared to LC-ESI-MS due to dilution of the CE eluent by the sheath liquid.

4.2 The Mass Analyzer

4.2.1 An Introduction to the Mass Analyzer

From the ion source, ions are electrostatically drawn toward the entrance of the mass spectrometer, after which they are electrically driven through a heated N_2 gas flushed sampling capillary [2]. Ions then typically pass through the central orifice of a cone shaped metal skimmer and are accelerated toward the mass analyzer via an ion guide (group of small parallel metal rods) and focusing lenses under increasing vacuum. The ion guide efficiently transmits ions above a certain mass limit via the application of a radio frequency (RF) voltage. Whereas focusing lenses, at varying degrees of voltage, are used to concentrate and improve the shape of the beam of ions moving toward the mass analyzer. Focusing lenses can also be used to extract ions from the ion source [50]. It should be noted, however, that instrument manufacturers will use different configurations of ion guides and focusing lenses, with the latter being placed anywhere along the flight path of a beam of ions before the mass analyzer. In summary, the combination of the skimmer, ion guide and focusing lenses improves the transmission of ions from the source to the mass analyzer and to the detector (Fig. 12) [51–56].

To avoid unnecessary collisions with and/or interferences from air molecules, a vacuum gradient is required to remove as much of the atmosphere as possible. This gradient utilizes less powerful roughing pumps (e.g., oil-free scroll or rotary vane, which requires oil) between the ion source and early stages of the ion guide to transition from atmospheric pressure to low vacuum. From there, more powerful turbo molecular vacuum pumps are used to create a region of very low pressure in the mass analyzer. The extent of the higher vacuum pump power will largely depend on the type of mass analyzer used. Higher resolution mass analyzers such as the orbitrap and Fourier transform ion cyclotron resonance (FT-ICR-MS) require the most powerful turbo molecular vacuum pumps, whereas quadrupoles require the least powerful.

The mass analyzer is the component of a mass spectrometer that resolves/separates rapidly moving ions based on their mass-to-charge ratios (m/z), thereby enabling the accurate measurement of the masses of ionized biomolecules within a given sample. Information gained from MS measurements can be used to characterize the mass of a charged molecule's elemental composition and

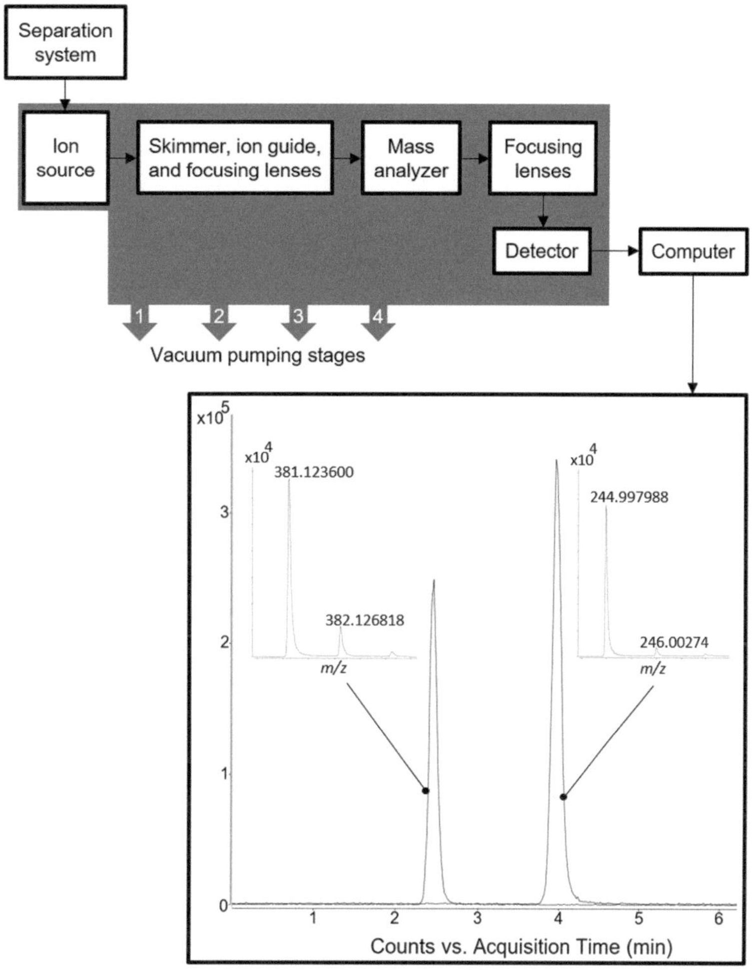

Fig. 12 MS system. As ions enter the MS system, their transmission to the mass analyzer is made more efficient as they pass through a skimmer, ion guide, and focusing lenses. After ions are separated in the mass analyzer (via their *m/z* ratios) they are transmitted through focusing lenses and then detected. The output of the detector is used to produce a plot of mass spectral peak intensity (or counts) versus *m/z*, which is then used to generate a mass chromatogram or mass electropherogram

elucidate its chemical structure. The mass of the most abundant isotope, i.e., the monoisotopic mass, is used for *m/z* measurements. Mass analyzers such as time-of-flight (TOF), orbitrap, and FT-ICR-MS can provide exact *m/z* measurements beyond the fourth decimal place primarily because they are able to distinguish mass spectral peaks of compounds with the same nominal mass (i.e., isobaric). A mass analyzer's ability to resolve two mass spectral peaks is referred to as the resolving power (RP). $RP = m/\Delta m$, where m is the mass and Δm is the difference between two adjacent mass spectral peaks (i.e., resolution, R) [57]. Consequently, mass accuracy is the difference between the theoretical *m/z* and measured *m/z* of a compound and is normally referred to as mass error (in ppm or mDa). The more accurate the *m/z* measurement, the easier the identification process becomes, as compound identity can be determined with higher degrees of confidence.

After ions have been separated by the mass analyzer, they are transferred to the electron multiplier (EM) for detection. The EM amplifies the ion signal, thereby improving MS sensitivity. The detector signal is sent to a computer that generates a mass spectrum, which is a plot of intensity (or counts) versus *m/z* (Fig. 12). A mass spectrum (at different points in time) is then used to produce a mass chromatogram or mass electropherogram, which are plots of mass spectral peak intensity versus time (the area under which can be quantified). Therefore, each data point is generated from a scan of the mass spectrum. LC and CE metabolomics experiments are generally conducted within an *m/z* range of 50–1000 since most metabolites have generally low molecular weights. GC-MS experiments, however, typically use a narrower mass range with a lower *m/z* limit as compounds are required to be volatile. Finally, an attractive feature of MS is its ability to produce molecular fragmentation, whether in-source or collision cell, to provide extra information on compound structure. This improves the selectivity and compound structural elucidation, which is especially important when identifying unknown compounds.

Of all the mass analyzers used in metabolomics studies, the quadrupole is the most robust and easiest to use and, when combined with a hexapole and an additional quadrupole, the resulting triple quadrupole (QQQ) achieves the highest sensitivities and quantitative precision of any mass spectrometer. TOF, however, provides the highest scan speeds of any mass analyzer, which is ideal for measuring efficient chromatographic and electrophoretic separations. The ion accumulation prowess of the ion trap allows this mass analyzer to attain the greatest scan sensitivities. FT-ICR-MS achieves the best mass spectral resolution and resolving power (with the orbitrap in second place in these categories, followed by TOF, and, lastly, the ion trap and quadrupole, which have similar resolving powers) since it can separate ions in the order of kilometers (i.e., the longer the distance ions travel, the better they separate from each other). While it is important to highlight the best features of these mass analyzers, it should be noted that at present the most popular mass analyzer for metabolomics work is TOF (Fig. 13) (i.e., popularity of TOF $<$ quadrupole $<$ orbitrap $<$ ion trap $<$ FT-ICR-MS). This is due to the all-round ability of this mass analyzer at a very reasonable cost. The typical cost of mass analyzers are as follows: FT-ICR-MS ($\geq$\$1000,000) $>$ Orbitrap (can be from $\sim$\$500,000 to \$ 1000,000) $>$ QTOF (typically \$500,000) $>$ QQQ $\approx$ TOF $\approx$ ion trap (can be up to \$300,000 or more, especially when considering hybrid systems) $>$ quadrupole (can be up to \$150,000).

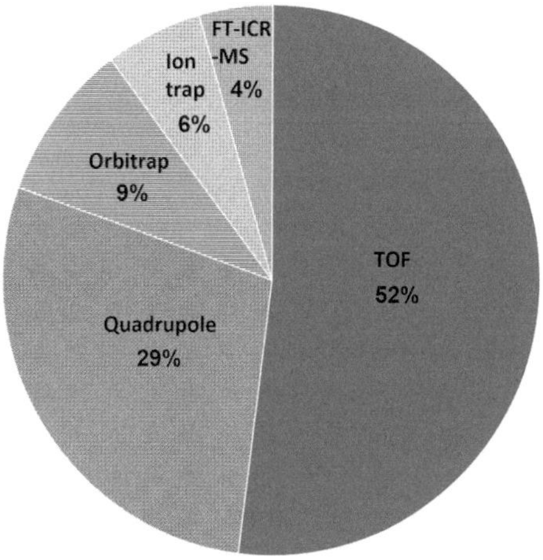

Fig. 13 Popularity of mass analyzers used in metabolomics work from 2000–2017. The popularity was based on the number of publications obtained from Web of Science. The search criteria are as follows: (1) metabolomics and time of flight and years 2000–2017, (2) metabolomics and quadrupole and years 2000–2017, (3) metabolomics and orbitrap and years 2000–2017, (4) metabolomics and ion trap and years 2000–2017, and (5) metabolomics and Fourier transform ion cyclotron resonance and years 2000–2017

4.2.2 The Quadrupole

The single quadrupole mass analyzer consists of four parallel cylindrical or hyperbolic rods equally spaced around a central axis that ions pass through [40, 51, 57]. A set of two rods are horizontally parallel to each other and the other set are vertically parallel (in a rhomboid-like configuration); with one set of rods acting as positive electrodes and the other as negative electrodes. The voltages applied to opposing sets of rods have both direct current (dc) and RF components. These voltages are increased at a constant ratio [40, 51, 57]. As positive ions pass through the pair of positive rods during the dc component of voltage, they stay in the center of the quad, due to electrostatic repulsion, and move toward the detector. When a negative RF voltage applied to these rods reaches a value greater than that of the positive dc voltage, ions then change direction and move toward the rods via electrostatic attraction. When the polarity of the RF voltage applied to these rods is changed to positive, the ions converge in the center and move through the quadrupole. Heavier ions with greater momentums toward the detector, whose trajectories are not easily perturbed, are transmitted through the quadrupole. Conversely, lighter ions collide with the rods and are not transmitted. Therefore, the pair of positive rods form a high mass filter that allows heavier ions with stable trajectories to pass through the quadrupole [40, 51, 57].

Concurrently, as positive ions move toward the pair of negative rods during the dc component of voltage, the addition of an applied

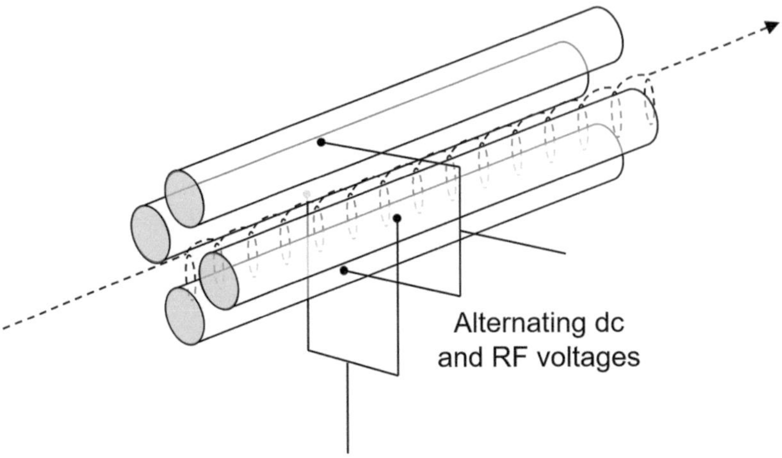

Fig. 14 The quadrupole. Electrically connected opposing sets of metallic rods have both dc and RF potentials applied to them. These voltages alternate, leading to high and low mass filters that create a region at the center of the rods where only ions of certain *m/z* ratios have stable spiral-like trajectories and are transmitted to the detector

positive RF voltage with a value greater than that of the dc voltage causes ions to converge in the center of the rods. Hence, lighter ions of molecular weights below a certain *m/z* value are transmitted through the quadrupole whereas heavier ions above a certain *m/z* value, and whose momentums cannot be perturbed, collide with the rods. Therefore, the pair of negative rods act as a low mass filter that allows lighter ions with stable trajectories to pass through the quadrupole [40, 51, 57].

When combined, both mass filters create a region at the center of the quadrupole that only ions of specific *m/z* values exhibiting stable trajectories can enter. In this region, ions move in a stable spiral-like trajectory from the inlet to the outlet end of the quadrupole (Fig. 14).

To scan a mass spectrum, both voltages are increased at the same time from zero to some maximum value while their ratio is maintained constant. A scan is, therefore, achieved in sequential steps, from the lowest to the highest *m/z* ratio. When scanning across the full *m/z* range, less time is spent on individual ions, thereby reducing signal-to-noise ratio and the sensitivity of the mass spectrometer. The scan mode does, however, provide abundant information for qualitative analysis since it covers the entire *m/z* range.

In the most sensitive operating mode of the quadrupole, selected ion monitoring (SIM), the mass analyzer is programmed to allow the passage of ions with specific *m/z* ratios (typically ± 0.5 *m/z*) through to the detector. In this way, the mass spectrometer spends more time collecting information on preselected ions rather than scanning across a wide mass range. This leads to a significant reduction in background noise and, by extension, an increase in signal-to-noise ratio and, therefore, sensitivity. Each SIM transition (i.e., precursor ion selection) can be made at a predetermined retention time. Consequently, data is only acquired when metabolites are eluted from the LC column, allowing more metabolites to be measured at higher sensitivities. SIM is ideally suited to targeted metabolite analyses.

Fragmentation patterns of metabolites can vary considerably during tandem MS (i.e., MS/MS) experiments due to the chemical diversity within the metabolome. Fragmentations are predicated on the presence of individual functional groups, bond type and position, as well as molecular structure. Quadrupole fragmentation experiments are generally conducted using a quadrupole (Q1), a hexapole (Q2), and a quadrupole (Q3) in series. The hexapole is the collision cell where molecular fragmentation occurs and has only an RF potential applied to it (which is responsible for ion transport from the inlet to the outlet end of this device). In addition to this, the hexapole provides a higher ion transmission efficiency than the quadrupole and is, therefore, used as a collision cell to minimize low mass product ion losses [40, 51, 57]. A neutral buffer gas is introduced to the hexapole to reduce (by dampening) the cross-sectional area of the beam of ions that transit the device for efficient ion transmission or for collision induced dissociation (CID), which is the most commonly used fragmentation process for quadrupole-based instruments. In CID, the precursor ion is kinetically excited (via an increased electric field potential), causing the conversion of a part of its translational energy to internal energy during collision with the buffer gas. The resulting internal energy is enough to cause bond breakage and product ion (i.e., fragment ion) formation [58]. CID fragmentation events of singly charged ions generate both charged and neutral fragments, with only the former being detected. CID generates mostly even electron species when coupled to ESI [58]. This enables the use of the nitrogen rule, which states that a protonated compound with an even number of electrons and an odd nominal mass contains an even number of nitrogen atoms, but a compound with an even nominal mass contains an odd number of nitrogen atoms [58]. For a more comprehensive understanding of the principle of CID, the reader is kindly referred to the following source of information [58].

There are five major modes of quadrupole-based MS/MS operation and in these modes Q1 and Q3 are either used for scanning or SIM: (1) For product ion scan, Q1 operates in the SIM mode by selecting the precursor ions of interest, which are fragmented in Q2 and the fragment (i.e., product) ions scanned in Q3 before detection. This mode is useful for structural elucidation studies. (2) During a precursor ion scan, Q1 scans for all the available masses, which are fragmented in Q2, from which a single product ion is selected the via SIM in Q3 before detection. Precursor ion scans are typically used for screening experiments where the identified compounds all yield the same product ion. (3) During selected reaction monitoring (SRM), Q1 operates in the SIM mode by selecting the ion(s) of interest, which is fragmented in Q2, and a product ion selected via SIM in Q3 prior to detection. (4) Multiple reaction monitoring (MRM) is similar to SRM except that multiple product ions are selected via SIM in Q3. In this way, even greater specificity is achieved than for SRM. Both SRM and MRM are ideal for quantitative metabolomics experiments as they significantly increase sensitivity (via increased signal-to-noise ratio from precursor and product ion selections) and selectivity. In dynamic MRM, each transition (i.e., precursor ion → production ion) is accompanied by a retention time. Thus, data is only acquired when metabolites are eluted from the LC column, which allows more metabolites to be measured at higher sensitivities. (5) For neutral loss experiments, Q1 and Q3 are scanned together with the latter being offset by neutral loss (e.g., loss of H_2O, CO, NH_3, etc.). Thus, after fragmentation in Q2, only product ions exhibiting specific neutral losses are detected. Neutral loss scans are typically used for screening experiments where the identified compounds all yield a loss of a neutral functional group [40, 51, 57, 58].

4.2.3 The Ion Trap

The 3D ion trap is essentially an ion storage mass analyzer device that is a closed version of the quadrupole [51]. Its major components are a central ring electrode with two convex shaped end-cap electrodes either side of it. At the center of the end-caps, are entry and exit orifices through which ions are transmitted [51, 57]. Once ions enter the ion trap, the end-caps are grounded, and a high voltage RF potential is applied to the ring electrode. This creates an oscillating potential difference between the ring and end-cap electrodes, with the resulting oscillating electric field being able to trap ions of a certain mass range. This field induces an oscillatory harmonic motion in the ions. In addition to this, helium gas removes energy from the beam of ions entering the ion trap. Together, the oscillating electric fields and helium gas focus and confine the ions to the center of the ion trap [59].

Ions in the trap are separated in two ways. Firstly, increasing the RF voltage causes the orbits of heavier ions to become stabilized while those of lighter ions become destabilized, creating a high mass filter like that of the quadrupole. Secondly, lower RF voltages stabilize the orbits of lighter ions creating a low mass filter. Thus, the m/z ratio of an ion is proportional to the applied RF voltage and a mass range is scanned as the amplitude of the RF voltage is ramped [57, 60]. To eject ions from the trap, a small RF voltage is applied to the end-caps. When the resonance frequency of the end-caps matches that of a trapped ion, the ion absorbs energy from the applied field causing its oscillatory motion to increase, eventually

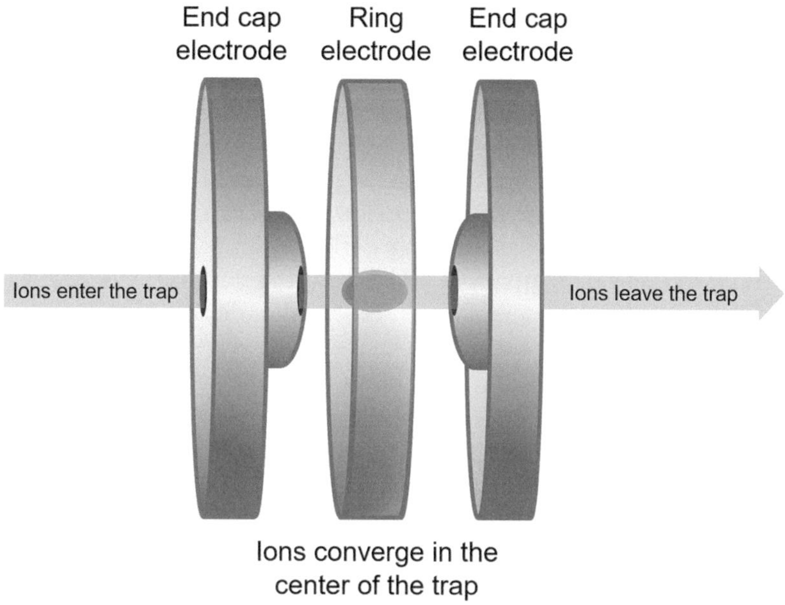

Fig. 15 The ion trap. As ions enter the mass analyzer they are subjected to oscillating RF electric fields applied to ring electrodes and helium gas cause ions to converge in the center of the trap. Here, ions rotate in specific trajectories according to their *m/z* ratios until an RF potential is applied to the end caps, causing them to rotate in higher orbits. Ions are ejected from the trap as they reach unstable orbits and are subsequently detected

leading to its displacement from the center of the trap [60]. Therefore, an ion's trajectory is increased toward the end cap electrodes until it reaches an unstable orbit and is ejected from the trap via the exit end-cap and is subsequently detected (Fig. 15) [60].

The ion trap accumulates ions of a wide mass range for an extended period of time, leading to improved signal-to-noise ratio and the production of full scan spectra at high sensitivity. However, when too many ions are close to each other, electrostatic repulsion occurs, causing space-charge effects and significant ion losses. This can lead to incorrect *m/z* assignments and reduced mass spectral resolution. To mitigate these effects, a maximum number of stored ions that will not compromise the mass analyzer's performance are allowed to enter the ion trap. This fixed number of ions is maintained by applying gain controls [51].

The ion trap can operate in a mode similar to SIM to improve the signal-to-noise ratio of a measurement [57]. In this case, an ion of a certain *m/z* ratio can be accumulated in the trap while all other ions are ejected before it is detected. The ion trap can also conduct MS/MS and MS^n (i.e., indefinite fragmentation) experiments without the need of a collision cell. In this mode of operation, the kinetic energies of selected ions are increased to cause CID with helium gas. The resulting fragment ions are then sequentially ejected from the trap and transmitted to the detector.

The linear ion trap can operate in both quadrupole mass filtering and ion trap accumulation modes. It consists of two ring electrodes at either side of a central quadrupole [59]. When ions enter the mass analyzer, they are retained and begin to accumulate as the end electrodes are turned on and a dampening gas is infused into the trap. While in the trap, a supplemental resonance excitation voltage can also be applied to induce the formation of fragments via precursor ion collision with the dampening gas [59]. Ions can then be sequentially released from the trap to the detector through the combined application of scanning dc and RF voltages on the quadrupoles and a supplemental resonance ejection voltage on the trapping electrodes. Since the linear ion trap has a greater trapping capacity than its 3D counterpart, it can accumulate many more ions, and improve the signal-to-noise ratio and overall sensitivity of MS measurements [59].

4.2.4 The Orbitrap

The orbitrap is comprised of two coaxial axisymmetric electrodes, with an inner spindle-shaped electrode contained in an outer barrel-shaped electrode. A constant electric potential is applied between the outer and inner electrodes. Since the surfaces of the electrodes are not truly parallel to each other, the electric field between them varies depending on the position on the longitudinal *z*-axis of the orbitrap (Fig. 16) [51]. The highest and lowest electric

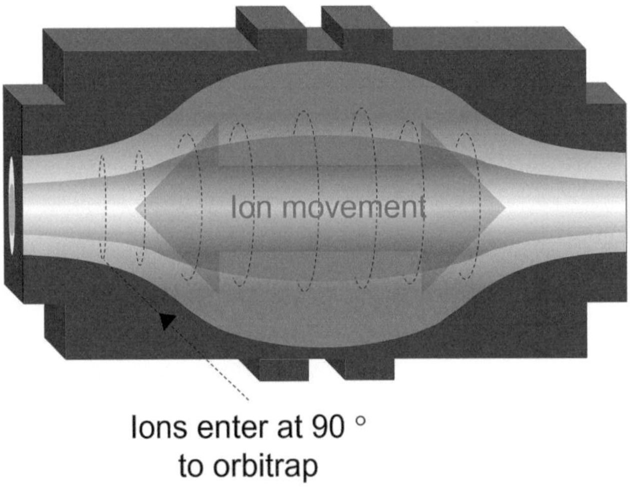

Fig. 16 The orbitrap. A constant electric potential is applied between the outer and inner electrodes. The highest and lowest electric field strengths are achieved at the ends and the center, respectively. The packet of ions is injected (at 90 ° to the center of the orbitrap) with a forward kinetic energy that matches the opposing potential energy from the radial electric field emanating from the interior of the orbitrap, causing them to follow circular orbits. They also oscillate along the central axis due to the different electric field strengths at the center and the ends of the orbitrap. The frequency of ion oscillation along the central axis is detected as a time-domain signal using image current detection that is subsequently Fourier-transformed into mass spectra

field strengths are achieved at points of least (at the ends) and greatest (at the center) separation between the two electrodes, respectively. A group of ions is injected (in tight packets) at 90° to the center of the orbitrap (on the longitudinal *z*-axis) with a forward kinetic energy that matches the opposing potential energy from the radial electric field between the inner surface of the outer electrode and outer surface of the inner electrode (the space between these two electrodes is referred to as orbitrap cavity) [51, 61]. This causes the ions to follow circular orbits (of similar radii) around the outer surface of the inner electrode. Ions also oscillate along the longitudinal *z*-axis of the inner electrode due to the different electric field strengths exhibited from the center to the ends of the orbitrap; with electric field strength increasing uniformly from the center to the ends (because the separation of electrode surfaces is greatest at the center of the orbitrap). Moreover, vectors of the electric fields become less radially parallel in the orbitrap cavity the further away they are from the center of the orbitrap. As ions migrate from the center to the ends, they decelerate and eventually lose the kinetic energy they had at the point of injection. As a result, ions stop moving and are accelerated back to the center of the orbitrap. The frequency of ion oscillation along the longitudinal *z*-axis is mass dependent. Such oscillations are detected as a time-domain signal using image current detection that is subsequently Fourier-transformed into mass spectra (i.e., frequency data to a mass spectrum).

The success of the orbitrap is in part due to efficient coordinated ion injection via the C-trap, which is essentially a curved linear ion-trap that is positioned after the ion optics and before the orbitrap. To begin with, an RF potential is applied to the C-trap, causing ions of a wide *m/z* range to be confined to its center (as they move back and forth through it). Ions subsequently lose kinetic energy through collisions with N_2 gas molecules in the C-trap, which serves to cool them down. The RF is then turned off and the DC potential turned on, causing the ions to accelerate out of the C-trap and focus into a tight packet (with a small kinetic energy distribution) before being injected into the orbitrap. The process of injection is coordinated with the ramping of the potential on the inner electrode of the orbitrap, leading to efficient ion transfer and trapping of the ion packet in the electric field between the outer and inner electrodes [51].

The orbitrap boasts resolving powers of >100,000 and achieves high mass accuracies without the need to continuously recalibrate the *m/z* axis with reference mass compounds; achieving similar performance to FT-ICR-MS. It is a compact, simple, robust mass analyzer that requires very little maintenance. The orbitrap has a higher trapping capacity than the 3D ion trap and, consequently, does not suffer from space charge effects as much. It can also perform elaborate fragmentation experiments as it is often coupled to a quadrupole and/or an ion trap, in addition to the C-trap. For example, parallel reaction monitoring, which is a new MS/MS method for the orbitrap, utilizes high resolution accurate mass spectrometry (HRAMS) provided by the orbitrap instead of a third quadrupole in an arrangement that is similar to QQQ, leading to high resolution accurate mass measurements of precursor and fragment ions. A major disadvantage of the orbitrap, however, is its rather low scan rate when compared to accurate mass TOF mass analyzers. For further information on the orbitrap, the reader is kindly referred to other source materials [51, 61–64].

4.2.5 FT-ICR-MS

FT-ICR-MS is comprised of a cubic cell that is housed in a strong magnetic field. The cube is made up of two receiver plates, two trapping plates, and two transmitter plates. As ions are pulsed into the FT-ICR-MS cell, they experience a strong magnetic field (generated from a superconducting high field magnet, typically >4.7 T), causing them to move in circular orbits (cyclotron motion) at 90° to the field. Electric potentials applied to the trapping plates, which are also at 90° to the magnetic field, keep the ions inside the cell, while RF voltages applied to the transmitter plates excite ions to higher orbits of rotation. As ions get closer to the receiver plates, the frequency of their motions is detected as an induced alternating current called the "image current" (which is a current that flows into a conductor to balance the change induced by an approaching packet of ions of opposing charge) [51, 57]. Excitation of each *m/z* is achieved when the frequency of the RF pulse matches (i.e., RF is resonant with) the cyclotron frequency of the ion packet [65]. Once the RF loses resonance, ions drop back to lower orbits and the next *m/z* ion packet is excited. In this way, a series of stepped RF frequencies is used to excite all ions within a given *m/z* range, leading to the production of a frequency versus time plot, which is Fourier-transformed into a frequency versus intensity spectrum. This is then converted into a mass versus intensity spectrum. Thus, the detected frequency is a sum of all the different frequencies of ions of various masses. Moreover, the frequency of an ion's motion is inversely proportional to its *m/z*, and the magnitude of the generated image current is related to the abundance of ions within the packet (Fig. 17) [57]. After detection, an RF voltage is applied to eject ions before the next group of ions enter the FT-ICR-MS cell. The entire trapping, excitation, and detection cycle takes <10 ms [57]. Stronger magnetic fields bring about higher frequencies of ion motion (increasing the difference in frequencies between two ions) and, therefore, higher mass spectral resolution. Like the orbitrap, the FT-ICR-MS has very low scan rates. It is also the costliest MS to purchase and maintain (partly due to the continual liquid He and N_2 cooling requirement of the FT

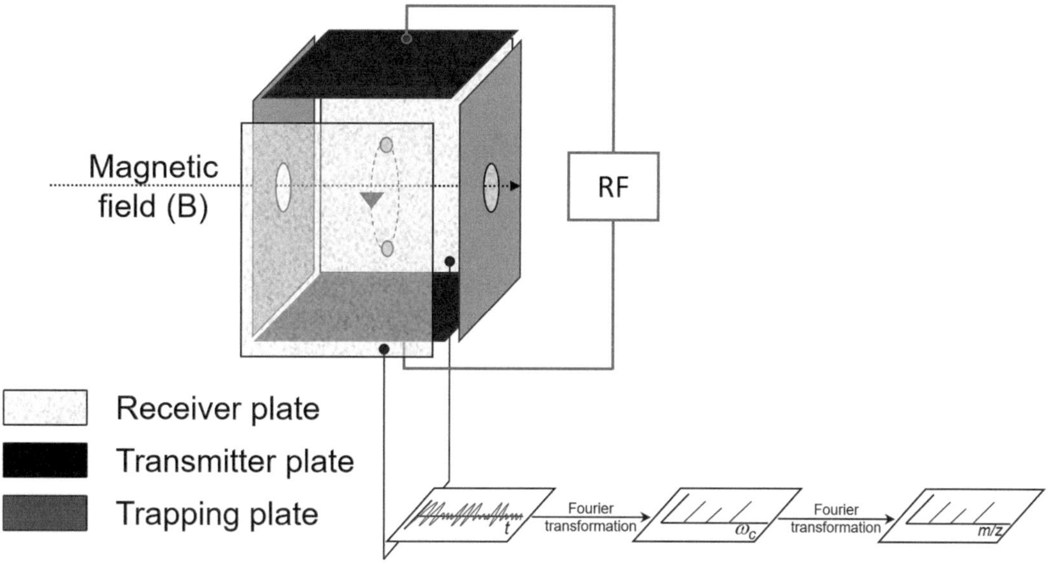

Fig. 17 FT-ICR-MS. When ions enter the FT-ICR-MS cell, they rotate as they experience a strong magnetic field that causes them to move in circular orbits at 90 ° to the field. Electric potentials applied to the trapping plates keep the ions inside the FT-ICR-MS cell while the RF voltage applied to the receiver plate plates excite them to higher orbits of rotation. As ions get closer to the receiver plates, the frequency of their motions is detected as the image current, which is Fourier-transformed into mass spectra

magnet). Consequently, TOFs and orbitraps are the most popular high resolution accurate mass analyzers used for metabolomic applications today (Fig. 13); and higher resolution orbitraps are beginning to take over the high-end HRAMS market that was once dominated by FT-ICR-MS.

MS/MS measurements can be achieved in the FT cell by the following steps: (1) retain ions of interest after ejecting all other ions via resonance excitation, (2) introduce a pulse of collision gas and apply voltage to the transmitter plate, (3) vary the amplitude of the resonance frequency of the voltage to increase ion oscillation and subsequent collisions with background gas molecules, (4) generate fragment ions from resulting collisions, and (5) apply RF voltages to the transmitter plate to excite ions to higher orbits to detect their frequencies of motion. While MS/MS experiments are very well executed in the FT cell, modern FT-ICR-MS instruments are equipped with quadrupole and/or iron traps for these purposes (with high resolution accurate mass measurements of precursor and fragment ions provided by the FT-ICR mass analyzer). For further information on FT-ICR-MS, the reader is kindly referred to other source materials [51, 57, 59, 65].

4.2.6 The TOF Mass Analyzer

Modern TOF mass analyzers consist of a long 1–2 m length flight tube under a high vacuum where ions with the same initial kinetic energy travel at different velocities that are proportional to their *m/*

z ratios and inversely proportional to their masses [2, 43]. The *m/z* ratios of ions are calculated by measuring the times they take to reach a certain distance [43]. The TOF process begins when ions are subjected to a pulsed electric field, at the entrance of the mass analyzer, causing them to accelerate through the flight tube. Ions that are lighter and multiply charged travel faster than their heavier singly charged counterparts; and after they have traveled a certain distance, they enter a region containing a homogenous electric field, known as the reflectron [51]. Here, ions are repelled toward the detector. Energetic ions with higher forward kinetic energies penetrate the retarding reflectron region more deeply and, once reflected, arrive at the detector at the same time as ions of lower initial kinetic energies of the same *m/z* (Fig. 18) [43, 57, 59]. Thus, the reflectron region minimizes variations (from varying spatial and

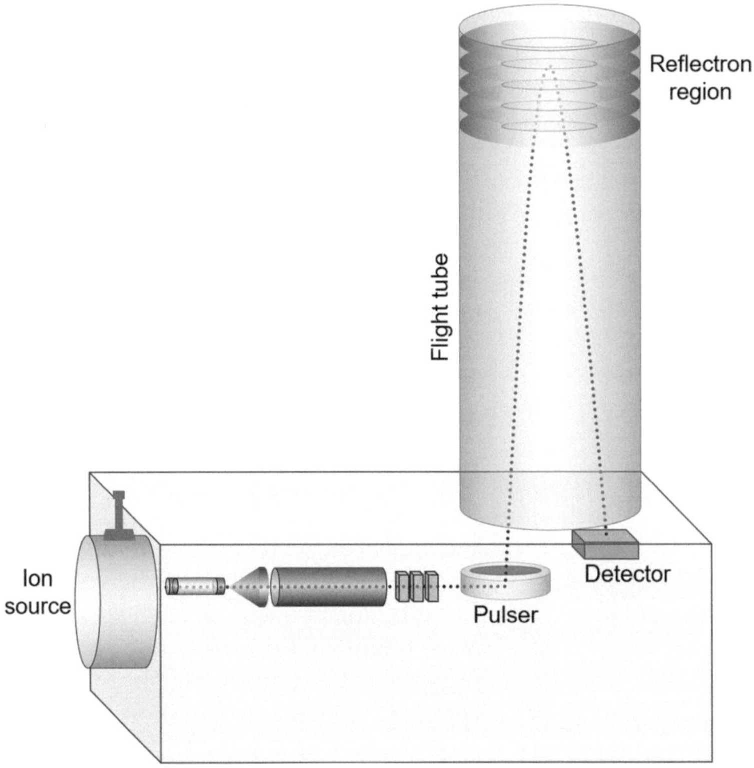

Fig. 18 The TOF mass analyzer. As ions enter the base of the flight tube, they are accelerated upwards (at the same kinetic energy) by a pulsed electric field. As ions reach the Reflectron region, they experience a homogenous electric field that repels them based on their forward kinetic energies. This adjusts any variations in flight time (i.e., from within the same molecules), allowing the reflected ions to arrive at the detector at the same time. The *m/z* ratios of ions are calculated by measuring the times they take to reach a certain distance, with lighter multiply charged ions reaching the detector before their heavier singly charged counterparts

kinetic energy distributions) in the flight times of ions, thereby improving mass resolution and accuracy [43]. The reflectron doubles the overall flight path of ions, giving them more time to separate and increasing the resolution of the TOF mass analyzer [43].

The movement of ions from the pulser region to the detector that produces a mass spectrum is referred to as a transient [43, 57, 59]. Each TOF-MS measurement (i.e., a data point) is a summation of thousands of transients. The more the transients that are summed, the more accurate the measurement and the greater the signal-to-noise ratio. TOF-MS produces the highest scan rates of any mass spectrometer, since ion flight times are very short (μs) and many thousands of transients take place per measurement. In fact, the inability of current detection/data systems to capture all transient information puts a limit on TOF-MS scan rates.

A major disadvantage of the TOF mass analyzer is that the flight tube is susceptible to unwanted expansion and contraction upon environmental temperature variations. Consequently, ion flight times can vary, which often leads to reduced mass accuracy. Modern TOF mass analyzers try to mitigate this issue in two ways. Firstly, the flight tube is constructed from metal alloys with low thermal expansion coefficients to reduce the physical effects of environmental temperature change. Secondly, reference compounds are continuously infused into the MS to recalibrate the TOF-MS *m/z* axis after a certain number of mass spectra have been averaged. Taken together, these approaches improve mass accuracy and compound identification.

The high acquisition rate, high sensitivity, high mass resolution and mass accuracy of TOF make it ideally suited to metabolomics analyses. When combined with either quadrupole or ion trap mass analyzers, the resulting quadrupole time-of-flight mass spectrometer (QTOF-MS) or ion trap time-of-flight mass spectrometer (ITTOF-MS) can conduct fragmentation experiments with accurate mass measurements, further improving the structural elucidation capabilities of TOF. For further information on TOF-MS, the reader is kindly referred to other source materials [40, 51, 59].

4.3 Detectors

The EM detects the presence of ion signals emerging from the mass analyzer. As ions strike the surface of the EM, they cause secondary emissions of electrons to be released from atoms in the surface layer [66], thereby amplifying the MS signal. Such amplifications can be $>10^6$. The number of secondary electrons released depends on the type of incident ion, its energy (smaller ions with higher kinetic energies and velocities elicit greater detector responses), and the EM surface material [66]. The EM is composed of a dynode, i.e., an electrode that permits secondary electron emissions under vacuum. The most common EM forms are the discrete-dynode (DD) and the continuous dynode (CD, also known as the channel electron

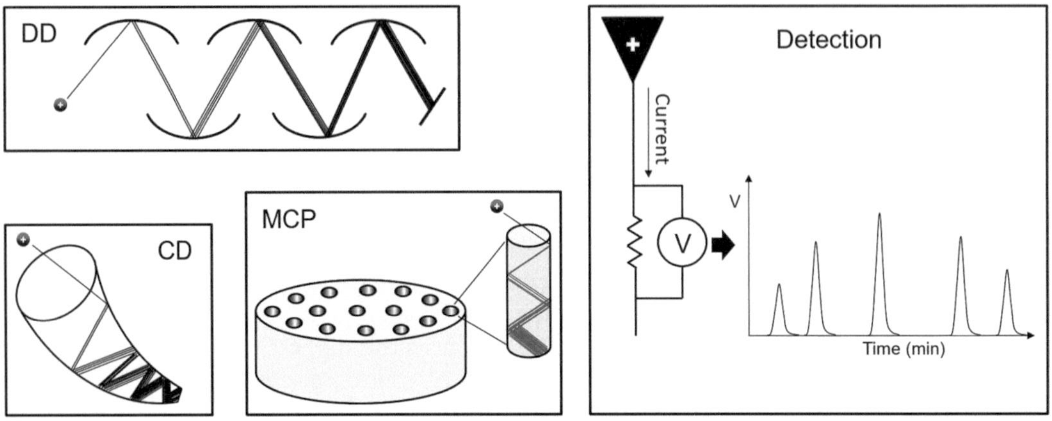

Fig. 19 Electron multiplier (EM) detectors. The discrete dynode (DD) contains a series of dynodes, with the first being a conversion dynode that converts ions to electrons. The resulting secondary electrons lead to a cascade of electron emissions, significantly amplifying the MS signal. When ions strike the inner surface of the horn-shaped continuous dynode (CD) secondary electrons are emitted, leading to a cascade of electron emissions with each successive strike, thereby amplifying the MS signal. As ions strike the interior of an MCP channel at an angle that is slightly off-axis, electrons are emitted in a similar way. The summation of electron emissions from all channels greatly multiplies the MS signal. During the detection phase, electrons collide with a metal anode, leading to a measurable current that is recorded

multiplier). The DD contains between 12 and 24 dynodes in series [66], with each successive dynode having a higher electric potential than its predecessor. Thus, when a beam of positive incident ions from the mass analyzer is focused on to a conversion dynode (that converts ions to electrons), secondary electrons are emitted, which are electrostatically drawn and accelerated to a second dynode causing the release of even more secondary electrons. The number of secondary electron emissions increases with each successive dynode [51] and the emitted electrons are eventually collected by the anode where current is measured (Fig. 19) [22]. The current is used to generate a plot of intensity (e.g., voltage) versus time. DD technology allows dynodes to incorporate different surface materials to achieve optimal performance and improve operational lifetime [66]. The wide circular entrance of the curved horn-shaped CD is adjusted at an angle to ensure that the beam of ions from the mass analyzer is readily captured and focused on its inner surface, where electrons are emitted and, subsequently, multiplied after each successive hit. The multiplication increases exponentially (due to a greater number of electron surface hits) toward the end of the CD, as the space between the inner surface of the walls becomes narrower. From there, electrons are directed to the anode for collection (Fig. 19). The curved design of the CD prevents positive ions from causing erroneous signals (due to secondary ionization of residual gas molecules). With prolonged usage, however, the surfaces of the dynodes become contaminated, resulting in decreased secondary electron emission and multiplier gain (i.e., amplification factor). As a result, the high voltage applied to electron multipliers is often increased over time to maintain the requisite gain. High-energy conversion dynodes (HED) can be used to increase sensitivity (which is especially important for high mass ions). Electrons generated from the HED are then directed to a detector for further amplification. HEDs, however, shorten the lifetime of the EMs. For the detection of negatively charged ions, a conversion dynode with a positive potential is normally required before the detector. As negative ions hit the surface of the former, newly liberated positive ions are accelerated into the EM, leading to signal amplification [22].

Microchannel plate (MCP) based detectors are popular among TOF-MS systems. The MCP detector consists of a thin resistive plate containing many microscopic cylindrical holes (typically millions of 10 μm channels) [43, 51, 67]. The wide planar area allows for the detection of many ions at the same time, which can improve detection limits and provide spatial resolution (as a few channels will result in the detection of a single ion) [51]. The MCP's working principle is as follows: Ions of sufficient energy strike the interior of the channels at an angle that is slightly off-axis, causing emissions of electrons (Fig. 19). Therefore, a single channel acts as an EM. MCPs can be coupled to photomultipliers, where electrons leaving the MCP are directed to a scintillator leading to the emission of photons that are concentrated via optical lenses onto a photomultiplier tube (PMT) [51]. The PMT increases the number of photons emitted and uses them to generate an electric signal (with the number of photons produced being proportional to the resulting electric signal). The scintillator provides good sensitivity, long lifetime, and robust detection. Modern MCPs can consist of a dual plate arrangement. The use of two MCPs tilted at an angle to each other increases the overall gain, while reducing the possibility of ion feedback [51, 68]. The gap between the MCPs can also increase the spatial range of emissions across multiple channels. Electrons leaving the back MCP can be accelerated to a metal anode where they collide, leading to a measurable current that is recorded.

The ion signal produced from the detector (e.g., the input voltage) is converted into a digital measurement (i.e., numeric), which is recorded and then interpreted by the data system. The ion signal is normally digitized by either time-to-digital conversion (TDC) or analog-to-digital conversion (ADC) [43]. TDC converts intermittent ion signal pulses into a digital representation of their time indices and does not record the intensity (height or area) of the ion signal. TDC triggers when the ion signal reaches a certain threshold and, therefore, provides the same response irrespective of whether one or more ions have contributed to the signal. Further, because its peak widths are independent of the input signal pulse width, it can achieve higher mass resolution at low *m/z*. MS peak heights are obtained by averaging. TDC is normally used in experiments that yield low ion currents [43]. ADC records the time and amplitude of ion signals and works by converting a continuous input of ion signals into digital values. The ion signals are sampled at very short intervals (e.g., ≤ns, which translates to frequency, i.e., cycles per second, in gigahertz (GHz)), with the output signal intensity being translated into a digital value during each sampling event. ADC begins to convert the signal arriving from the detector as soon as the acquisition system initiates the detection process. System noise can be reduced by setting a value close to zero as a threshold that represents no ion signal [51, 68]. Under this threshold, no signal is recorded. The advantage of ADC is its ability to record multiple ion signals from ions of the same mass arriving at the detector; with each successive packet of ions adding to the previous values in memory [43]. Such an approach provides an accurate representation of the detector output signal regardless of the intensity of the input ion current. This results in a wider dynamic range.

It should, however, be noted that for mass analyzers that generate image currents such as FT-ICR, ions do not collide with the detector but merely pass near detector plates for indirect detection. The frequency of ion motion that generates an image current is responsible for detection [51]. In this regard, the FT-ICR is the detector. The signal-to-noise ratio of FT-ICR-MS measurements is improved by averaging many cycles before transforming and storing data [57].

4.4 Emerging New Technologies

Recently, Link et al. demonstrated real-time metabolite profiling of microbial metabolism by directly injecting bacterial and yeast cells (with organic mobile phase) into a HRAMS system [69]. In this work, approximately 300 metabolites were profiled at 15–30 s intervals over the course of several hours. The technique was able to observe subtle shifts in metabolism such as the discovery of switch-like feedback inhibition in amino acid biosynthesis [69]. Emerging new technologies such as microfluidics also offer the possibility of real-time metabolomic measurements. A microfluidic-based inline small molecule extraction system allowed continuous metabolite profiling of biological systems using HRAMS [70]. This technology delivers high-throughput screening by providing an almost instantaneous profile of microbial metabolism, enabling researchers to monitor dynamic metabolic processes [71]. By directly interfacing microbial culture with MS, these real-time metabolomics systems eliminate the need for laborious sample preparation steps as they seamlessly transition from sampling, to metabolite extraction, and MS detection [70]. The lack of separation significantly reduces the quantitative capability of these techniques (as isomers are not easily resolved) and makes them exclusive to expensive HRAMS systems. However, given time and optimization, these technologies could be routinely applied to microbiological screening investigations.

Mass spectrometry imaging (MSI) allows the visualization of spatially distributed metabolites and their abundances in biological systems. MSI can directly analyze one dimensional metabolite data, via the exact *m/z* and/or MS/MS spectrum, without extraction and separation processes [72]. MSI enables the precise location of metabolites (amino acids, lipids, carbohydrates, etc. [71]) on specific regions and compartments, permitting the molecular mapping of biological surfaces. Much of this work has been pioneered on mammalian and plant tissue studies via MALDI [72–74]. MSI can be used to characterize biofilms, which are crucial to microbial survival since most microorganisms reside in them for nutritional benefits and protection from other organisms and environmental change [71, 75]. However, like real-time metabolomics, the lack of separation in MSI reduces its ability to quantify metabolites and resolve isomers. Furthermore, improvements in ionization efficiency and the use of matrices that reduce matrix interference are necessary to provide a wide coverage of metabolites [72]. Despite these draw backs, MSI is still a valuable diagnostic tool.

Interestingly, the combination of microfluidics, MSI (e.g., MALDI), and other detection methods (e.g., UV/vis) can be used for single cell metabolic profiling and imaging studies [76]. Recently, three dimensional secondary ion mass spectrometry imaging was used to investigate single cellular metabolite uptake [77]. These technologies have the potential to provide information on the metabolism of individual cells and cell-cell interactions and offer levels of specificity that have yet to be achieved in many microbiological applications. Single cell metabolic profiling would be most advantageous to microbial community studies and strain optimization.

The array-based technique nanostructure-initiator MS (NIMS) is a tool for spatially defined mass analysis and uses initiator molecules that are embedded in nanostructured surfaces to ionize adsorbed metabolites via MALDI [78–80]. NIMS was used to image acoustically printed bacteria colonies [78], with identified ions shown to localized to different regions between and within colonies, as well as to the agar support [78]. While NIMS could clearly be used for many applications, especially those requiring extensive screening of biological samples, its potential use in characterizing biofilm formation is very intriguing [75].

4.5 Data Analysis

Before going through the process of data analysis, we must first consider the steps that precede it. The design of an experiment can affect many components of a metabolomics study workflow. It is, therefore, important to define the type of study being conducted as this will determine the effectiveness of the experimental design. The process of data mining is generally performed with either hypothesis-driven or data-driven approaches in mind. The former focuses on answering a predetermined question (with some prior knowledge) and the latter tries to identify patterns, trends, or associations in an experiment whose observations are yet to be discovered [81]. While data-driven investigations can be used to form new hypotheses, either approach can lead to the other [81]. Minimizing variance in metabolomics experiments is especially important to the design process considering the composition of the metabolome can be rapidly altered in response to environmental or genetic perturbations. A study could be biased if these sources of variability (e.g., growth media, carbon source, genotype, etc.) are not controlled, since the metabolic changes observed can come from multiple sources of variation. Controlling sources of variation ensures that the observations made are due solely to the test parameter. Further, defining the number of samples used in a biological study guarantees that a suitable number of replicates will be used to provide an answer to a biological question with a high degree of confidence. Fewer replicates are required in situations where most sources of variability are controlled. The design of a robust scientific study leads to the accurate reporting of biological results, which should be validated in future hypothesis-testing studies.

The last hurdle the field of MS-based microbial metabolomics must overcome is achieving rapid, accurate, and comprehensive data analysis. At present, there are numerous computational tools that automate the data curation process, but one of the major challenges is to reliably detect relevant mass spectral signals in the raw data format and to correctly identify and annotate the underlying spectral features of the components of the metabolome [82]. However, the chemical complexity and diverse abundances of metabolites makes it difficult to achieve this in a swift automated manner. Consequently, researchers often have to manually curate large complex data sets. But as the performance of MS-based data acquisition systems continue to improve, along with increased computational processing power, RAM management, and precision of data analysis algorithms, we should see greater reliability in automated compound extraction, identification, and analysis processes.

In the early 2000s, attempted metabolomics studies were conducted in a more comprehensive fashion [36]. During that time, the goal of many analytical research laboratories was to achieve global metabolite profiles by trying to quantify all known metabolites in research worthy biological systems. Since metabolites can play roles in more than one biosynthetic pathway, it was thought necessary to apply metabolomics measurements to entire metabolic networks. This laborious approach, however, was not sustainable (with analytical technologies of the time or even present-day technologies) and cost-effective for the majority of government sponsored research laboratories and private companies. Additionally, some studies revealed that only a small subset of metabolomics data was useful to the biological experiments conducted [36, 84]. As a result, research laboratories began to weigh the cost versus benefit of comprehensive global metabolomics studies. To date, there are only a few nonprofit/academic laboratories (e.g., Michigan Regional Comprehensive Metabolomics Resource Core, NIH Eastern Regional Comprehensive Metabolomics Resource Core, Fiehn laboratory at University of California Davis) and private companies (e.g., Metabolon and Human Metabolome Technologies) that specialize in comprehensive metabolomics analyses. Most research laboratories have adopted untargeted metabolomics and targeted metabolomics approaches (Fig. 20). The former seeks to capture all available metabolite information from a biological sample through the comprehensive analysis of the measurable analytes [83]. The resulting metabolic features are then treated with the appropriate statistical analyses to find differences or commonalities among sample types, leading to a reduced number of relevant metabolites identified and, therefore, a reduction in the number of purchased chemical standards. The information gained could then be used to identify diagnostic markers of a particular biological state, experimental condition, etc. This approach can also be used in the discovery of high value biochemical products or intermediates. A major disadvantage of untargeted metabolomics experiments, however, is the long data extraction, curation, and analysis processes—a lot of computational processing power and RAM must be dedicated to extracting possibly hundreds of metabolic features (in a single sample) from the spectral noise before conducting statistical analyses. Untargeted metabolomics analysis workflows require data acquisition systems with excellent resolution (i.e., in both separation and MS), reproducibility (to minimize analytical variation), and high sensitivity and wide dynamic range.

Targeted metabolomics analysis measures defined groups of known chemically characterized and biochemically annotated metabolites [83]. Targeted metabolomics provides a focused strategy involving the measurement of localized metabolism, thereby only highlighting aspects of metabolomics data that are meaningful to a research study. The metabolites targeted are usually from a biosynthetic pathway and/or closely related pathways. Thus, targeted metabolomics requires the purchase of fewer chemical standards compared to comprehensive metabolomics, leading to reduced experimental costs. Targeted metabolomics requires excellent analytical reproducibility to differentiate variances among biological samples and to reduce the number of replicates. Large-scale targeted metabolite profiling studies yield rich data sets as they can provide information on biosynthetic pathways (via intermediate, product, and cofactor analysis) and surrounding metabolism. These experiments are generally performed by HRAMS (e.g., TOF or orbitrap) using accurate mass full scan measurements or MS/MS experiments if greater selectivity is needed. Small-scale targeted metabolite profiling is normally restricted to the biosynthetic pathway or a class of compounds and can therefore be used for high-throughput screening experiments. Low-resolution mass spectrometers such as QQQ (e.g., in SRM mode) are typically used for small-scale routine analyses. Since targeted metabolomics requires prior knowledge of the target metabolites, it is generally used for quantitative metabolomics (i.e., metabonomics) experiments, which are often achieved through the application of external chemical standard-based calibration curves. Quantitative metabolomics enables researchers to compare the contribution of each individual metabolite to microbial metabolism. In these experiments, internal standards (e.g., compounds of similar chemical structures to the target metabolites) are commonly employed to reduce errors in peak area measurements. In addition to this, the technique of isotope dilution mass spectrometry further reduces these errors by providing fully labeled stable isotopes as internal standards for each target metabolite [84]. Results obtained from Targeted metabolomics experiments, can be used to determine biosynthetic pathway flux and bottlenecks, metabolite composition, and intra- and extracellular metabolite dynamics [83].

Once data has been acquired from the biological sample, metabolic features can be obtained in the following way: (1) perform a baseline correction after removing the background noise (reduces baseline variability), (2) set deconvolution parameters (optimizes extraction of mass spectra from raw mass spectrum data by highlighting chemical properties of desired species), (3) extract metabolic features (extracts mass spectral features based on deconvolution parameters), (4) align mass chromatographic or electrophoretic peaks (reduces retention time drift via the use of internal standards), and (5) conduct peak normalization (reduces peak area fluctuations via the use of internal standards and/or sample concentration information such as dry cell weight, cell density, etc.) (Fig. 20).

Once extracted, metabolite data can be exported (e.g., as a .csv file) and treated with the appropriate statistical analysis method. The extracted metabolite data is commonly treated with univariate or multivariate (three or more variables) statistical analysis methods for biological significance. The former analyzes a single variable in a data set separately by exploring the range of values and the trends within them. Univariate analysis, therefore, describes the pattern of summarized metabolite feature data in response to the variable [85]. Examples of univariate analyses are analysis of variance (ANOVA) and Student's t-test [85]. Multivariate analysis

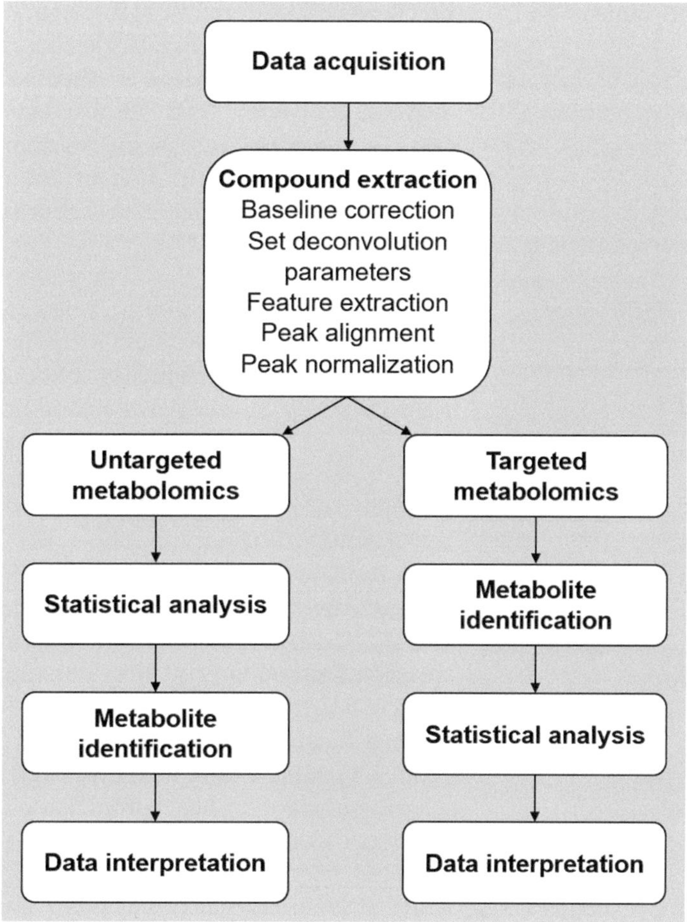

Fig. 20 Data analysis workflow. This workflow shows the typical steps for untargeted and targeted metabolomics experiments

simultaneously analyzes metabolic features and identifies relationships between them (e.g., principle component analysis, PCA, and partial least squares projection to latent structures, PLS) [86, 87]. Upon the completion of the previously described profiling experiment, accurate mass metabolic features are identified as they are searched against known MS databases and/or MS/MS spectral libraries (e.g., NIST, METLIN, Human Metabolome Database, and mzCloud). The use of retention times or retention time indices (i.e., the ratio of the analyte retention time to internal standard retention time) in metabolomics databases and spectral libraries improves the identification process (by significantly reducing the number of false-positive identifications). It should be noted that for targeted metabolomics experiments, the metabolite identification step can precede the statistical analysis step. In the interpretation phase of data analysis, the metabolite data is evaluated for biological relevance. Metabolite data can be correlated with pathway information via pathway analysis software (that utilize biological databases such as BioCyc/MetaCyc and KEGG) [88]. The use of statistical analysis methods, pathway analysis software, and knowledge of microbial physiology and metabolism, can help researchers to correctly interpret biological data (Fig. 20). For further information on data analysis, the reader is kindly referred to other source materials [71, 81, 82, 84–86, 89–91].

4.6 Applications

4.6.1 Metabolic Flux Analysis

Metabolomic technologies are being routinely applied to metabolic flux analysis (MFA) studies. MFA uses mathematical models to evaluate metabolite production and consumption in biological systems at metabolic steady state conditions [71]. Tracer-based mass isotopologue (i.e., chemical species that differ in isotopic composition) or mass isotopomer (i.e., isotopic isomers containing the same number of isotopic atoms that differ in the positions they are located) distributions allow MFA mathematical models to track the flow of mass isotope labels through sequential biochemical reactions [92]. MS/MS isotopomer analysis can provide positional ^{13}C-labeled fragments originating from a carbon back bone cleavage for more precise flux estimation [93]. Information derived from MFA studies are used to estimate intracellular fluxes by determining metabolite transport rates and balancing reaction networks [71]. ^{13}C-MFA allows fluxes through parallel pathways (e.g., glycolytic pathways) to be resolved and provides methods for studying inter-species metabolite exchange [94] and population dynamics [71] in microbial communities. At present, MFA studies are moving toward dynamic metabolic flux analysis (DMFA). DMFA is not limited to metabolic steady states as it estimates transient in vivo fluxes during the characteristic metabolic phases in fermentation experiments [95–97].

4.6.2 Systems and Synthetic Biology

Systems biology is an interdisciplinary biology-based research field that utilizes "omics" technologies and mathematical modeling to holistically characterize complex interactions within biological systems [98]. It links theory, computation, and experiment, all the while providing an iterative assessment of model parameters and variables [99, 100]. Integrated multi-omics data can be used as predictor variables for mathematical models that predict phenotypes and identify biomarkers [71, 101]. Systems biology relies on the development of analytical and computational technology to provide high quality data and modeling [71, 99, 100]. This is especially necessary for the metabolomic component of systems biology's analytical repertoire. A major focus of systems biology is the complete characterization and understanding of metabolic networks. Metabolomics data is integral to this since the metabolome, while being downstream from the proteome, can interact with and influence all preceding "omes" [102]. Metabolomics analysis can also be used to provide signaling pathway information, which can be incorporated into mathematical models, further strengthening them. The vast amount of knowledge obtained from systems biology experiments can be used in the development of synthetic biology tools, which facilitates the manipulation of complex biological systems, providing a deeper level of understanding [98].

Synthetic biology applies biology and engineering disciplines to design artificial biological systems, or redesign natural biological systems, to assemble biological components and modules for research and industrial applications centered on the environment and human health. While Genomics and transcriptomics have long been staples of synthetic biology research, the emergence of proteomics and metabolomics has illuminated this field of biology by directly correlating the outcome of genetic engineering experiments to phenotypic information. In this regard, metabolomics can be used by synthetic biologists to (1) discover, identify, and quantify engineered bioproducts, (2) test the effectiveness of industrial enzymes via kinetics studies, (3) assess the level of production of unwanted accumulated intermediates and by-products [103–106], (4) identify bottlenecks and modes of toxicity in engineered biosynthetic pathways [103–106], (5) assess the level of diversion of the primary carbon food source away from the engineered pathway (e.g., by correlating pathway information with central carbon metabolism) [106, 107], (6) assess the energetic cost of incorporating an engineered pathway (via assessment of the adenylate energy charge and reducing equivalents), (7) assess variation in engineered strains [107], and (8) assess the overall impact of genetic engineering experiments on cellular metabolism and physiology. In the biotechnology industry, information from such work can be used to improve the rational design of engineered strains for efficient conversion of carbon sources toward greater production of high-value bioproducts.

4.6.3 Environmental Microbial Communities

In recent years, there has been a growing emphasis on the study of environmental microbial communities due to their importance to biogeochemical cycles (i.e., the cycling of chemical substances through the biospheric and abiotic compartments of earth) [108, 109] and industrial applications. While microbial diversity appears to be related to ecosystem stability, very little is known, however, about the structure and functional diversity in these communities [108]. This is, in part, due to the difficulty in replicating environmental conditions in a laboratory setting, leading to recalcitrant microbial growth [110]. In addition, it is difficult for a single "omic" technique to provide all necessary information pertaining to new gene and pathway functions. Regarding the latter, the challenge for metabolomics is in the difficulty in determining the microbial origins of each metabolite, due to similarities in microbial metabolite compositions (perhaps single-cell approaches involving fluorescent activated cell sorting or microfluids can solve this problem). As a result, metabolomics is being used in the following ways: (1) to observe biological markers of metabolism or physiological states, (2) to observe total community metabolism through pooled meta-metabolomics, and (3) the use of either of the aforementioned approaches with other "omics" technologies in an integrated multi-omics approach. The employment of multi-omics technologies and establishing cultured isolates [110] in community studies is critical to obtaining an accurate depiction of the physiologies of microorganisms and the roles they play in their respective communities and the environment at large. A multi-omics approach that includes metabolomics allows us to link microbial community structure, dynamics, interactions, and function (s) [111], thereby providing answers to microbial diversity, output, resilience, and succession, as well as a deeper understanding of microbial response to abiotic and biotic stressors [112]. Knowledge from this research can be used to decode complex microbial interactions within communities and the environment and lead to the development of predictive mathematical models that address key environmental challenges.

The British Petroleum Deepwater Horizon marine oil spillage in the Gulf of Mexico is the largest of its kind in the petroleum industry and one of the worst man-made environmental disasters in history. What is fascinating about this oil spillage is the multitude of microorganisms found at the site that were implicated in the degradation of petroleum. It is known that Deep Sea sediments contain abundant and diverse microbial communities that may impact large-scale geochemical processes such as the biotransformation and degradation of petroleum from oil spills [113, 114]. In 2013, results from metagenomics and metabolomics analyses suggested that *Deltaproteobacteria* could be responsible for the degradation of petroleum following the Deepwater Horizon oil spill. This was presumably through anaerobic hydrocarbon metabolism since increased exposure to hydrocarbons enriched for *Deltaproteobacteria* at the site of the Deepwater Horizon oil spillage [114]. In 2015, Bargiela et al. [113] conducted shotgun metaproteomics and metabolomics analyses on microbial communities (archaea and bacteria) in crude oil polluted areas in the Mediterranean Sea. The study revealed that the majority of microbial metabolisms occurring in these areas were anaerobic (e.g., sulfate reducing bacteria and methanogens) and not directly related to petroleum degradation, unlike the communities found in Deepwater Horizon. These two major studies indicate that the biodegradation/biotransformation of petroleum is dependent on the microbial community present and is, therefore, site specific. In addition to the biotransformation of petroleum, integrated genomics and metabolomics, in conjunction with single cell analysis, have allowed researchers in the U.K. to evaluate the adaptation strategies of arctic snow algae by elucidating the function and ecology of green and red snow algal blooms, which play crucial roles in the maintenance of glacial ecosystems [115]. Similar approaches have also been used to evaluate harmful marine algal blooms [116]. Finally, integrated "omics" technologies (including metabolomics) have shed light on the impact that bacteriophages have on microbial communities and the environment [117]. This research could be used to identify biomarkers of the novel phage [117] and uncover mechanisms of host–bacteriophage association.

Microorganism coculture (i.e., the cultivation of two or more microorganisms under the same environmental growth condition) is being used to enhance chemical diversity for drug discovery [118]. These studies show that many biosynthetic gene clusters are silent under standard laboratory conditions. But, when grown in coculture, these cryptic biosynthetic pathways become activated, leading to the dynamic production of new bioactive metabolites [118]. Metabolomics technologies are being used to measure the production of these new chemicals, while providing information (in conjunction with other "omics" technologies) on microbial interactions. In addition to this, when microbial cocultures are used to produce high value bioproducts, they may reduce the metabolic burden (i.e., the cellular resources dedicated to foreign DNA maintenance and replication) exhibited during single culture growth, potentially increasing production. Metabolomics in combination with other "omics" technologies could be used to reveal the mechanisms behind metabolic burden, which would be of great benefit to the biotechnology industry.

4.6.4 Human Disease

Disease occurs when the normal state of a human being is impaired to such an extent that its vital functions are perturbed. But following microbial infection, only a few harmful microorganisms (e.g., <1% of bacteria) can strongly influence the development of certain human diseases [119]. The multiplication of these infectious pathogenic microbes in the human body can cause damage to an individual's vital functions or systems. Additionally, microbes may also contribute to the development of noninfectious human diseases such as cancers, diabetes, and heart disease. The economic cost of human disease can be substantial to an individual, a community, and a nation. Thus, it is imperative that resources are allocated to understanding the mechanisms behind human disease, especially in the case of microorganisms. In this regard, "omics" technologies have been heavily employed to answer these far-reaching biological questions.

Malaria is a mosquito-borne infectious disease caused by protozoans from the parasitic *Plasmodium* genus. It is estimated to give rise to >200 million cases and >400,000 deaths annually (World Health Organization [120]). In malaria research, metabolomics is being used to characterize host and *Plasmodium* parasite metabolism en route to identifying biomarkers of infection and new antimalarial compounds. Metabolomics was used to discover parasite-specific biomarkers such as 3-methylindole and succinylacetone from erythrocytes infected by the malaria parasite *Plasmodium falciparum*. Hierarchical cluster analysis of the metabolomic data was able to differentiate infected from noninfected culture supernatant [121]. Studies are currently underway to determine the presence of these and other molecules in malaria-infected humans and their associations with the parasite. A similar approach was recently used to associate parasitemia with *Plasmodium vivax* [122], another malaria parasite. Results from this study suggest that host response and/or parasite survival could be affected by metabolites involved in the degradation of heme and metabolism of certain lipids [122]. The authors postulate that their observations could potentially lead to the development of new antimalarial drugs [122].

Tuberculosis is an infectious disease caused by *Mycobacterium tuberculosis* and is generally considered to yield the highest global mortality rate for a single pathogenic bacterium. It is also the leading cause of deaths due to antimicrobial resistance in people with HIV (World Health Organization [120]). In tuberculosis research, metabolomics is being used to identify potential diagnostic biomarkers [123] and to characterize aspects of *Mycobacterium tuberculosis* metabolism [124] such as mycobacterial persistence, tuberculosis disease, drug resistance [125], drug efficacy [126], as well as drug discovery [127]. Metabolomics data also is helping researchers to understand *Mycobacterium tuberculosis* adaptation strategies [128] and host–pathogen interaction [129]. This information is allowing researchers to elucidate mechanisms of action of antituberculosis drugs, build more accurate experimental models of infection, and improve counter measures to *Mycobacterium tuberculosis* infection.

Cancer tissue is often reported to enter a different state of metabolism such as the "Warburg effect," where cancer cells rely on glycolysis for energy under aerobic or anaerobic conditions. Consequently, metabolites from the glycolytic pathways and tricarboxylic acid (TCA) cycle are being used to characterize cancerous tissue metabolism. Since it is thought that some microorganisms are responsible for certain types of cancers, metabolomic studies are being used to find diagnostic markers for the underlying mechanisms behind the microbial development of cancer. In this regard, Washio et al. [130] used metabolomics, via the measurement of glycolytic and TCA cycle metabolites, to reveal the metabolic associations between oral bacterial biofilms and oral squamous cell carcinomas [130]. Metabolomics has also linked bacterial biofilms to colon carcinogenesis via the identification of polyamine metabolites that were upregulated in host cancer and biofilm structures [131].

Microbial communities are known to exert numerous biological effects on the human gut, all of which can impact human health and disease. Meta-metabolomics, in conjunction with other "omics" technologies, is currently being used to study human gut microbiome structure, composition, and metabolism to unravel mechanisms that predispose individuals to disease [132, 133]. Metabolomics-based approaches are allowing researchers to identify the contribution of gut microbes to the host metabolome and to assess the potential of bacterial metabolites to influence disease outcomes [133]. Relationships between infection, immunity, altered metabolism, and bacterial metabolites are being used to understand the development of atherosclerosis [134]. Additionally, metabolomics is being used to determine the role of gut microbiota in obesity-related metabolic diseases such as nonalcoholic fatty liver disease, type-2 diabetes, and insulin resistance [135]. Further, researchers are now using metabolomics to identify biomarkers for early detection of diseases from dysbiosed microbiota. Metabolomics, combined with sequence-based (e.g., 16S rRNA) approaches, is also helping to decipher microbiota-mediated drug metabolism and interaction [136, 137]. This work is being used to identify drugs that are cometabolized by host and microbiota as well as the latter's influence on drug-induced toxicity and efficacy [136].

5 Conclusions

The complex nature of the metabolome has posed major challenges to metabolite sample preparation, data acquisition and analysis. Development in these key areas is crucial to providing more reliable metabolite data and advancing metabolomic centric research. For MS-based metabolomics experiments, HRAMS appears to be the way forward as it can provide both qualitative and quantitative data, with the targeted approach being favored by most research laboratories. The disadvantage of this approach, especially when low-resolution MS is employed, is the potential for significant metabolomic information to be missed. This is unlikely to occur with HRAMS, as it allows both untargeted and targeted metabolomics data to be collected simultaneously. When metabolomics is included in multi-omic studies, the combined information provides the most accurate representation of cellular metabolic activity and can be used to characterize metabolic networks in systems microbiology experiments. Metabolomics data can also be used to improve the rational design of engineered strains in synthetic biology experiments. Furthermore, work from microbial metabolomics experiments has made significant contributions to environmental and human related research. But while the scientific community is willing to put significant resources behind "omics" approaches, the quality of metabolomics data going forward will determine the role it will play in future microbiological research.

Acknowledgments

The authors would also like to acknowledge that this work was part of the DOE Joint BioEnergy Institute (http://www.jbei.org) supported by the US Department of Energy, Office of Science, Office of Biological and Environmental Research, through contract DE-AC02-05CH11231 between Lawrence Berkeley National Laboratory and the US Department of Energy.

References

1. Villas-Bôas SG, Roessner U, Hansen MAE et al (2007) Metabolome analysis: an introduction. John Wiley & Sons, Inc, Hoboken, NJ
2. Baidoo EEK, Benke PI, Keasling JD (2012) Mass spectrometry-based microbial metabolomics. In: Navid A (ed) Microbial systems biology: methods and protocols. Springer, New York, NY, pp 215–278
3. Murzin AG, Brenner SE, Hubbard T et al (1995) SCOP: a structural classification of proteins database for the Investigation of Sequences and Structures. J Mol Biol 247:536–540
4. Stryer L (1995) Biochemistry. W.H. Freeman & Company, New York, NY
5. Lodish H, Berk A, Zipursky L et al (2000) Molecular cell biology. W.H. Freeman & Company, New York
6. Pugh BF (2000) Control of gene expression through regulation of the TATA-binding protein. Gene 255:1–14
7. de KW, van DK (1992) A method for the determination of changes of glycolytic metabolites in yeast on a subsecond time scale using extraction at neutral pH. Anal Biochem 204:118–123
8. Rabinowitz JD, Kimball E (2007) Acidic acetonitrile for cellular metabolome extraction from Escherichia coli. Anal Chem 79:6167–6173
9. da Luz JA, Hans E, Zeng AP (2014) Automated fast filtration and on-filter quenching improve the intracellular metabolite analysis of microorganisms. Eng Life Sci 14:135–142
10. Kell DB, Brown M, Davey HM et al (2005) Metabolic footprinting and systems biology: the medium is the message. Nat Rev Microbiol 3(7):557–565
11. Pinu FR, Villas-Boas SG, Aggio R (2017) Analysis of intracellular metabolites from microorganisms: quenching and extraction protocols. Metabolites 7:E53
12. Breil C, Abert Vian M, Zemb T et al (2017) "Bligh and dyer" and Folch methods for solid–liquid–liquid extraction of lipids from microorganisms. Comprehension of solvatation mechanisms and towards substitution with alternative solvents. Int J Mol Sci 18:1–21
13. Bligh EG, Dyer WJ (1959) Can J Biochem Physiol 37:911–917
14. Folch J, Lees M, Sloane Stanley GH (1957) A simple method for the isolation and purification of total lipides from animal tissues. J Biol Chem 226:497–509
15. Axelsson M, Gentili F (2014) A single-step method for rapid extraction of total lipids from green microalgae. PLoS One 9:17–20
16. Oikawa A, Fujita N, Horie R et al (2011) Solid-phase extraction for metabolomic analysis of high-salinity samples by capillary electrophoresis-mass spectrometry. J Sep Sci 34:1063–1068
17. Johnson WM, Kido Soule MC, Kujawinski EB (2017) Extraction efficiency and quantification of dissolved metabolites in targeted marine metabolomics. Limnol Oceanogr Methods 15:417–428
18. Mousavi F, Bojko B, Pawliszyn J (2015) Development of high throughput 96-blade

solid phase microextraction-liquid chromatography-mass spectrometry protocol for metabolomics. Anal Chim Acta 892:95–104

19. Wang Z, Zhu H, Huang G (2017) Ion suppression effect in DESI mass spectrometry and ESI mass spectrometry. Rapid Commun Mass Spectrom 31(23):1957–1962
20. Buszewski B, Noga S (2012) Hydrophilic interaction liquid chromatography (HILIC)—a powerful separation technique. Anal Bioanal Chem 402:231–247
21. Baker DR (1995) Capillary electrophoresis. John Wiley & Sons, Inc, New York
22. Harris DC (2003) Quantitative chemical analysis. W. H. Freeman and Company, New York
23. Snyder LR, Kirkland JJ, Dolan JW (2010) Introduction to modern liquid chromatography. Wiley, Hoboken, NJ
24. Przybyciel M, Industries ES, Berlin W et al (2002) Phase collapse. ES Industries, West Berlin, NJ
25. Bajad SU, Lu W, Kimball EH et al (2006) Separation and quantitation of water soluble cellular metabolites by hydrophilic interaction chromatography-tandem mass spectrometry. J Chromatogr A 1125:76–88
26. ChromAcademy (2014) The Theory of HPLC. Chromatographic parameters. E-learning. Anal Chem Commun 1:23
27. Chawla G, Ranjan C (2016) Principle, instrumentation, and applications of UPLC: A novel technique of liquid chromatography. Open Chem J 3:1–16
28. Desai TK, Mahajan AA, Thaker A (2012) Ultra performance liquid chromatography: a step ahead to HPLC. Int J Pharm Rev Res 2:61–68
29. Lake R (2007) Easy transfer of HPLC methods to UPLC. Restek Advantage 4:10–11
30. Cunliffe JM, Maloney TD (2007) Fused-core particle technology as an alternative to sub-2-microm particles to achieve high separation efficiency with low backpressure. J Sep Sci 30:3104–3109
31. Abrahim A, Al-Sayah M, Skrdla P et al (2010) Practical comparison of 2.7 μm fused-core silica particles and porous sub-2 μm particles for fast separations in pharmaceutical process development. J Pharm Biomed Anal 51:131–137
32. Hübschmann HJ (2008) Handbook of GC/MS: fundamentals and applications. Wiley-VCH Verlag GmbH & Co. KGaA, Weinheim
33. Halket JM, Waterman D, Przyborowska AM et al (2005) Chemical derivatization and mass spectral libraries in metabolic profiling by GC/MS and LC/MS/MS. J Exp Bot 56:219–243
34. Halket J, Zaikin V (2003) Review: derivatization in mass spectrometry—1. Silylation. Eur J Mass Spectrom 9:1
35. Kanani H, Chrysanthopoulos PK, Klapa MI (2008) Standardizing GC-MS metabolomics. J Chromatogr B Anal Technol Biomed Life Sci 871:191–201
36. Soga T, Ohashi Y, Ueno Y et al (2003) Quantitative metabolome analysis using capillary electrophoresis mass spectrometry. J Proteome Res 2:488–494
37. Baidoo EEK, Benke PI, Neusüss C et al (2008) Capillary electrophoresis-Fourier transform ion cyclotron resonance mass spectrometry for the identification of cationic metabolites via a pH-mediated stacking-transient isotachophoretic method. Anal Chem 80:3112–3122
38. Soga T, Ueno Y, Naraoka H et al (2002) Pressure-assisted capillary electrophoresis electrospray ionization mass spectrometry for analysis of multivalent anions. Anal Chem 74:6224–6229
39. Harada K, Fukusaki E, Kobayashi A (2006) Pressure-assisted capillary electrophoresis mass spectrometry using combination of polarity reversion and electroosmotic flow for metabolomics anion analysis. J Biosci Bioeng 101:403–409
40. Hoffmann E d, Stroobant V (2002) Mass spectrometry: principles and applications. Wiley, Chichester
41. Gabellca V, De PE (2005) Internal energy and fragmentation of ions produced in electrospray sources. Mass Spectrom Rev 24:566–587
42. Skoog DA, Holler FJ, Nieman TA (1998) Principles of instrumental analysis. Brooks Cole, Pacific Grove, CA
43. Fjieldsted J (2011) Time-of-flight mass spectrometry technical overview this overview describes. Agilent Technologies, Santa Clara, CA
44. Stewart II (1999) Electrospray mass spectrometry: a tool for elemental speciation. Spectrochim Acta B Atom Spectrosc 54:1649–1695
45. Wilhelm O, Mädler L, Pratsinis SE (2003) Electrospray evaporation and deposition. J Aerosol Sci 34:815–836
46. Bruins a P (1998) Mechanistic aspects of electrospray ionization. J Chromatogr A 798:345–357

47. Smith JN, Flagan RC, Beauchamp JL (2002) Droplet evaporation and discharge dynamics in electrospray ionization. J Phys Chem A 106:9957–9967
48. Banks JF (1997) Review Recent advances in capillary electrophoresis/electrospray/mass spectrometry. Electrophoresis 18:2255–2266
49. Von BA, Nicholson G, Bayer E (2001) Recent advances in capillary electrophoresis/electrospray-mass spectrometry. Electrophoresis 22:1251–1266
50. Park CJ, Ahn JR (2005) A closed ion source with a cylindrical repeller for sensitivity enhancement in quadrupole mass spectrometry. Rev Sci Instrum 76:044101
51. Watson JT, Sparkman OD (2007) (First published 20 June 2008) Introduction to mass spectrometry: instrumentation, applications and strategies for data interpretation. John Wiley & Sons, Inc., Chichester. https://doi.org/10.1002/9780470516898, Print ISBN: 9780470516348, Online ISBN: 9780470516898
52. Hinterberger F (2006) Ion optics with electrostatic lenses. CAS, Cern Accel Sch Small Accel, Geneva, pp 27–44
53. Birkinshaw K, Hirst DM, Jarrold MF (1978) The focusing of an ion beam from a quadrupole mass filter using an electrostatic octopole lens. J Phys E 11:1037–1040
54. Zhang R, Lei W, Molina LT et al (2000) Ion transmission and ion/molecule separation using an electrostatic ion guide in chemistry ionization mass spectrometry. Int J Mass Spectr 194(1):B1–B2
55. Orloff J (2009) Handbook of charged particle optics. CRC Press, Boca Raton, FL
56. Limbach PA, Marshall AG, Wang M (1993) An electrostatic ion guide for efficient transmission of low energy externally formed ions into a Fourier transform ion cyclotron resonance mass spectrometer. Int J Mass Spectrom Ion Process 125:135–143
57. Willoughby R, Sheehan E, Mitrovich S (1998) A global view of LC/MS: how to solve your most challenging analytical. Global View Publishing, Pittsburg
58. Johnson AR, Carlson EE (2015) Collision-induced dissociation mass spectrometry: a powerful tool for natural product structure elucidation. Anal Chem 87:10668–10678
59. McMaster MC (2005) LC/MS: a practical users guide. John Wiley & Sons, Inc, Hoboken, NJ
60. Jonscher KR, Yates JR (1997) The quadrupole ion trap mass spectrometer – a small solution to a big challenge. Anal Biochem 224:1–15
61. Hu Q, Noll RJ, Li H et al (2005) The Orbitrap: a new mass spectrometer. J Mass Spectrom 40:430–443
62. Perry RH, Cooks RG, Noll RJ (2008) Orbitrap mass spectrometry: instrumentation, ion motion and applications. Mass Spectrom Rev 27:661–699
63. Scigelova M, Makarov A (2006) Orbitrap mass analyzer – overview and applications in proteomics. Proteomics 1:16–21
64. Eliuk S, Makarov A (2015) Evolution of Orbitrap mass spectrometry instrumentation. Annu Rev Anal Chem 8:61–80
65. Marshall AG, Hendrickson CL, GS J (1998) Fourier transform ion cyclotron resonance mass spectrometry: a primer. Mass SpectromRev 17:1–35
66. Products C (2008) Electron multipliers for mass spectrometry. Restek, Bellefonte, PA, pp 1–4
67. Ladislas Wiza J (1979) Microchannel plate detectors. Nucl Instrum Methods 162:587–601
68. Barnstedt J. Microchannel plate detectors. http://www.uni-tuebingen.de/fileadmin/Uni_Tuebingen/Fakultaeten/MathePhysik/Institute/IAAT/AIT/Lehrveranstaltungen/F-Praktikum/Dokumente/VersuchsAnleitungMCP_english.pdf
69. Link H, Fuhrer T, Gerosa L et al (2015) Real-time metabolome profiling of the metabolic switch between starvation and growth. Nat Methods 12:1091–1097
70. Heinemann J, Noon B, Mohigmi MJ et al (2014) Real-time digitization of metabolomics patterns from a living system using mass spectrometry. J Am Soc Mass Spectrom 25:1755–1762
71. Aretz I, Meierhofer D (2016) Advantages and pitfalls of mass spectrometry based metabolome profiling in systems biology. Int J Mol Sci 17:E632
72. Fujimura Y, Miura D (2014) MALDI mass spectrometry imaging for visualizing in situ metabolism of endogenous metabolites and dietary phytochemicals. Metabolites 4:319–346
73. Shroff R, Schramm K, Jeschke V et al (2015) Quantification of plant surface metabolites by matrix-assisted laser desorption-ionization mass spectrometry imaging: glucosinolates on Arabidopsis thaliana leaves. Plant J 81:961–972
74. Zaima N, Hayasaka T, Goto-Inoue N et al (2010) Matrix-assisted laser desorption/

ionization imaging mass spectrometry. Int J Mol Sci 11:5040–5055

75. Dunham SJB, Ellis JF, Li B et al (2017) Mass spectrometry imaging of complex microbial communities. Acc Chem Res 50:96–104
76. Svatos A (2011) Single-cell metabolomics comes of age new developments in mass spectrometry profiling and imaging. Anal Chem 83:5037–5044
77. Passarelli MK, Newman CF, Marshall PS et al (2015) Single-cell analysis: visualizing pharmaceutical and metabolite uptake in cells with label-free 3D mass spectrometry imaging. Anal Chem 87:6696–6702
78. Louie KB, Bowen BP, Cheng X et al (2013) "Replica-extraction-transfer" nanostructure-initiator mass spectrometry imaging of acoustically printed bacteria. Anal Chem 85:10856–10862
79. Northen TR, Yanes O, Northen MT et al (2007) Clathrate nanostructures for mass spectrometry. Nature 449:1033–1036
80. Woo HK, Northen TR, Yanes O et al (2008) Nanostructure-initiator mass spectrometry: a protocol for preparing and applying NIMS surfaces for high-sensitivity mass analysis. Nat Protoc 3:1341–1349
81. Banimustafa AH, Hardy NW (2012) A strategy for selecting data mining techniques in metabolomics. Methods Mol Biol 860:317–333
82. Baran R (2017) Untargeted metabolomics suffers from incomplete raw data processing. Metabolomics 13:107–110
83. Roberts LD, Souza AL, Gerszten RE et al (2012) Targeted metabolomics. Curr Protoc Mol Biol 1:1–24
84. Wu L, Mashego MR, Van DJC et al (2005) Quantitative analysis of the microbial metabolome by isotope dilution mass spectrometry using uniformly 13C-labeled cell extracts as internal standards. Anal Biochem 336:164–171
85. Saccenti E, Hoefsloot HCJ, Smilde AK et al (2014) Reflections on univariate and multivariate analysis of metabolomics data. Metabolomics 10:361–374
86. Worley B, Powers R (2013) Multivariate analysis in metabolomics. Curr Metabolomics 1:92–107
87. Gromski PS, Muhamadali H, Ellis DI et al (2015) A tutorial review: metabolomics and partial least squares-discriminant analysis – a marriage of convenience or a shotgun wedding. Anal Chim Acta 879:10–23
88. Warth B, Spangler S, Fang M et al (2017) Exposome-scale investigations guided by global metabolomics, pathway analysis, and cognitive computing. Anal Chem 2017:acs.analchem.7b02759
89. Cai Y, Weng K, Guo Y et al (2015) An integrated targeted metabolomic platform for high-throughput metabolite profiling and automated data processing. Metabolomics 11:1575–1586
90. Schwahn K, Beleggia R, Omranian N et al (2017) Stoichiometric correlation analysis: principles of metabolic functionality from metabolomics data. Front Plant Sci 8:1–12
91. Robinson MD, De SDP, Keen W et al (2007) A dynamic programming approach for the alignment of signal peaks in multiple gas chromatography-mass spectrometry experiments. BMC Bioinformatics 8:419
92. Alves TC, Pongratz RL, Zhao X et al (2015) Integrated, step-wise, mass-isotopomeric flux analysis of the TCA cycle. Cell Metab 22:936–947
93. Kappelmann J, Klein B, Geilenkirchen P et al (2017) Comprehensive and accurate tracking of carbon origin of LC-tandem mass spectrometry collisional fragments for 13C-MFA. Anal Bioanal Chem 409:2309–2326
94. Gebreselassie NA, Antoniewicz MR (2015) ^{13}C-metabolic flux analysis of co-cultures: a novel approach. Metab Eng 31:132–139
95. Leighty RW, Antoniewicz MR (2011) Dynamic metabolic flux analysis (DMFA): a framework for determining fluxes at metabolic non-steady state. Metab Eng 13:745–755
96. Schumacher R, Wahl SA (2015) Effective estimation of dynamic metabolic fluxes using 13C labeling and piecewise affine approximation: from theory to practical applicability. Metabolites 5:697–719
97. Julien F, Georges R, Vande A et al (2016) Direct metabolic of dynamic metabolic of dynamic metabolic analysis dynamic metabolic flux of and overdetermined underdetermined and overdetermined underdetermined and overdetermined underdetermined and metabolic. Science 49:318–323
98. Liu D, Hoynes-O'Connor A, Zhang F (2013) Bridging the gap between systems biology and synthetic biology. Front Microbiol 4:1–8
99. Kell DB (2006) Metabolomics, modelling and machine learning in systems biology – towards an understanding of the languages of cells. FEBS J 273:873–894
100. O'Hagan S, Kell DB (2018) Analysing and navigating natural products space for generating small, diverse, but representative chemical libraries. Biotechnol J 13:1–11

101. Ritchie MD, Holzinger ER, Li R et al (2015) Methods of integrating data to uncover genotype-phenotype interactions. Nat Rev Genet 16:85–97
102. Trivedi DK, Hollywood KA, Goodacre R (2017) Metabolomics for the masses: The future of metabolomics in a personalized world. New Horizons Transl Med 3:294–305
103. George KW, Thompson MG, Kang A et al (2015) Metabolic engineering for the high-yield production of isoprenoid-based C_5 alcohols in *E. coli*. Sci Rep 5:11128
104. Zhou K, Zou R, Stephanopoulos G et al (2012) Metabolite profiling identified methylerythritol cyclodiphosphate efflux as a limiting step in microbial isoprenoid production. PLoS One 7:e47513
105. Zou R, Zhou K, Stephanopoulos G et al (2013) Combinatorial engineering of 1-deoxy-D-xylulose 5-phosphate pathway using cross-lapping in vitro assembly (CLIVA) method. PLoS One 8:e79557
106. George KW, Thompson M, Kim J et al (2018) Integrated analysis of isopentenyl pyrophosphate (IPP) toxicity in isoprenoid-producing Escherichia coli. Metab Eng 47:60–72
107. Brunk E, George KW, Alonso-Gutierrez J et al (2016) Characterizing strain variation in engineered *E. coli*. Cell Syst 2:335–346
108. Panizzon JP, Luiz H, Júnior P et al (2015) Microbial diversity: relevance and relationship between environmental conservation and human health. Braz Arch Biol Technol 58:137–145
109. Nazaries L, Pan Y, Bodrossy L et al (2013) Evidence of microbial regulation of biogeochemical cycles from a study on methane flux and land use change. Appl Environ Microbiol 79:4031–4040
110. Stewart EJ (2012) Growing unculturable bacteria. J Bacteriol 194:4151–4160
111. Vanwonterghem I, Jensen PD, Ho DP et al (2014) Linking microbial community structure, interactions and function in anaerobic digesters using new molecular techniques. Curr Opin Biotechnol 27:55–64
112. Beale DJ, Crosswell J, Karpe AV et al (2017) A multi-omics based ecological analysis of coastal marine sediments from Gladstone, in Australia's Central Queensland, and Heron Island, a nearby fringing platform reef. Sci Total Environ 609:842–853
113. Bargiela R, Herbst FA, Martínez-Martínez M et al (2015) Metaproteomics and metabolomics analyses of chronically petroleum-polluted sites reveal the importance of general anaerobic processes uncoupled with degradation. Proteomics 15:3508–3520
114. Kimes NE, Callaghan AV, Aktas DF et al (2013) Metagenomic analysis and metabolite profiling of deep-sea sediments from the Gulf of Mexico following the Deepwater Horizon oil spill. Front Microbiol 4:50
115. Lutz S, Anesio AM, Field K et al (2015) Integrated "Omics", targeted metabolite and single-cell analyses of arctic snow algae functionality and adaptability. Front Microbiol 6:1–17
116. Anderson DM, Cembella AD, Hallegraeff GM (2012) Progress in understanding harmful algal blooms: paradigm shifts and new technologies for research, monitoring, and management. Annu Rev Mar Sci 4:143–176
117. Parmar KM, Gaikwad SL, Dhakephalkar PK et al (2017) Intriguing interaction of bacteriophage-host association: an understanding in the era of omics. Front Microbiol 8:559
118. Bertrand S, Bohni N, Schnee S et al (2014) Metabolite induction via microorganism co-culture: a potential way to enhance chemical diversity for drug discovery. Biotechnol Adv 32:1180–1204
119. World Health Organization. Infectious diseases. http://www.who.int/topics/infectious_diseases/en/
120. World Health Organization. Malaria. http://www.who.int/mediacentre/factsheets/fs094/en/
121. Park YH, Shi YP, Liang B et al (2015) High-resolution metabolomics to discover potential parasite-specific biomarkers in a Plasmodium falciparum erythrocytic stage culture system. Malar J 14:122
122. Gardinassi LG, Cordy RJ, Lacerda MVG et al (2017) Metabolome-wide association study of peripheral parasitemia in Plasmodium vivax malaria. Int J Med Microbiol 307:533–541
123. Lau SK, Lam C-W, Curreem SO et al (2015) Identification of specific metabolites in culture supernatant of Mycobacterium tuberculosis using metabolomics: exploration of potential biomarkers. Emerg Microbes Infect 4:e6
124. Garay CD, Dreyfuss JM, Galagan JE (2015) Metabolic modeling predicts metabolite changes in Mycobacterium tuberculosis. BMC Syst Biol 9:57
125. Xu Y, Zhang Z, Sun Z (2015) Drug resistance to Mycobacterium tuberculosis: From the traditional Chinese view to modern systems biology. Crit Rev Microbiol 41:399–410

126. Lobritz MA, Belenky P, Porter CBM et al (2015) Antibiotic efficacy is linked to bacterial cellular respiration. Proc Natl Acad Sci 112:8173–8180
127. Warner DF, Arlehamn CSL, Lewinsohn D et al (2014) Mycobacterium tuberculosis metabolism. Metabolism 5:a021121
128. Luier L, Loots DT (2016) Tuberculosis metabolomics reveals adaptations of man and microbe in order to outcompete and survive. Metabolomics 12:1–9
129. Mason S, van FAMT, Solomons R et al (2016) A putative urinary biosignature for diagnosis and follow-up of tuberculous meningitis in children: outcome of a metabolomics study disclosing host–pathogen responses. Metabolomics 12:1–16
130. Washio J, Takahashi N (2016) Metabolomic studies of oral biofilm, oral cancer, and beyond. Int J Mol Sci 17:870
131. Johnson CH, Dejea CM, Edler D et al (2015) Metabolism links bacterial biofilms and colon carcinogenesis. Cell Metab 21:891–897
132. Shaffer M, Armstrong AJS, Phelan VV et al (2017) Microbiome and metabolome data integration provides insight into health and disease. *Transl Res* 189:51–64. https://doi.org/10.1016/j.trsl.2017.07.001
133. Martinez KB, Leone V, and Chang EB (2017) Microbial metabolites in health and disease: Navigating the unknown in search of function. *J Biol Chem* 292 (21):8553–8559. https://doi.org/10.1074/jbc.R116.752899
134. Li DY, Tang WHW (2017) Gut Microbiota and Atherosclerosis. Curr Atheroscler Rep 19(10):39. https://doi.org/10.1007/s11883-017-0675-9
135. He X, Ji G, Jia W et al (2016) Gut microbiota and nonalcoholic fatty liver disease: insights on mechanism and application of metabolomics. Int J Mol Sci 17:300
136. Li H, He J, Jia W (2016) The influence of gut microbiota on drug metabolism and toxicity. Expert Opin Drug Metab Toxicol 12:31–40
137. Nichols RG, Hume NE, Smith PB et al (2016) Omics approaches to probe microbiota and drug metabolism interactions. Chem Res Toxicol 29:1987–1997

Chapter 3

Metabolomics: A Microbial Physiology and Metabolism Perspective

Chijioke J. Joshua

Abstract

Metabolomics is valuable for studying microbial metabolism, which is often used to elucidate biological functions. Effective application of metabolomics is enhanced by fundamental understanding of microbial physiology and metabolism. This review briefly highlights important aspects of metabolism that are essential for designing and executing effective metabolic and metabolomics studies. The influence of microbial physiology and metabolism on growth, energy metabolism and regulation is briefly reviewed. The chapter also evaluates factors affecting metabolic prediction.

Key words Microbial physiology, Microbial metabolism, Microbial growth, Central carbon metabolism, Metabolic regulation, Experimental design, Metabolic pathways

1 Introduction

Metabolism is a complex sequence of reactions that direct biological functions in living things, and metabolomics enables simultaneous analysis of metabolic reactions and pathways involved in these processes. Metabolism encompasses fundamental processes such as DNA replication, RNA transcription (and reverse transcription), and translation [1, 2]. Conventional metabolic studies often focus on a specific or subset of a pathway due to the volume, heterogeneity, and interconnectivity of the metabolic reactions. Although the total set of metabolic reactions in an organism is encoded in its genetic materials (chromosomes or plasmids), they DO NOT occur simultaneously. Some metabolic reactions or pathways are continuously expressed, while others are only induced in response to intrinsic and extrinsic stimuli. The ability of an organism to coordinate the expression of metabolic reactions or pathways is termed "regulation" [3–5].

Metabolomics evaluates the abundance and distribution of metabolites in the cells or tissues of organisms at any given time [6]. Metabolites are generated from complex networks of catalyzed

Edward E.K. Baidoo (ed.), *Microbial Metabolomics: Methods and Protocols*, Methods in Molecular Biology, vol. 1859, https://doi.org/10.1007/978-1-4939-8757-3_3,

and spontaneous reactions that occur in discrete sequences within a pathway. Unlike genomics, transcriptomics and proteomics that rely on a finite set of metabolites or subunits, metabolomics evaluates the total set of metabolites in an organism. For instance, genomics and transcriptomics analyses rely on a set of four nucleotides respectively, while proteomics is based on 20 amino acid subunits [7]. Therefore, the effectiveness of metabolomics is often undermined by the absence of analytical strategies that can identify and quantifying the broad set of metabolites in an organism. Although these analytical techniques and strategies are essential for metabolic studies, they represent a fraction of the requirement for understanding microbial physiology and metabolism. Hence, this review highlights philosophical and scientific considerations for executing effective metabolic studies.

2 Microbial Growth

Microbial growth provides insight into the physiology and functions in an organism [8]. Microbial growth is influenced by levels of organic and inorganic nutrients, temperature, pH, oxygen, as well as the presence of other organisms. Microbial growth could be measured by directly counting individual cells or colonies on solid media that are derived from individual cells [9]. Growth can also be estimated indirectly by measuring the turbidity (optical density) of uniformly dispersed liquid cultures, the dry weight of cells from solid and liquid media or the diameter of colonies on solid medium. The importance of growth in Metabolic studies is often understated, but it is the primary cause of most inconsistencies in these studies. Therefore, microbial growth should be closely monitored for effective execution, analysis, and interpretation of data from metabolic studies.

Microbial growth studies could be carried out in liquid and solid media, but most metabolic studies are carried out in liquid media because of the ease of recovering the cells and metabolites for analysis. The primary mode of microbial growth in liquid media are batch or continuous cultures [10]. In continuous culture, the growth rate of an organism is kept constant by continuously removing a constant amount of spent culture (cells, residual medium, and metabolic by-products) and replacing it with an equal amount of fresh growth medium [10, 11]. In comparison, the nongaseous constituents of batch culture (cells, nutrients, and by-products) remain in the growth chamber throughout the growth period, thus making it a semi-closed system. But in certain instances, the depleted nutrients in batch cultures are replaced at intervals (fed-batch) without removing the spent medium [10, 11].

In general, the physiological state of microbial cells in continuous cultures remains fairly constant over a long period (steady

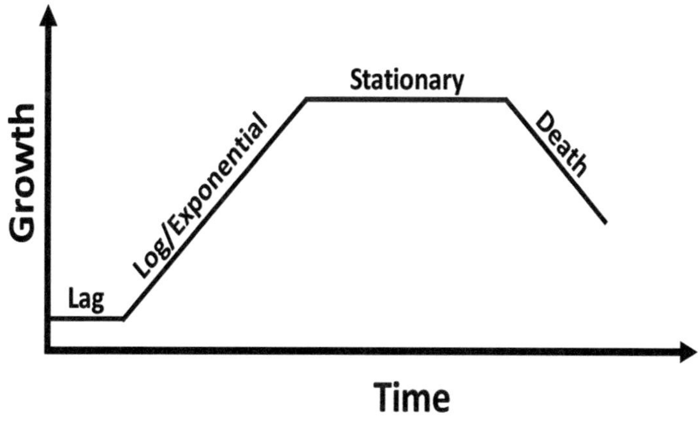

Fig. 1 Microbial growth curve in batch cultures highlighting lag, log/exponential, stationary, and death phases

state), whereas changes in the physiology of cells in batch cultures is very dynamic [12]. The physiological shifts in batch cultures correlates with the growth rate of the cells, which are influenced by dynamic changes in factors such as nutrients, secreted metabolites, pH, oxygen, and/or temperature during the growth period. The dynamic changes in the growth rate of an organism in batch cultures [8] results in four distinct primary growth phases termed lag, log or exponential, stationary, and death phases (Fig. 1). A profile of microbial growth phases in batch cultures over time [10, 11, 13] is called a growth curve. Microbial growth curve (Fig. 1) generally provides pictorial information on the physiology of an organism during metabolic studies. The growth curves from batch cultures are often comparable to the profiles observed during growth of most organisms in their natural habitat, especially in a static environment [14].

2.1 Lag Phases

Lag phase is the part of a growth curve that is characterized by absence or negligible growth. Although the significance of lag phase is often underestimated growth (Fig. 1), it is characterized by intense metabolic activities that are crucial for the viability of an organism. Some of the metabolic activities at lag phase include DNA, RNA, and protein syntheses that are critical for cell division and growth [10, 15]. The length of lag phase is influenced by the ability of the organism to rapidly adapt to its environment(s) or activate the expression of genes involved in the uptake and catabolism of substrates in the growth medium [10, 15]. The duration of lag phase of an organism is also influenced by the vigor of the cells, as well as the inoculum preparation strategy.

Lag phase is usually minimal or absent when a rapidly growing organism is transferred into the same growth medium and condition. However, the length of lag phase increases when the cells are subcultured into a different medium or when growth conditions such as temperature and pH are altered. To exit lag phase, the organism must overcome an intrinsic and extrinsic metabolic impediments or toxicity threshold. For example, an organism must express genes to overcome osmotic and temperature related shock each time it is inoculated into a growth medium from a cryo-stock. The expression of high levels of stress response genes might adversely affect the physiology of an organism, which might extend the lag phase or halt the growth of the organism. It is often a good practice to revitalize a cryo-stock of a microbial strain by inoculating the cells onto a solid medium before subculturing into a liquid medium to improve the growth vigor of the cell and reduce lag time. Alternatively, the organism should be inoculated into complex medium that is supplemented with the target substrate. The presence of the target substrate to the complex medium induces expression of genes involved in the metabolism of this substrate prior to subculturing into the appropriate growth medium.

2.2 Exponential Phase

The growth kinetics of an organism on any given substrate(s) is determined at exponential or log phase, which is characterized by rapid growth rate due to optimum nutrient supply and conditions [16]. Microbial cells divide at regular intervals called doubling or generation time (g) during exponential phase [8]. The maximum or specific growth rate (μ) of an organism on a given substrate is reached at mid-exponential phase under optimum growth conditions [11]. The rapid growth rate at exponential phase correlates with efficient metabolism, which occurs when an organism reaches metabolic homeostasis or steady-state metabolism [17]. The growth of an organism is considered "balanced" at steady state because metabolic machineries such as ribosomes, enzymes, intermediates, and cofactors are present at optimum levels [8]. Some organisms exhibit multiple exponential phases in batch cultures during growth on multiple substrates, a phenomenon called "diauxie" [8, 18]. Diauxie is triggered by a global regulatory mechanism called catabolic repression.

Although rapid growth rate enhances the ability of an organism to compete for scarce resources in its environment, it also increases the susceptibility of the species to adverse environmental conditions. Microbial cells are most susceptible to adverse environmental conditions or toxins at exponential phase due to increased permeability or fluidity of the cell membrane during cell division [19]. Membrane fluidity reduces cellular integrity, which results in loss of valuable intracellular resources or influx of harmful toxins from the environment.

Fast growing organisms are also susceptible to the adverse effect of their metabolic by-products, which are sometimes lethal. In general, fast growing organisms or primary colonizers dominate any given habitat or ecosystem when nutrient and environmental conditions are optimal. However, rapid metabolic activities of the primary colonizers often alter environmental factors such as pH or temperature, which are often toxic to organisms. The primary colonizers are subsequently succeeded by other species (secondary colonizers), which are also succeeded by tertiary and quaternary population. The dynamic effect of growth rate and microbial survival alters the natural or artificial ecosystems via a complex interspecies and intraspecies interaction, a phenomenon called "microbial succession." The vulnerability of microbial cells at exponential phase is the principle behind food preservation techniques such as "pasteurization" and "tyndallization," which are designed to eliminate pathogens and spoilage (or spore-forming) organisms respectively [20].

2.3 Stationary and Death Phases

Exponential growth phase is followed by stationary phase, which is a period of decreased or stalled growth due to nutrient depletion and/or accumulation of toxic metabolic by-products (Fig. 1). The unfavorable growth condition during stationary phase also results in the death and sometimes lysis of the cells. Growth and death rates of microbial cells are relatively at equilibrium during stationary phase [8, 10, 11, 19]. Stationary phase is sometimes followed by a death phase, when the death rate of the cells becomes significantly higher than growth rate. The physiological state of microbial cells at stationary and death phases are generally similar. The length of severity of each growth phase depends on the potency of the inducing factor(s), the ability of the organism to tolerate and mitigate adverse condition or restoration of optimum growth condition. Microbes are generally tolerant to onset of adverse conditions at stationary phase than exponential phase due to membrane integrity and ability to produce protective layers such as capsule and sheaths. Metabolic reactions at stationary phase are primarily directed toward maintaining or preserving the viability of the cells (maintenance metabolism). Stationary phase metabolism mostly involves recycling of intracellular and extracellular metabolites, and debris from dead cells [19, 21]. Some organisms can alternate between stationary and death phases for an extended period of time in batch cultures [21]. Most microbial cells can revert to exponential phases once optimum growth condition is restored [10].

2.4 Role of Growth Curve in Metabolic Studies

A growth curve is a valuable tool for trained microbial physiologists because it helps validate the conclusion reached from data derived from metabolic and metabolomics analyses. Growth curves highlight dynamic changes in the physiology of microbial cells in response to the conditions being evaluated, while also providing insights into the underlying metabolism. Carbon catabolite repression (CCR) was discovered by Jacques Monod from the growth curve of *Escherichia coli* cells grown on multiple sugars as co-carbon sources [8, 18, 22, 23]. The growth curve is also important in experimental design, because it helps identify the best sampling points and potential pitfalls while executing metabolic studies. Growth curve is also an effective quality and process control tool.

3 Microbial Nutrition and Metabolism

Metabolic reactions that build or breakdown complex organic molecules are called “anabolism” and “catabolism” respectively. In general, anabolic reactions consume energy, while catabolic reactions liberate the energy stored in chemical bonds of organic molecules that are broken in the process. Metabolic reactions such as respiration, DNA, RNA, protein, and fatty acid biosynthesis are essential because they are critical for growth and survival of an organism. Essential metabolic reactions are collectively called primary metabolism [24, 25], while nonessential metabolic reactions such as pigment and antibiotic production are termed secondary metabolism [24, 25]. Primary and secondary metabolisms often involve anabolic and catabolic reactions.

Growth condition and nutritional requirements of an organism are the first consideration in any metabolic studies. An organism requires a source of carbon and energy, as well as optimal levels of mineral (salts), oxygen, pH, and temperature to grow and function properly. Suboptimal levels of nourishments and environment conditions lead to imbalance growth or death in some cases. Some microbes require special nutrients such as vitamins, amino acids, or trace elements for growth in the laboratory. Microorganisms with special nutritional requirements are called “fastidious organisms” [26]. For instance, obligate intracellular pathogen such as *Coxiella burnetti* can only grow in cell or tissue cultures [26]. Although the nutritional requirements of fastidious organisms are encoded in their genetic materials, they are often empirically or serendipitously determined. Most unculturable organisms are fastidious because they require special nutrients that are absent in most general-purpose growth media.

Organisms that synthesize complex biological molecules from inorganic carbon sources are called “autotrophs,” while those that require organic sources of carbon are “heterotrophs.” Microbes are also grouped based on their source of metabolic energy. For instance, “phototrophs” and “chemotrophs” derive metabolic energy from light and chemical sources respectively. Phototrophs primarily obtain their energy from sunlight, but they can also derive energy from any natural or artificial source that emits photons. Chemotrophs are further divided into lithotrophs and organotrophs if they obtain energy from inorganic and organic molecules, respectively. Microorganisms also show great variation in their oxygen requirement. Obligate aerobes are organisms that strictly grow in the presence of oxygen, although some strict aerobes (microaerophiles) only require low levels of oxygen for optimal growth and function. In comparison, obligate anaerobes strictly grow in the absence of oxygen. Although molecular oxygen is toxic to most anaerobic organisms, some species called "aerotolerant anaerobes" can survive short-term exposure to oxygen. Furthermore, a subset of anaerobic organisms (facultative anaerobes) such as *E. coli*, can grow and function in the presence or absence of oxygen [27].

4 Energy Metabolism

Non-photosynthetic organisms derive metabolic energy from cellular respiration or fermentation via catabolic conversion of glucose molecules into water, carbon dioxide, and energy, often in the form of adenosine triphosphate (ATP) and heat [28, 29]. Cellular respiration and fermentation primarily differ in the amount of energy released, the mode of electron transfer and terminal electron acceptor. The energy yield from glucose during cellular respiration is at 15-fold higher relative to fermentation. Aerobic and anaerobic organisms carry out cellular respiration using different terminal electron acceptors. Also, some fermentative organisms exclusively derive metabolic energy from fermentation [29], while others utilize fermentation under suboptimal conditions. Cellular respiration and fermentation are accomplished via central carbon metabolic (CCM) pathways. The primary CCM pathways glycolysis and tricarboxylic acid (TCA) are critical to most primary metabolic processes (Fig. 2). Glycolysis occurs in the cytosol of all organisms, while TCA cycle takes place in the mitochondria of eukaryotes and in the cytosol and cell membrane of prokaryotes. Glycolysis and TCA cycle are critical in aerobic and anaerobic respiration, while fermentation only requires glycolysis. The ATP (energy) molecules are generated at the membrane electron transport chain during cellular respiration via a process known as oxidative phosphorylation [30–33]. Molecular oxygen is the terminal electron acceptor during aerobic respiration, while molecules such as sulfate, nitrate, and carbon dioxide are used as electron acceptors during anaerobic respiration [34]. The production of ATP occurs in the cytosol during fermentation via a process known as substrate-level phosphorylation [35]. Also, the terminal electron is transferred to an intermediate derived from the breakdown of glucose or other substrates during fermentation. In some fermentative organisms, the terminal electron is transferred to glucose-derived acetaldehyde and pyruvate molecules generating ethanol and lactic acid, respectively [36]. Glycolysis and TCA cycle mechanisms will be further examined.

4.1 Glycolysis

Glycolysis is the first step in glucose catabolism and it involves the cleavage of glucose molecule into two pyruvate molecules. Glycolysis is usually rapid during increased metabolism or nutrient supply. The primary glycolytic pathways are Embden–Meyerhof–Parnas (EMP) pathway, while the others are Entner–Doudoroff (ED) and pentose phosphate (PP) pathways [28]. The EMP

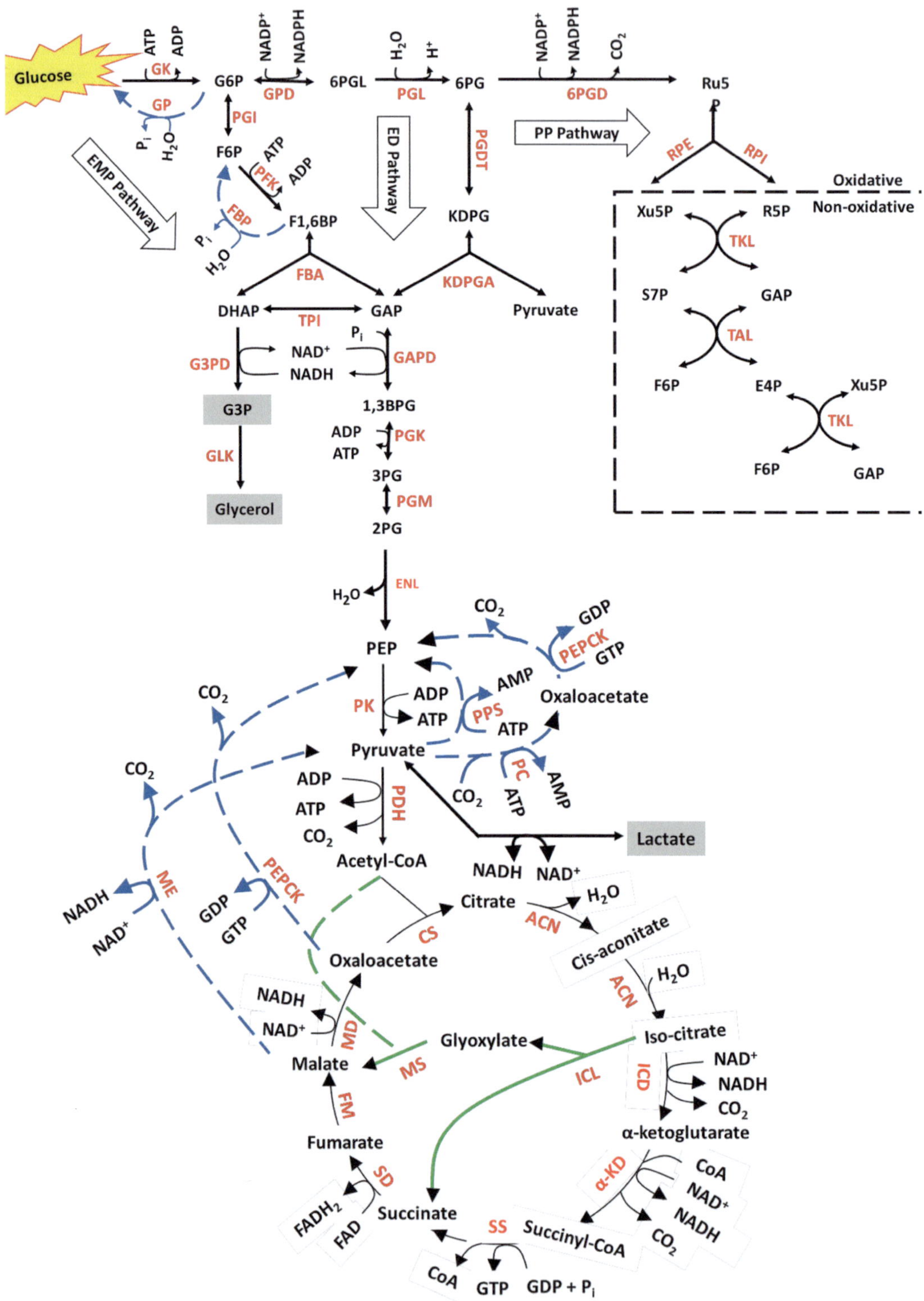

Fig. 2 Central carbon metabolism highlighting classical glycolytic pathways, tricarboxylic acid pathway, gluconeogenesis (blue lines), glyoxylate shunt (green lines), glycerol-3-phosphate and pyruvate–lactate redox shuttle pathways (gray boxes). Red abbreviations depict enzymes involved in each reaction. *See* Table 1 for description of abbreviations

Table 1
List of abbreviations in Fig. 2

Categories	Abbreviation	Description
Enzymes		
	6PGD	6-phosphogluconate dehydrogenase
	CAN	Aconitase
	CS	Citrate synthase
	ENL	Enolase
	FBA	Fructose bisphosphate aldolase
	FBP	Fructose-1,6-bisphosphatase
	FM	Fumarase
	G3PD	Glycerol-3-phosphate dehydrogenase
	GAPD	Glyceraldehyde-3-phosphate dehydrogenase
	GK	Glucokinase
	GLK	Glycerol kinase
	GP	Glucose-6-phophatase
	GPD	Glucose-6-phosphate dehydrogenase
	ICD	Isocitrate dehydrogenase
	ICL	Isocitrate lyase
	KDPGA	2-keto-3-deoxy-6-phosphogluconate aldolase
	LDH	Lactate dehydrogenase
	MD	Malate dehydrogenase
	ME	Malic enzyme
	MS	Malate synthase
	PC	Pyruvate carboxylase
	PDH	Pyruvate dehydrogenase
	PEPCK	Phosphoenolpyruvate carboxykinase
	PFK	Phosphofructokinase
	PGDT	6-phosphogluconate dehydratase
	PGI	Phosphoglucose isomerase
	PGK	Phosphoglycerate kinase
	PGL	6-Phosphoglucono-lactonase
	PGM	Phosphoglycerate mutase
	PK	Pyruvate kinase

(continued)

Table 1 (continued)

Categories	Abbreviation	Description
	PPS	Phosphoenolpyruvate synthetase
	RPE	D-riboluse-5-phosphate epimerase
	RPI	D-ribose-5-phosphate isomerase
	SD	Succinate dehydrogenase
	SS	Succinyl-CoA synthetase
	TAL	Transaldolase
	TKL	Transketolase
	TPI	Triose phosphate isomerase
	α-KD	α-ketoglutarate dehydrogenase
Metabolites		
	1,3 BPG	1,3-bisphospho-glycerate
	2PG	2-phosphoglycerate
	3PG	3-phosphoglycerate
	6PG	6-phosphogluconate
	6PGL	6-phosphogluconolactone
	ADP	Adenosine diphosphate
	ATP	Adenosine triphosphate
	DHAP	Dihydroxyacetone
	E4P	Erythrose-4-phosphate
	F1,6BP	Fructose-1,6-bisphosphate
	F6P	Fructose-6-phosphate
	FAD	Oxidized flavin adenine dinucleotide
	$FADH_2$	Reduced flavin adenine dinucleotide
	G3P	Glycerol-3-phosphate
	G6P	Glucose-6-phosphate
	GAP	Glyceraldehyde-3-phosphate
	GDP	Guanosine diphosphate
	GTP	Guanosine triphosphate
	KDPG	2-keto-3-deoxy-6-phosphogluconate
	NAD^+	Oxidized nicotinamide adenine dinucleotide
	NADH	Reduced nicotinamide adenine dinucleotide

(continued)

Table 1
(continued)

Categories	Abbreviation	Description
	$NADP^+$	Oxidized nicotinamide adenine dinucleotide phosphate
	NADPH	Reduced nicotinamide adenine dinucleotide phosphate
	PEP	Phosphoenolpyruvate
	P_i	Inorganic phosphate
	R5P	D-ribose-5-phosphate
	Ru5P	D-riboluse-5-phosphate
	S7P	D-sedoheptulose-7-phosphate
	Xu5P	D-xylulose-5-phosphate

pathway is the most studied glycolytic pathway, while PP pathway is primarily used to generate precursors such as phosphorylated NADH (NADPH) and ribose-5-phosphate and for various anabolic processes [35, 37–39].

The classic EMP pathway consists of ten enzyme-catalyzed reactions, which are divided into ATP-consuming and ATP-generating phases (Fig. 2). The first EMP reaction is the ATP-dependent and glucokinase catalyzed conversion of glucose to glucose-6-phosphate (G6P), which is mostly restricted to intracellular glucose catabolism [40]. In some organisms, the glucose is directly converted to G6P at the membrane during transport [22, 41] via the phosphoenolpyruvate (PEP) dependent phosphotransferase system (PTS). The G6P molecules are reversibly isomerized by phospho-glucose isomerase to fructose-6-phosphate (F6P), which is subsequently converted to fructose-1,6bisphosphate (F1,6BP) in a tightly regulated nonreversible reaction catalyzed by the ATP-dependent phosphofructokinase [28]. The rate of glycolysis is strongly influenced by the activity of phosphofructokinase. The activity of phosphofructokinase is inhibited by ATP and other molecules during rapid metabolism in nutrient-rich environment. The feedback inhibition of phosphofructokinase increases the intracellular level of F6P, which significantly reduces F-1,6-BP–F6P ratio in the cell. Therefore, low F1,6BP–F6P ratio generally signifies low-level or inhibition of glycolysis. The ATP-consuming phase of EMP pathway is terminated with the cleavage of F1,6BP into 3-carbon glyceraldehyde-3-phosphate (GAP) and dihydroxyacetone phosphate (DHAP) by fructose bisphosphate aldolase [35]. The DHAP molecules are reversibly converted to GAP by triose phosphate isomerase (Fig. 2).

The ATP-generating phase starts with the synthesis of 1,3-bisphospho-glycerate (1,3BPG) and nicotinamide adenine dinucleotide (NADH) molecules via glyceraldehyde-3-phosphate dehydrogenase mediated phosphorylation of GAP with an inorganic phosphate group [28]. An ATP molecule is produced from each 1,3BPG molecule via phosphoglycerate kinase mediated substrate-level phosphorylation reaction that also generates a 3-phosphoglycerate (3PG) molecule [35]. The 3PG molecules are sequentially converted to 2-phosphoglycerate (2PG) and phosphoenolpyruvate (PEP) by phosphoglycerate mutase and enolase, respectively (Fig. 2). A second ATP molecule is generated from each GAP molecule with the conversion of PEP to pyruvate by pyruvate kinase. In summary, a total of two ATP (net) and two NADH molecules are generated from the classic EMP pathway [28].

There are substantial difference between EMP pathway and the classic ED and PP pathways. The G6P molecule is converted to 6-phosphogluconolactone (6PGL) and NADPH by G6P dehydrogenase in classic ED and PP pathways (Fig. 2). The 6PGL is sequentially converted to 6-phosphogluconate (6PG) and 2-keto-3-deoxy-6-phosphogluconate (KDPG) by 6-phosphoglucono-lactonase and 6PG dehydratase respectively, in the ED pathway [28, 35, 42]. The KDPG is subsequently hydrolyzed to GAP and pyruvate by KDPG aldolase [28]. However, the 6PG molecule is oxidatively decarboxylated by 6PG dehydrogenase into riboluse-5-phosphate (Ru5P), NADPH, and CO_2 [28, 35]. The Ru5P is reversibly converted to xylulose-5-phosphate (Xu5P) and ribose-5-phosphate (R5P) by Ru5P epimerase and R5P isomerase respectively in the nonoxidative phase [28, 35]. The Xu5P and R5P are combined and converted to sedoheptulose-7-phosphate (S7P) and GAP by transketolase, and subsequently to erythrose-4-phosphate (E4P) and F6P by transaldolase [35]. In general, the ED pathway generates one net ATP, one NADH, and one NADPH per glucose molecule, while the PP pathway yields (net) one NADH, two NADPHs, and at least one ATP molecule [28, 35, 42].

There are various modifications of glycolytic pathways in several organisms. For instance, glycolysis is accomplished in most archaeal species via modified ED pathways that yield no net ATP [43, 44]. Archaeal species mostly generate their ATP via aerobic and anaerobic respiration (Fig. 2). However, some fermentative archaeal species such as *Pyrococcus furiosus* and *Thermococcus kodakarensis* generate ATP molecules via modified ADP-dependent EMP pathway [29, 45–47]. The ATP molecules from these species are generated from the conversion of PEP to pyruvate by ADP-dependent pyruvate kinase or AMP and inorganic phosphate-dependent PEP synthase [42, 47, 48]. Archaeal species also utilize NADPH and ferredoxin as reducing equivalents instead of NADH [43, 44].

4.2 Tricarboxylic Acid Cycle

TCA cycle is involved in complete oxidation and conversion of pyruvate from glycolysis into high energy molecules such as NADH (NADPH in archaea), reduced flavin adenine dinucleotide ($FADH_2$) and GTP [49]. The cycle also generates important intermediates that are consumed by various primary and secondary metabolic reactions [49]. TCA cycle starts with the condensation of oxaloacetate and acetyl-CoA into citrate molecule (Fig. 2) by citrate synthase [44, 49]. The acetyl-CoA molecule is mostly derived from decarboxylation of pyruvate by pyruvate dehydrogenase complex that also generates an NADH molecule [50]. The decarboxylation reaction occurs in the cytosol and mitochondrial lumen in prokaryotes and eukaryotes, respectively. The citrate molecule is sequentially converted to *cis*-aconitate, isocitrate, α-ketoglutarate, succinyl-CoA, succinate, fumarate, malate, and oxaloacetate via a series of enzyme catalyzed reactions that generate NADH (or NADPH in archaea), $FADH_2$, GTP, and CoA molecules (Fig. 2).

The NADH molecules are generated from sequential decarboxylation of isocitrate to α-ketoglutarate and succinyl-CoA and oxidation of malate to oxaloacetate, while the $FADH_2$ is produced from succinate dehydrogenase catalyzed conversion of succinate to fumarate (Fig. 2). The high energy NADH and $FADH_2$ molecules are converted to ATP at the electron transport chain (oxidative phosphorylation) found in the inner membrane of mitochondria or cytoplasmic membrane of prokaryotes by ATP synthase [30–33]. The high energy GTP and CoA molecules are produced from the conversion of succinyl-CoA to succinate by succinyl-CoA synthetase (Fig. 2). Most TCA cycle reactions are reversible. The reversed or reductive TCA cycle is used to synthesize ATP and complex organic molecules from water and carbon dioxide (carbon fixation) by autotrophic or photosynthetic organisms [51–54].

4.3 Gluconeogenesis

Although energy metabolism flows from glycolysis to TCA cycle, metabolic flux between these pathways is reversible. The flow of metabolism from TCA cycle toward glycolysis is known as gluconeogenesis (Fig. 2) and is primarily used to replenish the intracellular pool of glucose during fasting or when external supply of nutrient is low or depleted [44, 54, 55]. Also, gluconeogenesis is used for storing energy in the form of starch and glycogen [44] when there is excess supply of nutrients (feast). Gluconeogenesis is utilized to accomplishing various critical cellular functions such as glycosylation or biosynthesis of amino-sugars. The direction of metabolic flux through glycolysis and TCA cycle is mostly regulated at the nonreversible glycolytic steps [56]. Glucokinase, phosphofructokinase, and pyruvate kinase in classic EMP pathway are inactivated during gluconeogenesis, while the reversed reactions are carried out by different enzymes [28, 35]. For instance, the conversion of G6P to glucose during gluconeogenesis is mediated by glucose-6-phophatase, while the conversion of F1,6BP to F6P is catalyzed by fructose-1,6-bisphosphatase [55, 56]. Similarly, pyruvate is converted to PEP during gluconeogenesis by ATP-dependent phosphoenolpyruvate synthetase in most prokaryotes and *E. coli* [47, 57]. However, the conversion of pyruvate to PEP is accomplished via a two-step reaction during gluconeogenesis in most eukaryotes [55, 56]. First, an ATP-dependent pyruvate carboxylase converts one pyruvate and one CO_2 molecule to oxaloacetate [56], which is then converted to PEP and CO_2 by GTP-dependent phosphoenolpyruvate carboxykinase (PEPCK). Oxaloacetate from the TCA cycle is also converted to PEP and CO_2 by PEPCK during gluconeogenesis, while malate is converted to pyruvate by malic enzyme [44, 56, 58].

5 Metabolic Regulation

Microorganisms control onset and direction of metabolic reactions in accordance with their physiological need and to prevent unproductive metabolism. Metabolic regulation is also used to regenerate depleted metabolites or recycle accumulated metabolites that are sometimes toxic to the cells. Metabolism is regulated in most organisms at transcriptional (RNA) and translational (protein) levels [59, 60]. Some of the mechanisms employed in the control of cellular metabolism include substrate-level regulation, cataplerosis/anaplerosis, redox shuttles and metabolic shunts.

5.1 Substrate-Level Regulation

Substrate-level regulation involves interaction between enzymes and their products or substrates via allosteric association and covalent modifications. Substrate-level regulation controls the direction of metabolic flux through a pathway to prevent wasteful reactions or accumulation of toxic metabolites. Accumulation of a metabolite within a pathway is often an indication that the adjacent a reaction step is inefficient or that the enzyme(s) is temporarily inhibited, often via substrate-level regulation. The regulated reaction is usually the rate limiting step in most pathways. However, the accumulation of a given metabolite could also result from the simultaneous activation of multiple metabolic pathways that generate the molecule. For instance, the intracellular pool of GAP in rapidly growing cells could be produced from EMP, ED and nonoxidative phosphate pathways [35].

Some of the CCM enzymes regulated via substrate-level regulation include glucokinase, phosphofructokinase, pyruvate kinase, pyruvate dehydrogenase, citrate synthase, isocitrate dehydrogenase, and α-ketoglutarate dehydrogenase [61]. The CCM enzymes are broadly regulated by intermediates such as ATP, ADP, NADH, F1,6BP, acetyl-CoA, and citrate. Phosphofructokinase, pyruvate kinase, and pyruvate dehydrogenase are allosterically inhibited by high levels of ATP in cells. Also, glucokinase, pyruvate dehydrogenase, and α-ketoglutarate dehydrogenase are allosterically inhibited by high levels of G6P, acetyl-CoA, and succinyl-CoA, respectively, which are their immediate products [32]. Generally, high levels of ATP and CCM intermediates such as acetyl-CoA and citrate are correlated with rapid catabolism, whereas accumulation of ADP, AMP, and NAD^+ indicates low level of energy metabolism or nutrient depletion.

5.2 Cataplerosis and Anaplerosis

Metabolic intermediates are commonly exchanged between TCA cycle pathway and other pathways during imbalance metabolism. The exit of metabolites from the TCA cycle to other pathways is known as cataplerosis, while anaplerosis refers to the influx of metabolites from adjacent pathways to replenish the supply of TCA cycle intermediates [58]. The flow of metabolic intermediates between the TCA cycle and other pathways provides valuable information on the physiological or metabolic state of an organism. Cataplerotic reactions are generally induced during high rate of cellular metabolism such as during exponential growth. The excess TCA cycle intermediates are diverted to other pathways to maintain homeostasis and redox balance in the cell. Cataplerosis ensures a constant or balanced flow of intermediates through the TCA cycle during steady-state growth, which is characterized by a constant supply of acetyl-CoA from glycolysis. On the contrary, anaplerotic reactions are induced during low energy or maintenance metabolism, such as during stationary phase growth that is characterized by suboptimal rate of glycolysis. Cataplerotic and anaplerotic reactions are critical in amino acid metabolism. For example, most organisms employ cataplerotic/anaplerotic reactions in the interconversion of pyruvate and α-ketoglutarate to alanine and glutamate, respectively [58, 62].

5.3 Metabolic Redox Shuttles and Shunts

Metabolic shuttles and shunts are short pathways that bypass some reaction steps in a metabolic pathway. Metabolic shuttles and shunts are used to recycle cofactors or depleted metabolites during suboptimal metabolism to maintain redox balance and flow of critical intermediates through metabolic pathways. The intracellular levels of cofactors such as ATP, ADP, AMP, NADH, NAD^+, and CoA are critical for optimum cellular metabolism. Imbalanced cofactor ratios adversely impact the physiological state of an organism, which is often mitigated via the use of metabolic shuttles and shunts. For instance, the NADH generated from glycolysis and the TCA cycle are oxidized to NAD^+ at the electron transport chain via oxidative phosphorylation. The generated NAD^+ molecules are rapidly converted to NADH via glycolytic and TCA cycle reactions during cellular respiration, resulting in low NADH–NAD^+ ratio. The NADH molecules are exclusively generated from glycolysis and decarboxylation of pyruvate during fermentation due to inactivation or absence of the TCA cycle in fermentative organisms [62, 63]. However, NADH in the cell during imbalanced metabolism resulting in high NADH/NAD^+ ratio. Most organisms regenerate depleted NAD^+ molecules using metabolic shuttles such as pyruvate–lactate and glycerol-3-phosphate shuttles (Fig. 2). The pyruvate–lactate shuttle is a reversible lactate dehydrogenase-catalyzed conversion of pyruvate to lactic acid that is coupled to the oxidation of NADH to NAD^+ [62, 64, 65]. The lactic acid generated from the one-step enzyme catalyzed reaction is excreted to prevent the reverse reaction [62, 64, 65]. Similarly, the glycerol-3-phosphate shuttle is mostly used by eukaryotic cells to regenerate NAD^+ from the NADH generated by GAP dehydrogenase catalyzed glycolysis reaction [66, 67]. The glycerol-3-phosphate (G3P) shuttle involves sequential conversion of GAP to DHAP and G3P by triose phosphate isomerase and glycerol-3-phosphate dehydrogenase, respectively [66]. The conversion of DHAP to G3P is coupled to the oxidation of NADH to NAD^+, which helps restore redox balance in the cell. The G3P produced by the shuttle is converted to glycerol by glycerol kinase and subsequently excreted or diverted to fatty acid metabolism [66, 68]. Microorganisms utilize other metabolic shuttles such as the malate–aspartate shuttle to restore and maintain redox and metabolite balance in their cells [69].

Metabolic shunts are used by organisms to bypass reaction steps within a metabolic pathway that is not essential under the prevailing physiological condition. The most studied metabolic shunt is the glyoxylate shunt (Fig. 2), which is very similar to the TCA cycle, except that it bypasses the decarboxylation reactions between isocitrate and succinate, thus eliminating biosynthesis of α-ketoglutarate and succinyl-CoA [63, 70]. The isocitrate molecule is directly cleaved into succinate and glyoxylate by isocitrate lyase [63, 71]. The glyoxylate molecule combines with acetyl-CoA to form malate in a reaction catalyzed by malate synthase [63, 71]. The malate and succinate generated can be consumed by the TCA cycle for energy metabolism [71], but they are generally utilized for cataplerotic reactions. Glyoxylate shunt is also used to divert excess acetyl-CoA and TCA cycle intermediates toward anabolic processes such as gluconeogenesis [71, 72]. Some of the excess acetyl-CoA molecules are converted to acetate and excreted to restor redox balance in the cells during high flux of carbon through CCM [71, 72]. The glyoxylate shunt is mostly activated in response to imbalanced cellular metabolism, especially during stationary phase growth. The glyoxylate shunt is also used to generate TCA cycle intermediates for energy metabolism (anaplerosis) during microbial growth on less ideal substrates such as ethanol, acetate and fatty acids [71–73].

6 Microbial Physiology and Metabolic Engineering

Metabolic engineering is increasingly used by a wide range of industries to improve their manufacturing processes and product(s) yield. Metabolic engineering broadly refers to the process of introducing novel metabolic pathways into an organism or modifying an existing pathway. Metabolic engineering is analogous to the mechanism of viral infection because it diverts the metabolic machinery of the host toward producing molecule(s) or metabolite(s) of interest. The novel metabolic capability is introduced into an organism by inserting extraneous genetic materials using molecular or synthetic biology tools.

Metabolic engineering mostly involves secondary metabolism because primary metabolism is tightly regulated. The efficiency of a metabolic engineering process is determined by the relative yield of the desired product(s). However, uncontrolled production of an engineered molecule might be detrimental to the host organism. Biosynthesis of engineered products largely depends on the availability of metabolic precursors such as acetyl-CoA, which are derived from primary metabolism. Overproduction of an engineered molecule(s) could lead to the depletion of critical primary metabolites that are required for proper metabolic and physiological functions in an organism. For instance, heterologous production of limonene, a cyclic terpene in *E. coli* via the mevalonate pathway requires constant supply of acetyl-CoA and ATP from CCM [74]. Furthermore, stationary phase is generally the ideal stage for heterologous production of a bioproduct because of its characteristic low rate of primary metabolism, compared to exponential growth phase.

Metabolic engineering often generates products and intermediates that are toxic to the host cells. The toxic metabolites often trigger global stress responses that adversely impacts the physiology and growth of the host organism, sometimes resulting the death of the host. The primary objective of metabolic engineering is to the increase the yield or titer of the target products. But this objective is sometimes achieved by elucidating the cause of metabolic toxicity and the host response mechanism. Elucidating a metabolic stress response mechanism is challenging because it requires broad evaluation of the metabolic processes in an organism and deep understanding of the physiology. Therefore, metabolic engineering studies must always incorporate a strategy for evaluating the impact of metabolic intermediates on the physiology of the host organism.

6.1 The Acetyl-CoA "Myth"

Acetyl-CoA is a critical primary metabolite because of its role as a precursor in several metabolic pathways (native and engineered). The level of acetyl-CoA is generally low or unchanged during balanced metabolism. Acetyl-CoA level is tightly regulated in the cell because of its global influence on the physiology of an organism. However, low yield of desired products is often attributed to low level of acetyl-CoA in metabolic engineering studies. Some metabolic engineering studies incorporate strategies for increasing the intracellular pool of acetyl-CoA in their experimental designs. But a high intracellular level of acetyl-CoA is physiologically undesirable because it triggers regulatory responses that results in the excretion of excess acetyl-CoA in the form of molecules such as acetate, glycerol, and succinate to maintain cellular homeostasis. The loss of the excreted metabolites could be attributed to the low yield or conversion efficiency by many engineered strains. Although some organisms can consume the excreted metabolites, this second effort does not negate the adverse effect on the conversion efficiency of the substrate.

Increasing the intracellular level of acetyl-CoA by itself does not improve the efficiency of a pathway or yield of the desired product (s). In fact, an increased intracellular pool of acetyl-CoA is a strong indication that the downstream pathways, including heterologous pathways, are not operating optimally. The accumulation of acetyl-CoA is often associated with slower growth rate due to the existence of a metabolic bottleneck triggered by the heterologous pathway. However, some metabolic bottlenecks affect pathways that are not directly linked to the heterologous pathway, and hence cannot be deciphered by monitoring targeted metabolites. The intracellular level of acetyl-CoA is expected to be low when there is strong pull or efficient conversion by the heterologous pathway, although the acetyl-CoA could be diverted to other secondary pathways when the heterologous pathway is inefficient. Therefore, it is always important to include well-designed physiological experiments in the design of metabolic engineering studies. Physiological studies will help detect bottlenecks and side reactions that divert resources from the target products.

7 Experimental Design and Metabolism Prediction

Although metabolomics data is sufficient for predicting metabolism, it does not provide complete information on the physiology of an organism. Recent advances in information and computational technologies has increasingly shifted the focus of microbial physiology studies to metabolic prediction. But predicting metabolism is a complex collaborative process that requires contribution from multiple scientific disciplines such as microbiology, analytical chemistry, computer science, and statistics. The complexity of metabolic prediction is due to interconnectivity of metabolic pathways, as well as the ability of organisms to rapidly alter their metabolism in response to internal and external stimuli. The fluidity of metabolism sometimes undermines reproducibility of the data generated from these studies, which has adverse impact on the veracity of scientific conclusions and any associated metabolic prediction models. Therefore, an effective metabolic prediction study requires a combination of data from metabolomics, proteomics, and transcriptomics analyses.

Experimental design is the framework or blueprint of a scientific study that employs a priori and empirical knowledge from all aspects of the research subject. The design of a study for predicting microbial physiology and underlying metabolism is influenced by the study objective. An effective metabolic prediction experiment must have a clearly defined objective, hypothesis, analytical techniques and methods, as well as a strategy for data analysis and interpretation. For instance, the design of a study for evaluating the mechanism of glucose catabolism in *E. coli* would be markedly different from one that focuses on the role of flagella in chemotaxis in the organism. The effectiveness of a metabolic prediction model also depends on the quality of the data generated, as well as the availability of robust data analysis and computational tools. Furthermore, a strong metabolic prediction model requires large sets of data from well-designed metabolic studies that have been independently validated.

In general, the quality of the data generated from metabolic studies is influenced by the execution of the experiments. Most flawed metabolic prediction models are often attributed to poor execution of the experiments, which often result in erroneous scientific conclusions. The flaws of metabolic prediction models can also be attributed to computational errors and poor data interpretation. Computational errors commonly result from invalid scientific assumptions, which are often attributed to limited knowledge of metabolism. Invalid data analysis and interpretation due to computational errors such as weak algorithms can be readily corrected once detected. But it is practically impossible to correct error resulting from poorly generated experimental data without altering the experimental design. The challenge of developing effective metabolic prediction models is exacerbated by the knowledge gap among the diverse experts involved in these studies. Therefore, it is critical to involve each of these experts in the design of the experiments to minimize the impact of the knowledge gap.

7.1 Executing Studies for Predicting Microbial Metabolism

The significance of growth in microbial metabolic studies is broadly understated and often considered trivial due to its relative simplicity. But microbial growth is the most important and consequential aspect of metabolic studies because it provides valuable information on the physiology of an organism. Irrespective of the study objective, the first and critical step in the execution of a well-designed metabolic study is to establish a valid and reproducible growth profile for the organism(s) under the test condition. In fact, it is always a good practice to conduct preliminary studies to optimize the growth condition to limit experimental errors. Preliminary studies also help optimize the strategy for sample collection and preparation, as well as identify potential pitfalls that might adversely impact the studies.

The strategy for evaluating microbial growth depends on the type of organism, growth condition and study objective. For instance, optical density measurement is the preferred method for growth evaluation for species such as *E. coli* that produce uniformly dispersed cells in liquid cultures. However, optical density is not ideal for evaluating growth of *Streptomyces* and fungal species that often produce cells that aggregate into particulate masses in liquid cultures. Also, complete growth curve is sometimes required, but often growth measurement at sampling points is sufficient. Proper execution of microbial growth, sample collection and handling, is often critical because the design of metabolic studies could be easily altered at these steps. For instance, the growth rate and physiology of a fast-growing organism is temporarily altered each time agitation of culture is stopped for sampling. The altered growth rate is due to change in the rate of cellular metabolism resulting from changes in temperature and gas exchange rate. These sudden metabolic changes, which can cumulatively alter the physiology of an organism are often responsible for most errors in metabolic studies.

In general, the most consequential errors in metabolic studies are those that are not noticed during execution of the experiments. Unnoticed errors are important in metabolic studies because they insidiously alter the design of an experiment and scientific conclusion reached from the data generated from these studies. A valid conclusion that reflects the actual physiology of the organisms under an erroneously altered experiment might be rejected because it disagrees with the expected outcome based on the original experimental design. Therefore, it is very important that the execution of metabolic studies be supervised by a trained microbial physiologist to maintain the integrity and design of the experiment. Similarly, the sample collection and preparation should be executed in collaboration with analytical chemists to ensure that the data generated reflect the true metabolic state in the organism.

The sampling points and frequency in metabolic studies depends on study objective and cultural characteristics of the organism. For instance, mid-exponential growth phase is the preferred sampling point for establishing the profile of the metabolic processes in an organism during growth on a given carbon source. In contrast, late-exponential or early stationary phase are the preferred phases for evaluating cellular metabolism during the heterologous production of a targeted metabolite by an engineered strain. Inducing a heterologous pathway at stationary phase improves conversion efficiency by minimizing the diversion of the substrate (s) toward cell growth. The quality of sample collection and processing is also carried out rapidly to minimize metabolite degradation, which could adversely impact the quality and accuracy of the data generated.

8 Conclusion

Physiological studies and metabolic prediction require effective strategy for quantitative metabolite analysis and the accuracy of metabolomics makes it an ideal analytical tool for these studies. Metabolomics provides a snapshot of metabolism at any given moment in an organism, which is further enhanced by the addition of data from other analytical strategies such as transcriptomics, proteomics, and biochemical analyses. This comprehensive analytical approach is very critical for predicting microbial metabolism because of interconnectivity of metabolic processes and rapid physiological changes in these organisms in response to fluctuating environments. In addition to reproducible analytical tools, metabolic prediction requires fundamental understanding of metabolism and physiology, which is critical for experimental design and data interpretation. Furthermore, improving the throughput of microbial growth, sample preparation, and analysis will enhance the utility of metabolomics as a routine tool for metabolic studies.

Acknowledgments

This work was part of the DOE Joint BioEnergy Institute (http://www.jbei.org) supported by the US Department of Energy, Office of Science, Office of Biological and Environmental Research, through contract DE-AC02-05CH11231 between Lawrence Berkeley National Laboratory and the US Department of Energy.

References

1. Crick F (1970) Central dogma of molecular biology. Nature 227:561–563
2. Schreiber SL (2005) Small molecules: the missing link in the central dogma. Nat Chem Biol 1:64–66
3. Metallo CM, Vander Heiden MG (2013) Understanding metabolic regulation and its influence on cell physiology. Mol Cell 49:388–398
4. Marson A, Levine SS, Cole MF, Frampton GM, Brambrink T, Johnstone S, Guenther MG, Johnston WK, Wernig M, Newman J, Calabrese JM, Dennis LM, Volkert TL, Gupta S, Love J, Hannett N, Sharp PA, Bartel DP, Jaenisch R, Young RA (2008) Connecting microRNA genes to the core transcriptional regulatory circuitry of embryonic stem cells. Cell 134:521–533
5. Reik W (2007) Stability and flexibility of epigenetic gene regulation in mammalian development. Nature 447:425–432
6. Rochfort S (2005) Metabolomics reviewed: a new "omics" platform technology for systems biology and implications for natural products research. J Nat Prod 68:1813–1820
7. McLean TI (2013) "Eco-omics": a review of the application of genomics, transcriptomics, and proteomics for the study of the ecology of harmful algae. Microb Ecol 65:901–915
8. Monod J (1949) The growth of bacterial cultures. Annu Rev Microbiol 3:371–394

9. Larsson U, Hörstedt P, Normark S (1979) Frequency of dividing cells, a new approach to the determination of bacterial growth rates in aquatic environments. Appl Environ Microbiol 37:805–812
10. Maier RM, Pepper I, Gerba C (2000) Bacterial growth. Environ Microbiol 2000:43–59
11. Peleg M, Corradini MG (2011) Microbial growth curves: what the models tell us and what they cannot. Crit Rev Food Sci Nutr 51:917–945
12. Herbert D, Elsworth R, Telling R (1956) The continuous culture of bacteria; a theoretical and experimental study. Microbiology 14:601–622
13. Joshua CJ, Dahl R, Benke PI, Keasling JD (2011) Absence of diauxie during simultaneous utilization of glucose and xylose by Sulfolobus acidocaldarius. J Bacteriol 193:1293–1301
14. Baranyi J, Roberts TA (1994) A dynamic approach to predicting bacterial growth in food. Int J Food Microbiol 23:277–294
15. Rolfe MD, Rice CJ, Lucchini S, Pin C, Thompson A, Cameron ADS, Alston M, Stringer MF, Betts RP, Baranyi J, Peck MW, Hinton JCD (2012) Lag phase is a distinct growth phase that prepares bacteria for exponential growth and involves transient metal accumulation. J Bacteriol 194:686–701
16. Joshua CJ, Perez LD, Keasling JD (2013) Functional characterization of the origin of replication of pRN1 from Sulfolobus islandicus REN1H1. PLoS One 8:e84664
17. Saldanha AJ, Brauer MJ, Botstein D (2004) Nutritional homeostasis in batch and steady-state culture of yeast. Mol Biol Cell 15:4089–4104
18. Monod J (1942) Recherches sur la croissance des cultures bactâeriennes. Hermann, Paris
19. Kolter R, Siegele DA, Tormo A (1993) The stationary phase of the bacterial life cycle. Annu Rev Microbiol 47:855–874
20. Brown JV, Wiles R, Prentice GA (1979) The effect of a modified tyndallization process upon the sporeforming bacteria of milk and cream. Int J Dairy Technol 32:109–112
21. Finkel SE (2006) Long-term survival during stationary phase: evolution and the GASP phenotype. Nat Rev Microbiol 4:113–120
22. Gorke B, Stulke J (2008) Carbon catabolite repression in bacteria: many ways to make the most out of nutrients. Nat Rev Microbiol 6:613–624
23. Magasanik B (1961) Catabolite repression. Cold Spring Harb Symp Quant Biol 26:249–256
24. Sudarsan S, Dethlefsen S, Blank LM, Siemann-Herzberg M, Schmid A (2014) The functional structure of central carbon metabolism in Pseudomonas putida KT2440. Appl Environ Microbiol 80:5292–5303
25. Sulpice R, McKeown PC (2015) Moving toward a comprehensive map of central plant metabolism. Annu Rev Plant Biol 66:187–210
26. Brouqui P, Raoult D (2001) Endocarditis due to rare and fastidious bacteria. Clin Microbiol Rev 14:177–207
27. Gray CT, Wimpenny JWT, Hughes D, Mossman MR (1966) Regulation of metabolism in facultative bacteria: 1. Structural and functional changes in Escherichia coli associated with shifts between the aerobic and anaerobic states. Biochim Biophys Acta 117:22–32
28. Chen X, Schreiber K, Appel J, Makowka A, Fähnrich B, Roettger M, Hajirezaei MR, Sönnichsen FD, Schönheit P, Martin WF (2016) The Entner–Doudoroff pathway is an overlooked glycolytic route in cyanobacteria and plants. Proc Natl Acad Sci 113:5441–5446
29. Kengen SW, de Bok FA, van Loo ND, Dijkema C, Stams AJ, de Vos WM (1994) Evidence for the operation of a novel Embden-Meyerhof pathway that involves ADP-dependent kinases during sugar fermentation by Pyrococcus furiosus. J Biol Chem 269:17537–17541
30. Gautheron DC (1984) Mitochondrial oxidative phosphorylation and respiratory chain: review. J Inherit Metab Dis 7:57–61
31. Chaban Y, Boekema EJ, Dudkina NV (2014) Structures of mitochondrial oxidative phosphorylation supercomplexes and mechanisms for their stabilisation. Biochim Biophys Acta 1837:418–426
32. Fernie AR, Carrari F, Sweetlove LJ (2004) Respiratory metabolism: glycolysis, the TCA cycle and mitochondrial electron transport. Curr Opin Plant Biol 7:254–261
33. Gel'man NS, Lukoyanova MA, Ostrovskii DN (1967) Oxidative phosphorylation in bacteria. In: Gel'man NS, Lukoyanova MA, Ostrovskii DN (eds) Respiration and phosphorylation of bacteria. Springer, Boston, MA, pp 161–192. https://doi.org/10.1007/978-1-4899-5526-5_5
34. Saraste M (1999) Oxidative Phosphorylation at the fin de siècle. Science 283:1488
35. Wamelink MMC, Struys EA, Jakobs C (2008) The biochemistry, metabolism and inherited defects of the pentose phosphate pathway: a review. J Inherit Metab Dis 31:703–717

36. Clark DP (1989) The fermentation pathways of *Escherichia coli*. FEMS Microbiol Rev 5:223–234
37. Spaans SK, Weusthuis RA, van der Oost J, Kengen SWM (2015) NADPH-generating systems in bacteria and archaea. Front Microbiol 6:742
38. Fuhrer T, Sauer U (2009) Different biochemical mechanisms ensure network-wide balancing of reducing equivalents in microbial metabolism. J Bacteriol 191:2112–2121
39. Horecker BL (2002) The pentose phosphate pathway. J Biol Chem 277:47965–47971
40. Lenzen S (2014) A fresh view of glycolysis and glucokinase regulation: history and current status. J Biol Chem 289:12189–12194
41. Martin SA, Russell JB (1986) Phosphoenolpyruvate-dependent phosphorylation of hexoses by ruminal bacteria: evidence for the phosphotransferase transport system. Appl Environ Microbiol 52:1348–1352
42. Verhees CH, Kengen SWM, Tuininga JE, Schut GJ, Adams MWW, De Vos WM, Van der Oost J (2004) The unique features of glycolytic pathways in Archaea. Biochem J 377:819–822
43. Ahmed H, Ettema TJG, Tjaden B, Geerling ACM, van der Oost J, Siebers B (2005) The semi-phosphorylative EntnerDoudoroff pathway in hyperthermophilic archaea: a re-evaluation. Biochem J 390:529–540
44. Snijders APL, Walther J, Peter S, Kinnman I, de Vos MG, van de Werken HJ, SJJ B, van der Oost J, Wright PC (2006) Reconstruction of central carbon metabolism in *Sulfolobus solfataricus* using a two-dimensional gel electrophoresis map, stable isotope labelling and DNA microarray analysis. Proteomics 6:1518–1529
45. Kengen SWM (2017) Pyrococcus furiosus, 30 years on. J Microbial Biotechnol 10:1441–1444
46. Kengen SM, Stams AJM, de Vos WM (1996) Sugar metabolism of hyperthermophiles. FEMS Microbiol Rev 18:119–137
47. Imanaka H, Yamatsu A, Fukui T, Atomi H, Imanaka T (2006) Phosphoenolpyruvate synthase plays an essential role for glycolysis in the modified Embden-Meyerhof pathway in Thermococcus kodakarensis. Mol Microbiol 61:898–909
48. Sakuraba H, Utsumi EMI, Schreier HJ, Ohshima T (2001) Transcriptional regulation of phosphoenolpyruvate synthase by maltose in the hyperthermophilic Archaeon, Pyrococcus furiosus. Soc Biotechnol 92:108–113
49. Akram M (2014) Citric acid cycle and role of its intermediates in metabolism. Cell Biochem Biophys 68:475–478
50. Sugden MC, Holness MJ (2003) Recent advances in mechanisms regulating glucose oxidation at the level of the pyruvate dehydrogenase complex by PDKs. Am J Physiol Endocrinol Metab 284:E855
51. Berg IA (2011) Ecological aspects of the distribution of different autotrophic CO2 fixation pathways. Appl Environ Microbiol 77:1925–1936
52. Hügler M, Wirsen CO, Fuchs G, Taylor CD, Sievert SM (2005) Evidence for autotrophic CO2 fixation via the reductive tricarboxylic acid cycle by members of the ε subdivision of Proteobacteria. J Bacteriol 187:3020–3027
53. Tang K-H, Blankenship RE (2010) Both forward and reverse TCA cycles operate in green sulfur bacteria. J Biol Chem 285:35848–35854
54. Buchanan BB, Arnon DI (1990) A reverse KREBS cycle in photosynthesis: consensus at last. Photosynth Res 24:47–53
55. Hers HG, Hue L (1983) Gluconeogenesis and related aspects of glycolysis. Annu Rev Biochem 52:617–653
56. Pilkis SJ, Granner DK (1992) Molecular physiology of the regulation of hepatic gluconeogenesis and glycolysis. Annu Rev Physiol 54:885–909
57. Chulavatnatol M, Atkinson DE (1973) Phosphoenolpyruvate synthetase from *Escherichia coli*: effects of adenylate energy charge and modifier concentrations. J Biol Chem 248:2712–2715
58. Owen OE, Kalhan SC, Hanson RW (2002) The key role of anaplerosis and cataplerosis for citric acid cycle function. J Biol Chem 277:30409–30412
59. Jablonsky J, Papacek S, Hagemann M (2016) Different strategies of metabolic regulation in cyanobacteria: from transcriptional to biochemical control. Sci Rep 6:33024
60. Shimizu K (2013) Metabolic regulation of a bacterial cell system with emphasis on Escherichia coli metabolism. ISRN Biochem 2013:47
61. Chubukov V, Gerosa L, Kochanowski K, Sauer U (2014) Coordination of microbial metabolism. Nat Rev Microbiol 12:327–340
62. Gray LR, Tompkins SC, Taylor EB (2014) Regulation of pyruvate metabolism and human disease. Cell Mol Life Sci 71:2577–2604
63. Erkut C, Gade VR, Laxman S, Kurzchalia TV (2016) The glyoxylate shunt is essential for desiccation tolerance in C elegans and budding yeast. Elife 5:e13614
64. Cerdán S, Rodrigues TB, Sierra A, Benito M, Fonseca LL, Fonseca CP, García-Martín ML

(2006) The redox switch/redox coupling hypothesis. Neurochem Int 48:523–530
65. Draoui N, Feron O (2011) Lactate shuttles at a glance: from physiological paradigms to anti-cancer treatments. Dis Model Mech 4:727–732
66. Larsson C, Påhlman IL, Ansell R, Rigoulet M, Adler L, Gustafsson L (1998) The importance of the glycerol 3-phosphate shuttle during aerobic growth of Saccharomyces cerevisiae. Yeast (Chichester, Engl) 14:347–357
67. Henningsen BM, Hon S, Covalla SF, Sonu C, Argyros DA, Barrett TF, Wiswall E, Froehlich AC, Zelle RM (2015) Increasing anaerobic acetate consumption and ethanol yields in Saccharomyces cerevisiae with NADPH-specific alcohol dehydrogenase. Appl Environ Microbiol 81:8108–8117
68. Hagopian K, Ramsey JJ, Weindruch R (2008) Enzymes of glycerol and glyceraldehyde metabolism in mouse liver: effects of caloric restriction and age on activities. Biosci Rep 28:107–115
69. Cheeseman AJ, Clark JB (1988) Influence of the malate-aspartate shuttle on oxidative metabolism in synaptosomes. J Neurochem 50:1559–1565
70. Ahn S, Jung J, Jang I-A, Madsen EL, Park W (2016) Role of glyoxylate shunt in oxidative stress response. J Biol Chem 291:11928–11938
71. Dunn MF, Ramírez-Trujillo JA, Hernández-Lucas I (2009) Major roles of isocitrate lyase and malate synthase in bacterial and fungal pathogenesis. Microbiology 155:3166–3175
72. Wolfe AJ (2005) The acetate switch. Microbiol Mol Biol Rev 69:12–50
73. Maloy SR, Bohlander M, Nunn WD (1980) Elevated levels of glyoxylate shunt enzymes in Escherichia coli strains constitutive for fatty acid degradation. J Bacteriol 143:720–725
74. Alonso-Gutierrez J, Chan R, Batth TS, Adams PD, Keasling JD, Petzold CJ, Lee TS (2013) Metabolic engineering of Escherichia coli for limonene and perillyl alcohol production. Metab Eng 19:33–41

Part II

Metabolite Sample Preparation

Chapter 4

Untargeted Soil Metabolomics Using Liquid Chromatography–Mass Spectrometry and Gas Chromatography–Mass Spectrometry

Tami L. Swenson and Trent R. Northen

Abstract

The molecular composition of soil organic matter (SOM) sets the foundation for terrestrial microbial community structures and carbon cycling dynamics. However, the specific chemical constituents of SOM are underexplored. In this chapter we present a protocol for the extraction of small molecule metabolites from soil followed by compound detection and identification using liquid chromatography–mass spectrometry and gas chromatography–mass spectrometry. There are options within the protocol to assess either the extracellular pool of metabolites or the total pool (including intracellular) and either polar or nonpolar metabolites, depending on the reader's research interests. These methods can be followed individually for a more targeted analysis or all methods can be combined to obtain a more comprehensive understanding of SOM metabolite composition (such as amino acids, nucleobases, organic acids, fatty acids, carbohydrates, secondary metabolites, and antibiotics).

Key words Soil organic matter, Metabolite extraction, Liquid chromatography-mass spectrometry, Gas chromatography–mass spectrometry, Metabolomics

1 Introduction

Soils play an essential role in the global carbon cycle [1]. Despite the importance of soils, the molecular composition of soil organic matter (SOM) is still largely unknown. This information can be used to understand and predict the roles of terrestrial ecosystems in climate change through for example understanding the lability of SOM and its connection to microbial activities. Fortunately, many recent advances in analytical techniques have begun to shed light on SOM molecular structure and composition. One of the biggest insights has been through the use of imaging technology which completely shifted our understanding of SOM structure. Historically, SOM was thought to be composed of large polymeric humic substances [2], but now SOM is increasingly recognized to be largely composed of a variety of plant- and microbe-originating

Edward E.K. Baidoo (ed.), *Microbial Metabolomics: Methods and Protocols*, Methods in Molecular Biology, vol. 1859, https://doi.org/10.1007/978-1-4939-8757-3_4,

small molecules [3]. Although bulk soils may contain thousands of compounds, what may be of the most interest for microbiologists are the water-soluble metabolites (<1000 Da) that are exchanged within soil microbial communities (i.e., the water-extractable exo-metabolite pool). Ultimately, it is the composition of this bioavailable and non-mineral associated [4] metabolite pool that shapes microbial communities and therefore the biological activity (carbon cycling) of soil [5]. Other fractions that may also provide critical insights into soil ecosystems include the intracellular microbial metabolites (which may be released into the environment during cell lysis) and nonpolar metabolites involved in microbial defenses (e.g., secondary metabolites and antibiotics).

A variety of techniques currently exist to characterize the small molecule composition of soils including capillary electrophoresis–mass spectrometry [6], gas chromatography-mass spectrometry (GC-MS) [7, 8], nuclear magnetic resonance (NMR) [9], Fourier transform ion cyclotron resonance mass spectrometry [10] and high performance liquid chromatography [11]. Of these, we focus on widely available liquid chromatography-mass spectrometry (LC-MS) and GC-MS approaches. GC-MS is a very common and relatively inexpensive method, but requires sample derivatization to increase volatility of analytes and has some limitations in the range of detectable metabolites. Another frequently used method that is becoming useful for soil metabolomics is LC-MS [12]. This technique yields thousands of chemical features that can be identified by comparison with authentic chemical standards relative to retention time, *m/z* values and fragmentation spectra. LC-MS also provides a path to identification of novel compounds, typically in conjunction with isolation and NMR structural analysis [13]. LC-MS and GC-MS approaches can either be used individually or samples can be fractionated and analyzed using a complementary set of methods. For example, for this protocol, GC-MS is used to identify carbohydrates and small organic acids while LC-MS is used for the analysis of larger and more diverse sets of polar and nonpolar metabolites.

In this chapter, we present a protocol to extract and analyze the small molecule composition of soil, either extracellular or total (intracellular + extracellular) (Fig. 1). Within the protocol, there are options to extract either polar metabolites (e.g., amino acids, nucleobases, organic acids) or nonpolar metabolites (e.g., secondary metabolites, antibiotics) followed by metabolite detection using either LC-MS or GC-MS, depending on the reader's instrument accessibility and scientific goals. There are many LC column options that are used to separate different classes of molecules, but here we present two common and broad chromatographies: hydrophilic interaction liquid chromatography (i.e., HILIC, to detect polar metabolites) and reversed-phase (i.e., RP, to detect nonpolar metabolites). The following details are specific to the instruments and therefore it is recommended that the reader consult instrument manuals for method optimization. Data analysis

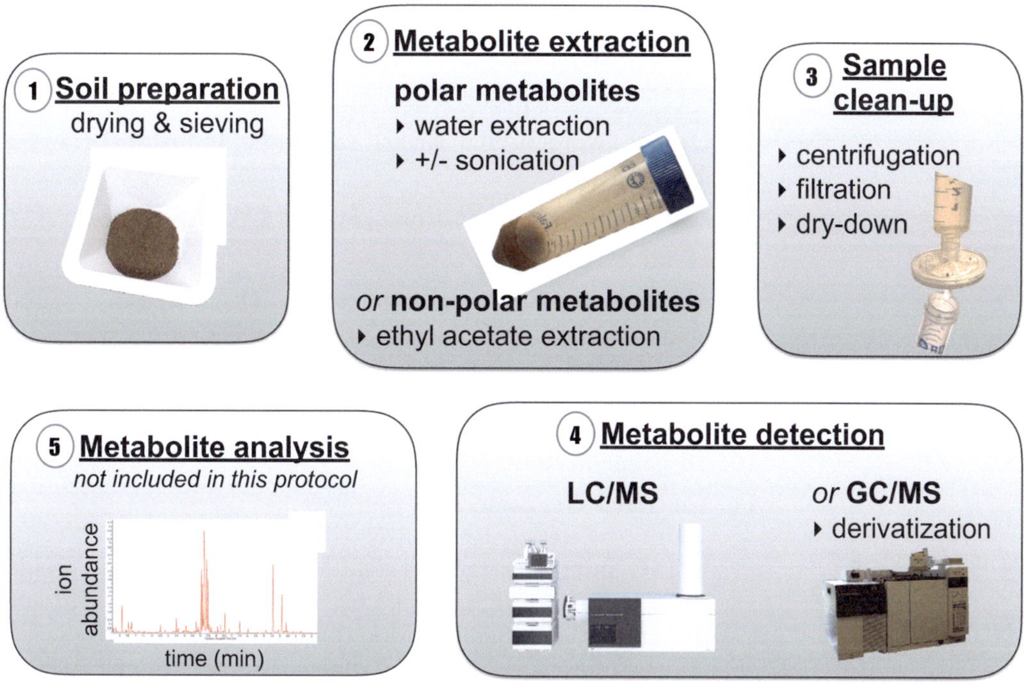

Fig. 1 General workflow for the extraction and detection of soil metabolites. The major steps (included in this protocol) involved in the extraction and detection of polar and nonpolar metabolites are shown

techniques are beyond the scope of this chapter. While GC-MS data analysis tends to be more straightforward due to the availability of electron impact mass spectral libraries (e.g., the FiehnLib [14]), LC-MS data analysis may be more challenging. However, there are many choices of software and readily accessible web-based tools (e.g., Metabolite Atlas [15]) to facilitate LC-MS feature detection and annotation.

2 Materials

Diligently follow all appropriate waste disposal regulations when disposing of waste materials.

2.1 Soil Preparation (Drying and Sieving)

1. 70% ethanol.
2. Labconco FreeZone 2.5 lyophilizer.
3. 50 mL polypropylene conical tubes.
4. −80 °C freezer.
5. 18-gauge needles.
6. AS 200 sieve shaker, sieve (2 mm mesh), collection pan (Retsch).
7. Balance.
8. Weigh boats and spatulas.

2.2 Soil Extraction (Polar Metabolites)

1. LC-MS grade water (JT Baker), kept at 4 °C.
2. Serological pipettes (10 mL) and pipettor.
3. Q125 QSonica probe sonicator.
4. Orbital-Genie orbital shaker (Scientific Industries), kept at 4 °C.
5. Eppendorf 5810R centrifuge, kept at 4 °C.
6. 10 mL syringes.
7. 32 mm 0.45 μm syringe filters (Pall Acrodisc Supor membrane).
8. 15 mL polypropylene conical tubes.
9. −80 °C freezer.
10. 18-gauge needles.
11. Labconco FreeZone 2.5 lyophilizer.

2.3 HILIC LC-MS

2.3.1 Chemicals

1. HILIC internal standards solution: 4-(3,3-dimethyl-ureido) benzoic acid (Sigma-Aldrich), d_4-lysine (Sigma-Aldrich), d_5-benzoic acid (Sigma-Aldrich), 9-anthracene carboxylic acid (Sigma-Aldrich), ^{13}C-^{15}N-L-phenylalanine (Sigma-Aldrich) each at 5 μg/mL in LC-MS grade methanol.
2. Mobile phase A: 5 mM ammonium acetate in water.
3. Mobile phase B: 95% acetonitrile, 5% 100 mM ammonium acetate in water.

2.3.2 Equipment

1. Vortexer.
2. 1.5 mL Eppendorf tubes.
3. Eppendorf 5430R centrifuge.
4. 0.22 μm centrifugal membranes (Ultrafree-MC-GV PVDF, Millipore).
5. 2 mL screw cap glass vials (Agilent).
6. 250 μL glass vial inserts with polymer feet (Agilent).
7. Vial screw caps with septa (Agilent).
8. ZIC®-HILIC column (100 mm × 2.1 mm, 3.5 μm, 200 Å) (Millipore).
9. Agilent 1290 UHPLC.
10. Thermo QExactive MS.

2.4 GC-MS

2.4.1 Chemicals

1. LC-MS grade methanol.
2. GC-MS internal standard solution: d_{27}-myristic acid at 50 μg/mL in LC-MS grade methanol.
3. O-methoxyamine hydrochloride (Supelco).
4. Pyridine.

5. N-methyl-N-trimethylsilyltrifluoroacetamide (MSTFA) with 1% trimethylchlorosilane (Restek).
6. Fatty acid methyl ester (FAME) marker solution: methylcaprylate (20 mg), methylperlargonate (20 mg), methylcaprate (20 mg), methyllaurate (20 mg), methylmyristate (20 mg), methylpalmitate (20 mg), methylstearate (10 mg), methyleicosanoate (10 mg), methyldocosanoate (10 mg), methyllignocerate (10 mg), methylhexacosanoate (10 mg), methyloctacosanoate (10 mg), methyltriacontanoate (10 mg), all in 25 mL chloroform.

2.4.2 Equipment

1. Vortexer.
2. 1.5 mL Eppendorf tubes.
3. Eppendorf 5430R centrifuge.
4. 0.22 μm centrifugal membranes (Ultrafree-MC-GV PVDF, Millipore).
5. Savant SpeedVac SPD111 (Thermo Scientific).
6. Glass serological pipette (1 mL) and pipettor.
7. Eppendorf thermomixer.
8. 2 mL screw cap glass vials (Agilent).
9. 250 μL glass vial inserts with polymer feet (Agilent).
10. Vial screw caps with septa (Agilent).
11. Agilent 7890 GC equipped with a 15 m length × 0.25 mm ID Rtx-5MS column with a 10 m Integra-Guard column (Restek).
12. Gerstel MultiPurpose Sampler.
13. Agilent 5977 single quadrupole MS.

2.5 Nonpolar Metabolite Extraction and RP LC-MS

2.5.1 Chemicals

1. Nonpolar extractant: 50% ethyl acetate, 50% water.
2. C18 internal standard solution: 2-Amino-3-bromo-5-methylbenzoic acid (Sigma-Aldrich) at 5 μg/mL in LC-MS grade methanol.
3. Mobile phase C: 0.1% formic acid in water.
4. Mobile phase D: 0.1% formic acid in acetonitrile.

2.5.2 Equipment

1. Glass serological pipettes (10 mL) and pipettor.
2. Orbital-Genie orbital shaker (Scientific Industries), kept at 4 °C.
3. Eppendorf 5810R centrifuge, kept at 4 °C.
4. Glass Pasteur pipettes.
5. 15 mL polypropylene conical tubes.
6. Savant SpeedVac SPD111 (Thermo Scientific).

7. Vortexer.
8. 1.5 mL Eppendorf tubes.
9. Eppendorf 5430R centrifuge.
10. 0.22 μm centrifugal membranes (Ultrafree-MC-GV PVDF, Millipore).
11. 2 mL screw cap glass vials (Agilent).
12. 250 μL glass vial inserts with polymer feet (Agilent).
13. Vial screw caps with septa (Agilent).
14. C18 column: Zorbax Eclipse Plus C18 (50 mm × 2.1 mm, 1.8 μm) (Agilent).
15. Agilent 1290 UHPLC.
16. Thermo QExactive MS.

3 Methods

The following protocol was established using a sandy loam soil. Appropriate adjustments may need to be made depending on the soil type and mineral content. It is highly recommended that the protocol be followed initially with a small set of pilot samples (e.g., a mixture of target compounds to determine recovery) prior to full-scale analysis. Extraction controls should also be included throughout the protocol (samples without soil) to account for extraction background. *Note that it is critical that researchers follow best practices as far as lab safety. It is beyond the scope of this protocol to describe safety protocols.*

3.1 Soil Drying

To ensure comparable extraction efficiency from sample-to-sample, soil should be dried prior to extraction. Lyophilization is a preferred method over heating (i.e., speedvac) to limit metabolite degradation. It is recommended to measure soil weight before and after drying to determine the percent water weight of original sample (*see* **Note 1**).

1. Place each soil sample in a container suitable for lyophilization such as a 50 mL polypropylene conical tube.
2. Freeze at −80 °C.
3. Poke holes in lids using 18-gauge needle and place samples in precooled lyophilizer. Turn on vacuum according to instrument protocol.
4. Allow soil to dry completely. This process may take anywhere from a few hours to a few days, depending on the amount of sample and its water content.

3.2 Soil Sieving

These steps may not need to be done for certain fine-textured soils (such as sand-based desert soils).

1. Prepare a 2 mm mesh sieve and a collection pan by spraying with 70% ethanol and wiping clean. Be sure everything is dry prior to the addition of soil samples.
2. Place soil in sieve, place over collection pan and sieve for approximately 2 min at medium vibration or until all the sample has passed through.
3. For each sample, place 2 g sieved soil (dry weight) in appropriately labeled 50 mL conical tube for extraction (*see* **Note 2**).

3.3 Polar Metabolite Extraction

These steps are for the extraction of mostly polar metabolites typically found in dissolved organic matter (e.g., amino acids, nucleobases, sugars). If the focus is on nonpolar metabolites, follow steps in Subheading 3.6 instead. Alternatively, at this time, soil samples can be divided into two fractions; one for polar metabolite extraction and one for nonpolar metabolites.

1. Add 8 mL of LC-MS grade water to each (2 g) sample (*see* **Note 3**).
2. If intracellular metabolites are to be analyzed, sonicate samples using a probe sonicator at 50% amplitude for 2 × 30 s.
3. Place samples on a refrigerated orbital shaker and shake at 200 rpm for 1 h (angle the tube rack approximately 45 degrees to allow for adequate sample movement and extraction).
4. Centrifuge samples at 3220 × *g* for 15 min (at 4 °C).
5. Decant supernatant into a 10 mL syringe fitted to a 0.45 μm (32 mm) filter disc and filter into 15 mL conical tubes (*see* **Note 4**).
6. Freeze extracts at −80 °C.
7. Poke holes in lids using 18-gauge needle and lyophilize extracts for 6–8 h or until dry.
8. Once samples are dry, proceed with resuspension according to Subheading 3.4 for LC-MS or Subheading 3.5 for GC-MS. To use both LC-MS and GC-MS, follow Subheading 3.4 until noted.

3.4 Polar Metabolite LC-MS Sample Preparation and Analysis

An Agilent 1290 UHPLC and Thermo QExactive MS is used in this protocol. Parameters and methods will need to be adjusted and optimized according to the specific instruments. Here, polar metabolites are separated using normal-phase chromatography and detected by MS.

1. Resuspend dried water extracts in 200 μL of HILIC internal standards solution (*see* **Notes 5** and **6**), making sure to recover sample that may have adhered to the sides of the vial during lyophilization. Vortex samples to get extracts into solution (*see* **Note 7**).

2. Transfer samples to 1.5 mL tubes.
3. Centrifuge samples at 5000 × *g* for 5 min to pellet insoluble salts.
4. Add supernatant to centrifugal membranes and filter by centrifuging at 5000 × *g* for 3 min.
5. At this point, **if performing both LC-MS and GC-MS**, remove an aliquot (e.g., 100 μL) for GC-MS and place into a new 1.5 mL tube. Add 100 μL of GC-MS internal standard solution. Proceed to Subheading 3.5, **step 5**.
6. Transfer samples to glass LC-MS vials with glass inserts, capping immediately.
7. For HILIC separation, use a ZIC®-HILIC column with the following LC parameters: Autosampler temperature of 4 °C, sample injection volume of 2 μL, and a column compartment temperature of 40 °C left/right.
8. Chromatographic separation of polar metabolites can be achieved by the gradient shown in Table 1 using mobile phases A and B.
9. For mass spectra, collect data in both negative and positive polarities with a mass range of 70–1050 *m/z* in centroid data mode. Acquire MS/MS fragmentation spectra using collision energies of 10–40 eV.
10. It is recommended that a known mixture of quality control standards be injected approximately every ten samples to check for consistent instrument performance (*m/z* accuracy and retention time shifting) throughout the sequence. Methanol blanks should also be included frequently to check for metabolite carryover.

Table 1
Recommended LC gradient for HILIC chromatography

Time (min)	A (%)	B (%)	flow (mL/min)
0.0	0	100	0.45
1.5	0	100	0.45
15.0	35	65	0.45
18.0	100	0	0.60
23.0	100	0	0.60
25.0	0	100	0.45
29.0	0	100	0.45
30.0	0	100	0.45

11. Data analysis can be done with the reader's software choice and may be targeted or untargeted. An example of LC-MS-detected polar metabolites extracted from biological soil crust (with and without sample sonication) is shown in Fig. 2.

3.5 Polar Metabolite GC–MS Sample Preparation and Analysis

For GC-MS, there are many derivatization procedures to detect different classes of metabolites, but here we report the protocol for a standard MSTFA derivatization to detect a broad range of metabolites.

1. Resuspend dried water extracts in 200 μL of methanol (*see* **Note 5**). Add 100 μL of GC-MS internal standard solution (*see* **Note 6**). Be sure to recover sample that may have adhered to the sides of the vial during lyophilization. Vortex samples to get extracts into solution (*see* **Note 7**).
2. Transfer samples to 1.5 mL tubes.
3. Centrifuge samples at 5000 × *g* for 5 min to pellet insoluble salts.
4. Add supernatant to centrifugal membranes and filter by centrifuging at 5000 × *g* for 3 min.
5. Dry samples using a speedvac for about 30 min (*see* **Note 8**).
6. Prepare a 40 mg/mL methoxamine solution in pyridine (using a glass pipette) and dissolve by shaking at 60 °C for 15 min on a thermomixer. Be sure to make an adequate amount (10 μL is needed per sample).
7. Add 10 μL methoxamine solution to each dried sample.
8. Shake samples in thermomixer at maximum speed at 30 °C for 1.5 h.
9. In the meantime, FAME markers can be prepared by adding 20 μL FAME marker solution to each bottle (1 mL) of MSTFA (*see* **Note 9**). FAMEs are included with samples to track GC column stability and for use as a retention time index [14].
10. Add 90 μL MSTFA/FAME standard mix to each sample, capping immediately.
11. Shake in thermomixer at maximum speed at 37 °C for 30 min.
12. Centrifuge at 5000 × *g* for 1 min to pellet any undissolved material.
13. Transfer liquid contents to glass GC-MS vials with glass inserts, capping immediately.
14. For this protocol and specific GC-MS, the following GC gradient is used: 50 °C initial oven temperature, ramp to 65 °C at 5 °C/min, hold for 0.2 min, ramp to 80 °C at 15 °C/min, hold for 0.2 min, ramp to 310 °C at 15 °C/min, hold for 12 min.

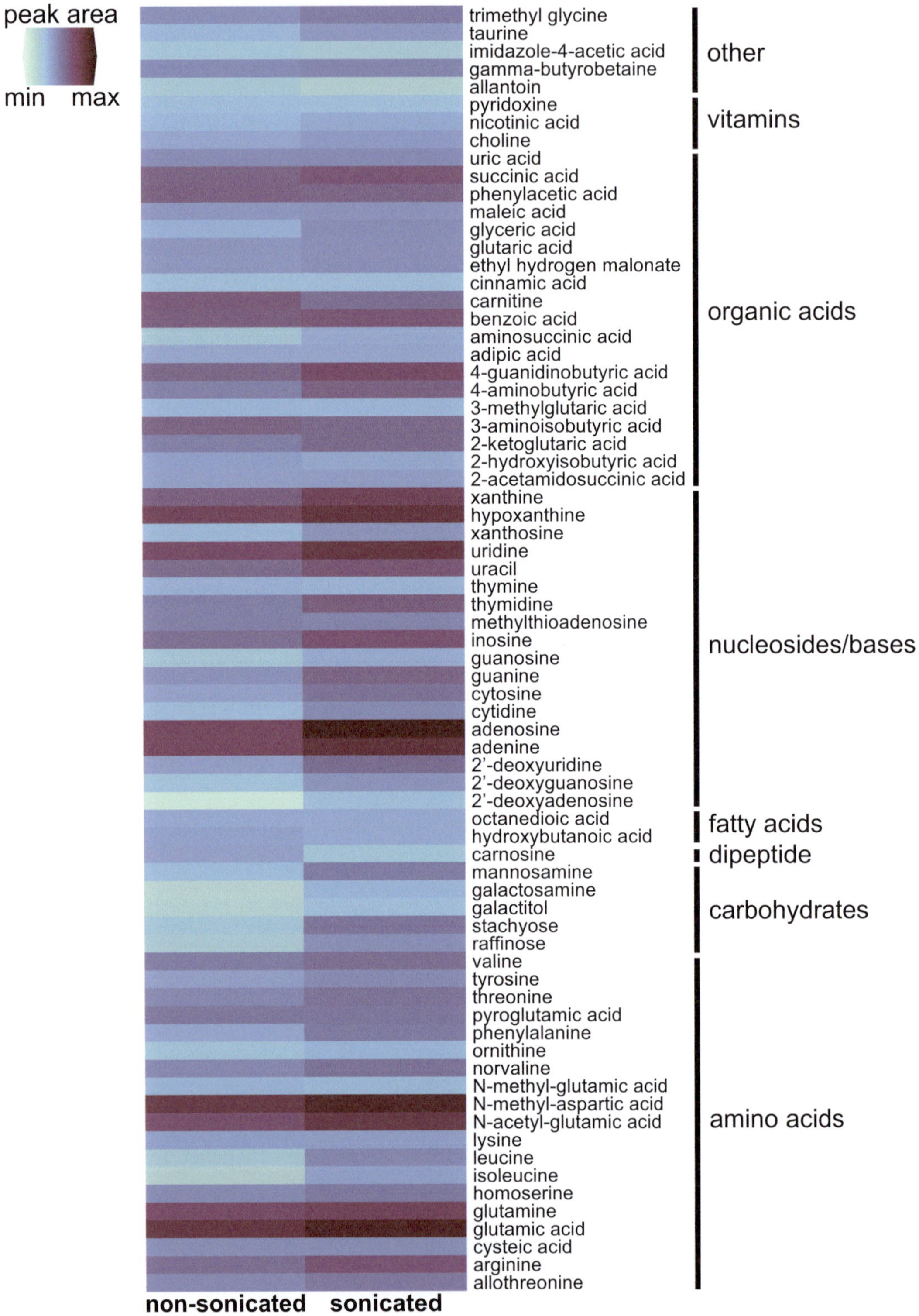

Fig. 2 Common soil metabolites detected by LC-MS. Metabolites shown were water-extracted from (nonsonicated and sonicated) biological soil crust. Data were collected in both negative and positive mode with an Agilent 1290 UHPLC and Thermo QExactive MS. Metabolites were identified using the Metabolite Atlas [15] and validated based on comparison with authentic standards (*m/z*, retention time and MS/MS fragmentation spectra). Peak areas were normalized from minimum-to-maximum within the dataset

15. MS parameters: transfer line: 250 °C, electron ionization: 70 eV, ion source: 230 °C, MS range from 50 to 700 *m/z* at an acquisition rate of 8 spectra/s.
16. As with LC-MS, it is recommended to include quality control standards throughout the run to ensure instrument consistency. Data analysis can be done using the reader's software choice.

3.6 Nonpolar Metabolite Extraction and LC-MS Sample Preparation and Analysis

These steps can be followed to extract nonpolar metabolites such as sterols, fatty acids, lipids and secondary metabolites (note that typical nonpolar extractants lyse most microbial cells). The same LC-MS system as indicated above is used for the detection of nonpolar metabolites. However, RP LC is performed for the separation of nonpolar metabolites.

1. Add 8 mL nonpolar extractant to each (2 g) sample using glass serological pipette.
2. Place samples on a refrigerated orbital shaker and shake at 200 rpm for 1 h (angle the tube rack approximately 45 degrees to allow for adequate sample movement and extraction).
3. Centrifuge samples at 3220 × *g* for 15 min (at 4 °C).
4. Using a glass Pasteur pipette, remove the upper ethyl acetate layer and place into 15 mL conical tubes.
5. The ethyl acetate layer containing nonpolar metabolites can be dried using a speedvac for approximately 1 h (time will depend on the exact instrument).
6. Resuspend dried ethyl acetate extracts in 200 μL of C18 internal standard solution (*see* **Notes 5**–7).
7. Transfer samples to 1.5 mL tubes.
8. Centrifuge samples at 5000 × *g* for 5 min to pellet insoluble salts.
9. Add samples to centrifugal membranes and filter by centrifuging at 5000 × *g* for 3 min.
10. Transfer samples to glass LC-MS vials with glass inserts, capping immediately.
11. For nonpolar metabolites, use a C18 column with the same LC parameters as indicated in Subheading 3.4, **step** 7.
12. Chromatographic separation of nonpolar metabolites can be achieved by setting the LC to a flow rate of 0.4 mL/min and using the gradient shown in Table 2 with mobile phases C and D.
13. For mass spectra, collect data in both negative and positive polarities with a mass range of 70–2050 *m/z* in centroid data mode. Acquire MS/MS fragmentation spectra using collision energies of 10–40 eV.

Table 2
Recommended LC gradient for reversed-phase (C18) chromatography

Time (min)	C (%)	D (%)
0.0	100	0
1.0	100	0
8.0	0	100
9.5	0	100
10.5	100	0
12.0	100	0
29.0	0	100
30.0	0	100

4 Notes

1. In order to determine the amount of soil to begin with, the reader will have to determine or approximate dry weight. This protocol uses 2 g of dry soil for each sample. For example, if the soil is expected to be 50% water weight, start with at least 4 g wet soil (although, if possible, it is recommended to start with more to account for sample loss during sieving and sample transfers).
2. If sample quantity is limited, the protocol can be scaled down. The extractant should be adjusted accordingly to maintain a 1:4 soil–extractant ratio. For example, for 0.5 g soil, use 2 mL extractant.
3. Throughout the extraction procedure, keep extractants and samples as cool as possible to limit continued microbial activity.
4. Filters can clog easily with some soils. In this case, the supernatant can be decanted into a 15 mL tube and recentrifuged at a higher speed followed by filtration. Additionally, a prefilter with a larger pore size (e.g., 3 μm) may be necessary.
5. Resuspension volume may vary depending on the type of soil (i.e., rich organic soils may require a larger volume) and instrument sensitivity. A good starting point is 200–500 μL. Instrument injection volumes can also be modified to adjust analyte signal intensity.
6. Internal standard composition can be modified depending on the instrument type and sensitivity. Typically, stable isotopes and/or xenobiotic compounds are used.

7. Some soil extracts are difficult to get into solution and may require a bath sonicator (for 5 min) or extensive vortexing.
8. All moisture must be removed from samples. Any water present will interfere with the derivatization reaction.
9. MSTFA is extremely volatile. The addition of FAMEs must be performed in a hood and done quickly to limit loss.

References

1. Lal R (2004) Soil carbon sequestration to mitigate climate change. Geoderma 123:1–22
2. Stevenson FJ (1994) Extraction, fractionation, and general chemical composition of soil organic matter. In: Humus chemistry genesis, composition, reactions. John Wiley & Sons, Inc., Hoboken, NJ, pp 24–58
3. Schmidt MWI, Torn MS, Abiven S et al (2011) Persistence of soil organic matter as an ecosystem property. Nature 478:49–56
4. Swenson TL, Bowen BP, Nico PS et al (2015) Competitive sorption of microbial metabolites on an iron oxide mineral. Soil Biol Biochem 90:34–41
5. Judd KE, Crump BC, Kling GW (2006) Variation in dissolved organic matter controls bacterial production and community composition. Ecology 87:2068–2079
6. Warren CR (2014) Response of osmolytes in soil to drying and rewetting. Soil Biol Biochem 70:22–32
7. Swenson TL, Jenkins S, Bowen BP et al (2015) Untargeted soil metabolomics methods for analysis of extractable organic matter. Soil Biol Biochem 80:189–198
8. Kakumanu ML, Cantrell CL, Williams MA (2013) Microbial community response to varying magnitudes of dessication in soil: a test of the osmolyte accumulation hypothesis. Soil Biol Biochem 57:644–653
9. Jones OAH, Sdepanian S, Lofts S et al (2014) Metabolomic analysis of soil communities can be used for pollution assessment. Environ Toxicol Chem 33:61–64
10. Roth V-N, Dittmar T, Gaupp R et al (2015) The molecular composition of dissolved organic matter in forest soils as a function of pH and temperature. PLoS One 10:1–23
11. Fischer H, Meyer A, Fischer K et al (2007) Carbohydrate and amino acid composition of dissolved organic matter leached from soil. Soil Biol Biochem 39:2926–2935
12. Baran R, Brodie EL, Mayberry-Lewis J et al (2015) Exometabolite niche partitioning among sympatric soil bacteria. Nat Commun 6:1–9
13. Sumner LW, Amberg A, Barrett D et al (2007) Proposed minimum reporting standards for chemical analysis. Metabolomics 3:211–221
14. Kind T, Wohlgemuth G, Lee DY et al (2009) FiehnLib: mass spectral and retention index libraries for metabolomics based on Quadrupole and time-of-flight gas chromatography/mass spectrometry. Anal Chem 81:10038–10048
15. Yao Y, Sun T, Wang T et al (2015) Analysis of metabolomics datasets with high-performance computing and metabolite atlases. Meta 5:431–442

Chapter 5

Fatty Acid Metabolome Extraction from Mycobacterial Cells for GC-MS Metabolomics Analysis

Ilse du Preez, Derylize Beukes, and Du Toit Loots

Abstract

Metabolomics is becoming an increasingly popular research tool for identifying new biomarkers, which can, among other applications, be applied to elucidate various microbial growth and virulence mechanisms. Since the lipid composition of numerous microorganisms are unique and characteristic of the particular species, and in many instances also associated with several of their growth and virulence features, we developed a method for extracting the total free fatty acid metabolome from mycobacterial cells, in order to better characterize these using a gas chromatography–mass spectrometry (GC-MS) metabolomics approach. The described method can be considered an optimized Bligh–Dyer approach, since it uses the traditional solvents; chloroform, methanol and water, in a ratio of 1:2:1. However, due to the robust cell walls associated with mycobacteria, and many other microorganisms, the method was adapted to include a step which allows for the physical disruption of the cells using a vibration mill, which dramatically increases the efficiency of this approach. Hereafter, the organic phase is collected, dried, and methylated (as a derivatization step), prior to GC-MS analyses.

Key words Fatty acid metabolome extraction, Gas chromatography–mass spectrometry, Metabolomics, Mycobacteria

1 Introduction

Metabolomics can be defined as the quantitative measurement of dynamic metabolic changes in living systems, in response to genetic modifications or physiological stimuli, including that caused by nutrients and/or drugs [1]. As opposed to global metabolomics (or total metabolome analysis), where the aim is to identify all metabolites in a specific sample, the purpose of a semitargeted metabolomics approach, is to analyze a predetermined fraction of the metabolome, for instance only the amino acids or lipids. This approach is typically applied when the number of compounds in an extract using the total metabolome approach could potentially overwhelm the dimensionality of the analytical instrument and, subsequently, lead to a loss of important information. For such

Edward E.K. Baidoo (ed.), *Microbial Metabolomics: Methods and Protocols*, Methods in Molecular Biology, vol. 1859, https://doi.org/10.1007/978-1-4939-8757-3_5, © Springer Science+Business Media, LLC, part of Springer Nature 2019

analyses, metabolite classes are selectively extracted using specific solvent combinations. Prior to gas chromatography–mass spectrometry (GC-MS) analyses, the extracted metabolites are derivatized in order to close off polar groups of highly polar compounds [2]. Due to the fact that so many microorganism species contain unique and characteristic fatty acids or fatty acid combinations, which may or may not be associated with various growth and virulence mechanisms, this metabolite class is of great interest to the microbial research community. For the purpose of extracting and analyzing the total fatty acid metabolome from bacteria with robust cell walls, such as *M. tuberculosis*, using a GC-MS metabolomics approach, we modified the well-known Bligh–Dyer fatty acid extraction procedure [3]. One of the modifications include the use of a vibration mill to disrupt the cell wall structure of the microbial cells, subsequently improving the extraction capacity and reducing the time required to achieve optimal extraction to a mere 5 h (which is a vast improvement on the 16 h required for completion of the original method). Additionally, the total solvent volume required for complete extraction of the fatty acid metabolome was reduced to 10 mL, significantly lowering the cost, complexity and environmental impact of the original method [4]. The application of this method to bacterial samples allowed for the differentiation of various *Mycobacterium* species based on their fatty acid metabolome profiles, using only 2500 cells per extraction and analysis [5]. Furthermore, when using this extraction procedure in the context of a metabolomics analysis, not only were the previously described fatty acids for mycobacterial samples detected, but also new fatty acid biomarkers never before described/associated with these species [4].

2 Materials

Warning: When this extraction procedure is used to extract the fatty acid metabolome of highly infectious organisms such as *M. tuberculosis*, all the necessary precautions to prevent infection should be adhered to, as set out by the World Health Organization's (WHO) TB Laboratory Biosafety Manual [6]. All the steps of this extraction method should be performed in a biosafety class II cabinet equipped with an exhaust system. As an additional biosafety measure, a N95 surgical mask (from 3M, St. Paul, USA) should be worn over the mouth and nose of the analyst. Also, it is essential that nitrile gloves (from Sigma-Aldrich, St. Louis, MO, USA) (as opposed to latex) are worn at all times, since tears and breaks to these gloves are clearly visible and they are additionally resistant to organic solvents, thereby protecting the analyst and preventing contamination to the sample. All organic solvents used should be ultrapure analytical grade (preferably, Burdick and Jackson brand).

All reagents can be prepared at room temperature and waste should be disposed of by following the appropriate biohazardous waste regulations.

2.1 Chemicals

1. Cultured mycobacterial cells (*see* **Note 1**).
2. Internal standard (phenyl butyric acid (0.104 μg/mL) made up in water) (*see* **Note 2**). Store at 4 °C until use.
3. Extraction solvent. Mix chloroform, methanol, and water in a 1:2:1 ratio (*see* **Note 3**). Store in a capped glass bottle (such as a Duran® Schott bottle) at 4 °C until use.
4. Chloroform.
5. 100% methanol.
6. Methanolic KOH (0.2 M) (VWR International Gmb/H, Darmstadt, Germany).
7. Hexane.
8. Glacial acetic acid (1 M). Add 6 mL glacial acetic acid (98%) to 94 mL water to make up 100 mL of 1 M glacial acetic acid.
9. Water.
10. External standard (methylated-C19 dissolved in trimethylpentane to a final concentration of 15.6 ng/mL) (*see* **Note 4**).

2.2 Equipment

1. MM 400 vibration mill (Retsch GmbH and co. KG, Haan, Germany), or similar apparatus.
2. Centrifuge with adaptors able to accommodate 2 mL microcentrifuge tubes (13,684 × *g*) as well as 10 mL culture tubes (849 × *g*).
3. Nitrogen gas solvent evaporator with heating blocks able to accommodate 2 mL microcentrifuge tubes as well as 10 mL culture tubes (Bonna-Agela Technologies Inc., Wilmington, USA).
4. Weighing balance able to accurately weigh 30 mg of wet cell mass, or 0.3 mg of lyophilized cell mass, depending on the samples being extracted.
5. Tungsten carbide beads (3 mm) (Retsch GmbH and co. KG, Haan, Germany).
6. Glass microliter syringes (1 mL and 100 μL, *see* **Note 5**).
7. Glass Pasteur pipettes with rubber teats.
8. PCR microcentrifuge tubes with screw on caps (2 mL) (*see* **Note 6**).
9. Screw cap glass culture tubes with PTFE-faced cap liners (*see* **Note 7**).
10. Clear glass GC vials (2.0 mL) and polypropylene screw tops with PTFE liners.

11. Pulled point conical glass inserts (250 µL) for GC vials.
12. Glass volumetric flasks (100 mL).
13. Disposable spatulas (from 3 M, St. Paul, USA).
14. Stainless steel tweezers.
15. Sample racks able to accommodate 2 mL centrifuge tubes and 10 mL culture tubes, respectively.

3 Methods

All the extraction steps (Fig. 1) can be performed at room temperature. A total of 24 samples can be extracted simultaneously (dependent on the number of sample vial positions and nitrogen gas nozzles on the drying block). If the sample cohort exceeds this number, we suggest that samples are randomly assigned to various batches and that these batches are extracted on consecutive days, in order to prevent a possible time dependent batch effect. Additionally, a quality control (QC) sample should be made up by combining a small and equal amount of all the collected bacterial samples. This stock sample can subsequently be aliquoted into several separate smaller quantities, equal to that of the mass of each experimental sample. One of these QC aliquots can then be extracted and analyzed with the experimental samples in each sample batch, and the generated data from these QC samples can be used to monitor

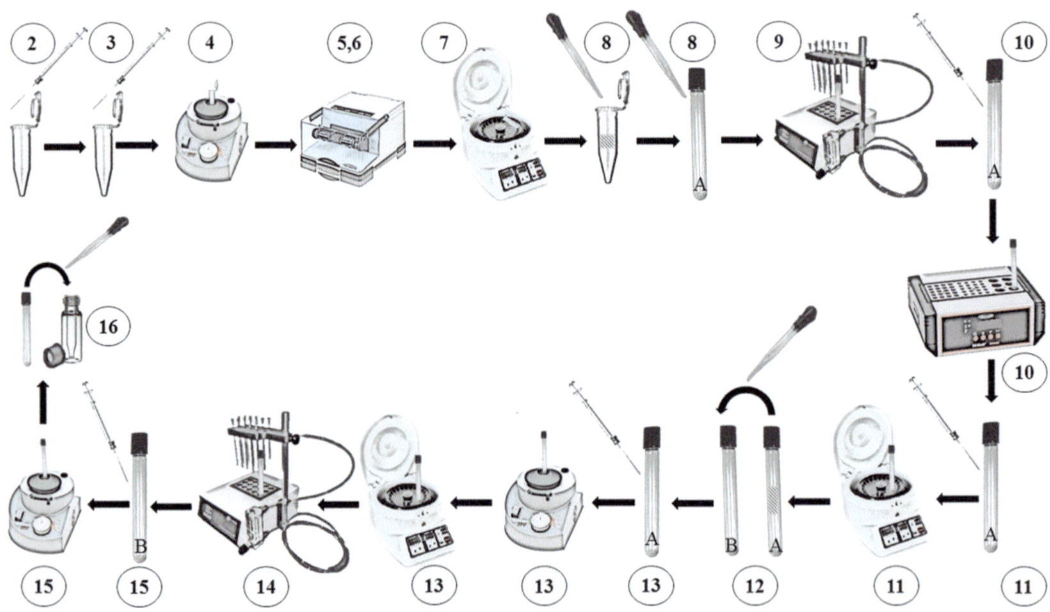

Fig. 1 Schematic illustration of the steps pertaining to the fatty acid extraction procedure. Encircled numbers correspond to those used for the steps described in Subheading 3

the repeatability of the extraction method and correct for any analytical drift.

1. Accurately weigh the required amount of sample material (*see* **Note 8**) and place in a microcentrifuge tube. Correctly label these sample tubes using a permanent marker and place into a sample rack.
2. Add 50 μL of the internal standard stock solution to each sample vial, using a 100 μL glass syringe (*see* **Note 9**).
3. Add 1 mL of the extraction solvent stock solution to each sample, using a 1 mL glass syringe.
4. Close the microcentrifuge tubes' caps securely and consecutively vortex each sample vial for approximately 5 s to ensure that the tubes' contents are sufficiently mixed.
5. Consecutively open each microcentrifuge tube and add one tungsten carbide bead to each sample using a clean, stainless steel tweezer (*see* **Note 10**), and close the cap securely again once this has been done.
6. Place the sealed microcentrifuge tubes into the vibration mill and allow them to shake for 2 min at 30 Hz. Rotate the sample holders (milling cups) so that the tubes at the lower end of the sample holder are now on top. Shake the samples for an additional 2 min at 30 Hz (*see* **Note 11**).
7. Remove the microcentrifuge tubes from the vibration mill and insert into the centrifuge vial holders in such a way that the rotor is properly balanced. Centrifuge for 5 min at 13,684 × *g* for phase separation to occur.
8. Carefully remove the organic acid phase (top layer) from the microcentrifuge tube using a glass Pasteur pipette with rubber teat, and transfer this layer to a clean culture tube (hereafter referred to as tube A) (*see* **Note 12**).
9. Place tube A (uncapped) in a solvent evaporator with a heat block (at 30 °C) and apply a gentle stream of nitrogen to break the surface tension, until all the solvent has evaporated completely and the culture tubes are completely dry.
10. Using 1 mL glass syringes, add 0.5 mL chloroform, 0.5 mL methanol, and 1 mL of methanolic KOH to the dried culture tube A, cap securely and incubate the sample mixture for 30 min at 60 °C (*see* **Note 13**).
11. After incubation, 2 mL hexane, 0.2 mL glacial acetic acid and 2 mL water should be added to each tube (A) using a 1 mL glass syringe, cap securely and centrifuge for 5 min at 849 × *g*.
12. Collect the organic phase (top layer), using a glass Pasteur pipette and rubber teat, and transfer this layer to a clean culture tube (hereafter referred to as tube B).

13. Add 2 mL hexane to the remaining water phase in tube A, cap, briefly vortex each tube and centrifuge for 5 min at 849 × *g*. Collect the organic phase (top layer) again (as done in **step 12**) and add this to the previously collected organic phase in tube B. Repeat this step three times, adding all collected organic phases accumulatively to tube B.
14. Place tube B (uncapped), containing all collected organic phases, in a solvent evaporator with a heat block (set to 30 °C) and apply a gentle stream of nitrogen, enough to break the surface tension of the contents of the tube, until the culture tubes are completely dry.
15. Add 50 μL of the external standard stock solution to the dried contents of tube B using a 100 μL glass syringe, cap and briefly vortex in order to resuspend the extracted fatty acids.
16. This extract can now be transferred to the appropriate insert in the GC vial, capped immediately and placed into the auto sampler tray of the GC-MS. We suggest that the chromatographic analyses of these samples commence within 24 h or less.
17. Although the focus of this document is the fatty acid extraction method, the GC-MS settings recommended for the analysis of these samples are described in **Notes 14** (GC) and **15** (MS).

4 Notes

1. The specific cell cultures or isolates required for the experiment should be obtained using conventional mycobacterial culturing or cell isolation methods. It should be kept in mind that for metabolomics analyses, all metabolic activities of these cells should be stopped/quenched at the time of sampling, while at the same time also ensuring no or minimal cell leakage. Various quenching methods, such as rapid freezing and acidic treatments, have been developed and investigated for metabolomics applications [7, 8], and should be applied when harvesting these bacteria. Additionally, all samples should be harvested in the exact same growth phase (preferably the mid-log phase), and the collected samples should be stored at −80 °C immediately after harvesting and quenching. If repeated analyses of the same sample are required, it is essential that smaller sample aliquots are prepared prior to extraction, since repeated freeze/thaw cycles may induce changes to the fatty acid metabolome. If cell cultures are grown at a different location as to where the metabolomics analyses will be performed, samples should be transported frozen on dry ice.

2. Ensure that the total volume of the internal standard stock solution (50 μL of which is required per sample) is adequate for the analyses of the whole sample cohort, the quality control samples and the repeated analyses which might be required. Making up a new batch of internal standard stock solution in the midst of an experiment may lead to a slight concentration variation which, in turn, can relate to an unnecessary batch effect, thereby influencing the overall outcome of the metabolomics analyses. The internal standard stock solution (100 mL) is made up as follows: (a) Weigh 10.4 mg phenyl butyric acid into a glass volumetric flask and fill the flask to 100 mL with water (Solution A); (b) transfer 1 mL of solution A to a new glass volumetric flask and fill the flask with water to a final volume of 100 mL (Solution B); (c) transfer 10 mL of Solution B into a third glass volumetric flask and fill up to 100 mL with water (Solution C). This is the final stock solution of internal standard used for the extraction procedure (0.104 μg phenyl butyric acid/mL). The glass volumetric flasks containing these mixtures should be closed tightly when not in use, and refrigerated at 4 °C.
3. Of this mixture, 1 mL is required per sample. We suggest making a stock solution with at least three times the required volume (1 mL × number of samples × 3). This will ensure that an adequate amount of the mixture is available for the total sample extractions and the washing steps described in Subheading 3. To prevent the highly volatile chloroform from evaporating, and thereby changing the ratio of the various solvents in the mixture, the glass bottle containing this mixture should be capped tightly when not in use, and refrigerated at 4 °C.
4. The external standard stock is made up as follows: (1) Weigh 15.6 mg methylated-C19 into a glass volumetric flask, and fill the flask to 100 mL with trimethylpentane (solution D); (2) dilute this solution by transferring 1 mL of solution D to a new 100 mL glass volumetric flask and make up to 100 mL with trimethylpentane (Solution E); (3) transfer 1 mL of Solution E to a third volumetric flask, and once again, make up to 100 mL with timethylpentane (Solution F). This is the final stock solution of external standard used for the extraction procedure (15.6 ng methylated-C19/mL). The glass volumetric flasks containing these mixtures should be closed tightly when not in use, and refrigerated at 4 °C.
5. We recommend using a Hamilton Microliter™ syringe (1 mL and 100 μL) or a similar glass syringe with a steel plunger. Conventional pipettes with polypropylene or polyethylene plastic tips should not be used for this extraction since they are only conditionally resistant to the organic solvents used,

and the dissolved plastics may interfere with the detected metabolite profiles.

6. The preferred tube to use for this extraction is the PCR clean, DNA LoBind Microcentrifuge Tube, from Eppendorf®. If these tubes are not available, microcentrifuge tubes which are resistant to acids and organic solvents should be used. It is imperative that the tubes used can properly be sealed, preferably with screw-on caps.
7. Do not use tubes with rubber cap liners when doing this extraction. We recommend the Kimax® screw cap glass culture tubes with PTFE-faced cap liners. These tubes and caps should be cleaned by washing with foam free soap, rinsing with distilled water, and drying in an oven at 100 °C. The dried tubes can then be rinsed with hexane prior to use.
8. For optimal extraction of fatty acids, this method requires 30 mg of wet cell mass, or 0.3 mg of lyophilized cell mass. We suggest that cells are weighed directly into the microcentrifuge tube which will be used for the extraction. This will minimize sample loss when transferring the sample from a weighing boat for example, to the microcentrifuge tube used for extraction. This step can also be done directly after harvesting and quenching of cells, before storing at −80 °C. It is recommended that individual disposable spatulas are used for weighing the culture samples to prevent intersample contamination.
9. Rinse the glass syringe three times by pumping with hexane to waste before use. Whilst in use, if the needle of the syringe happens to touch a sample tube, glove, fume hood, or any other surface other than that of the liquid that is being transferred, the needle should again be washed and rinsed with hexane immediately, in order to prevent contamination of subsequent samples. We find it best to dedicate individual syringes to each of the different solvents or solutions (internal standard, extraction solvents, and external standard) in order to minimize the possibility of contamination and corrosion of the syringes.
10. After use, beads can be washed with a foam free soap such as Sterigel™ from Quadex International, rinsed with distilled water and stored in a beaker filled with the extraction mixture. The tweezers can also be washed by rinsing it with the extraction mixture.
11. It is crucial to ensure that the microcentrifuge tubes are properly sealed before placing them into the vibration mill. In the case of a spillage (determined by fluid in the milling cup after shaking), all microcentrifuge tubes should be inspected to identify which tube is leaking. Discard the leaking sample

(this sample should then be re-extracted with the next sample batch) and spray the milling cup and milling chamber with an antituberculocidal (such as to the Q/T tuberculocidal spray disinfectant cleaner from Canberra Corporation, Toledo, OH, USA) to kill the infectious agents which may be present.

12. When removing the organic phase, slowly draw up and transfer only small volumes, to ensure that none of this enters the teat accidentally, which will contaminate the sample, and subsequently result in a total re-extraction of that particular sample from scratch. Also, be careful not to draw up any liquid from the lower water phase. We find it best to rather leave a small volume of the organic phase behind, for which the internal standard will compensate.
13. For the incubation step, culture tubes should preferably be in the upright position. If an oven with an appropriate heat rack is unavailable, sample mixtures can be incubated in the heating block of the solvent evaporator. Caps should also be checked to ensure the liners are intact, and that they properly seal the tube, in order to prevent evaporation and subsequently inadequate derivatization, which would require total re-extraction and derivatization of the sample.
14. Recommended GC settings include: DB-1MS (low polarity) capillary column (30 m × 0.25 mm i.d., 0.25 μm film thickness) as the primary column. In the case of a GC×GC system, a DB-17MS (medium polarity) capillary column (2 m × 0.1 mm i.d., 0.1 μm film thickness) is the recommended secondary column. An injection volume of 1 μL; splitless injection (1:10 split for GC×GC), injector temperature of 250 °C for the total run and helium as carrier gas at a constant flow of 1.2 mL/min, are recommended. The temperature program (primary oven) used should be: an initial GC oven temperature of 50 °C for 1 min followed by an initial increase in oven temperature of 20 °C/min to 160 °C, followed by an increase in oven temperature of 10 °C/min to 190 °C and then a further increase of 5 °C/min to a final temperature of 300 °C, at which the temperature should be maintained for 5 min. The recommended temperature program for the secondary oven, in the case of GC×GC, is identical to that of the primary column, only with an offset of +15 °C. Cryomodulation and a hot pulse of nitrogen gas of 0.7 s, every 3 s can be used to control the effluent emerging from the primary column onto the secondary column for GC×GC analyses.
15. With regards to the MS settings, we recommend that no mass spectra are recorded for the first 450 s of each run, which serves as a solvent delay. The transfer line is kept at a constant 270 °C and the ion source temperature at 200 °C, for the entire run.

For quadrupole-MS detectors, the recommended detector voltage is 1500 V and the filament bias −70 eV. Mass spectra can be collected at an acquisition rate of 200 spectra/s from 50 to 650 *m/z*. Although these parameters can be a good starting point, the optimum GC-MS conditions should, however, be determined before each study, and should be optimized according to the specific samples and instrument used.

Acknowledgments

The financial assistance from the National Research Foundation (NRF) of South Africa for this research is gratefully acknowledged (UID: 95269). The opinions expressed and conclusions derived are those of the authors and are not necessarily those of the NRF.

References

1. Collino S, Martin FJ, Rezzi S (2012) Clinical metabolomics paves the way towards future healthcare strategies. Br J Clin Pharmacol 75 (3):619–629
2. Dunn WB, Bailey NJC, Johnson HE (2005) Measuring the metabolome: current analytical technologies. Analyst 30:606–625
3. Bligh EG, Dyer WJ (1959) A rapid method of total lipid extraction and purification. Can J Biochem Physiol 37(8):911–917
4. Olivier I, Loots DT (2012) A comparison of two extraction methods for differentiating and characterizing various *Mycobacterium* species and *Pseudomonas aeruginosa* using GC-MS metabolomics. Afr J Microbiol Res 6(13):3169–3172
5. Du Preez I, Loots DT (2013) Detection limit for differentiating between various *Mycobacterium* species and *Pseudomonas aeruginosa* using GC-MS metabolomics: a comparison of two extraction methods. Afr J Microbiol Res 7 (9):797–801
6. World Health Organization (2012) Tuberculosis laboratory biosafety manual. http://www.who.int/tb/publications/2012/tb_biosafety/en/
7. Brown M, Dunn WB, Ellis DI, Goodacre R, Handl J, Knowles JD et al (2005) A metabolome pipeline: from concept to data to knowledge. Metabolomics 1(1):39–51
8. Mashego MR, Rumbold K, De Mey M, Vandamme E, Soetaert W, Heijnen JJ (2007) Microbial metabolomics: past, present and future methodologies. Biotechnol Lett 29 (1):1–16

Chapter 6

Total Metabolome Extraction from Mycobacterial Cells for GC-MS Metabolomics Analysis

Derylize Beukes, Ilse du Preez, and Du Toit Loots

Abstract

Over the past 10 years, the number of metabolomics based publications in the available scientific literature has exponentially grown, a large portion of which describing new biomarkers better elucidating microbial disease mechanisms and improved diagnostics and treatment thereof. Here, we describe a metabolomics method for extracting the total metabolome (all compounds present in the microbial cell irrespective of the compound class), for analysis in a single analytical run using only one analytical instrument. This method includes disruption of robust microbial cell walls, and the precipitation of proteins and cell debris using a combination of mechanical methods and solvents. These extracts are subsequently derivatized, in order to improve the volatility of polar compounds for efficient gas chromatography-mass spectrometry (GC-MS) analysis. This methodology can be applied to all microbes, including those with robust cell walls, such as *M. tuberculosis*. To date, the biomarkers identified using this approach have led to improved tuberculosis (TB) diagnostics, improved TB treatment approaches, and better understanding of host–microbe interactions and associated mycobacterial genomics.

Key words Metabolomics, Total metabolome extraction, GC-MS, Mycobacteria

1 Introduction

A metabolomics research approach is generally used to identify biomarkers characterizing a specific metabolic phenotype, related to an external or internal perturbation (e.g., a genetic mutation or disease state), and achieved by comparing the generated metabolome data, after extraction and analyses of the individual samples of two separate sample groups, with and without the perturbation. The general aim of a total metabolomics research experiment is to extract, identify, quantify and characterize all the metabolites, or small molecules (<1500 Da), such as metabolic intermediates, secondary metabolites, and other signaling compounds found in a sample, cell, organ, or organism [1]. The described total metabolomics research approach, subsequently allows for a holistic understanding of all possible physiological/metabolic abnormalities

Edward E.K. Baidoo (ed.), *Microbial Metabolomics: Methods and Protocols*, Methods in Molecular Biology, vol. 1859, https://doi.org/10.1007/978-1-4939-8757-3_6, © Springer Science+Business Media, LLC, part of Springer Nature 2019

which may occur due to the perturbation, since it attempts to visualize the metabolome in its entirety, and subsequently this approach is generally useful in the initial research stages or when new hypotheses generation is the aim of the investigation [2]. For this purpose, we developed a method that is capable of extracting and derivatizing metabolites belonging to all compound classes including among others: fatty acids, amino acids, alcohols, organic acids, monosaccharides, alkenes, alkanes, purines, and pyrimidines from mycobacterial cells, considering the guidelines proposed by "The Metabolomics Society." Considering the larger dimensionality of the total metabolome extract comparative to that achieved when extracting only subfractions of the metabolome, subsequent analyses using a 2D gas chromatograph coupled to a time-of-flight mass spectrometer (GC×GC-TOFMS) is preferable, due to the larger dimensionality, dynamic range and sensitivity this technique offers over that of more conventional GC-MS based analytical approaches (however, more traditional GC-MS systems could also be used by optimizing the temperature program). In a nutshell, the described method includes disruption of robust microbial cell walls, and the precipitation of proteins and cell debris using mechanical methods together with a combination of solvents, the latter of which are capable of solubilizing compounds from all compound classes and polarities. These extracts are then derivatized in order to close off polar groups of all polar compounds, making them suitable for subsequent GC-MS analyses. This method can be completed in less than 4 h, using only 1.0 mL of the solvent mixture per sample [3], and requires only 250 CFUs [4], making it a fast, cost effective, efficient and an environmentally friendly approach to total metabolome extraction.

Metabolomics, and the use of this extraction method in particular, has contributed substantially to our fundamental knowledge of *Mycobacterium tuberculosis* and other *Mycobacterium* species. Among other applications, this research strategy has been used to define the functionality of unknown genes [5, 6], to better describe virulence [7], drug resistance [8, 9], host–microbe interactions [10, 11], and to identify variations between various *Mycobacterium* species [3].

2 Materials

It is important to always adhere to the World Health Organization's (WHO) TB Laboratory Biosafety Manual [12], before attempting to work with infectious samples. All the described steps for this extraction method should be performed in a biosafety class II cabinet, equipped with an exhaust system, in order to prevent the analyst from inhaling harmful volatile organic solvents and from becoming infected with an infectious agent. As an added precaution, a N95 surgical mask (from 3 M, St. Paul, USA) covering the nose and mouth of the analyst must be worn, as well as nitrile gloves (from Sigma-Aldrich (St. Louis, MO, USA)) during the entire extraction procedure. All organic solvents used should be ultrapure analytical grade (such as the Burdick and Jackson brand from Honeywell International Inc., Muskegon, MI, USA) and water should be double distilled. Reagent mixtures can be prepared at room temperature. Waste disposal should adhere to biohazardous waste regulations.

2.1 Chemicals

1. Isolated mycobacteria cells (*see* **Note 1**).
2. Phosphate buffered saline (PBS) solution (*see* **Note 2**).
3. Internal standard (3-phenylbutyric acid (0.13125 mg/mL) made up in water) (*see* **Note 3**).
4. Chloroform.
5. Methanol.
6. Extraction solvent mixture (Prepare by mixing chloroform, methanol and water in a 1:3:1 ratio (*see* **Note 4**)). Store in a Duran® Schott glass bottle at room temperature.
7. Pyridine (Sigma-Aldrich, St. Louis, MO, USA).
8. Methoxyamine hydrochloride (15 mg/mL made up in pyridine) (Sigma-Aldrich, St. Louis, MO, USA) (*see* **Note 5**).
9. BSTFA with 1% TMCS (Sigma-Aldrich, St. Louis, MO, USA) (*see* **Note 6**).

2.2 Equipment

1. MM 400 vibration mill (Retsch GmbH and co. KG, Haan, Germany), or similar (*see* **Note 7**).
2. Microcentrifuge (*see* **Note 8**).
3. Tungsten carbide beads (3 mm) (Retsch GmbH and co. KG) (*see* **Note 9**).
4. Glass microliter syringe (100 μL) (Hamilton Company, Reno, USA) (*see* **Note 10**).
5. PCR clean microcentrifuge tubes (2.0 mL) (Eppendorf, Hamburg, Germany) (*see* **Note 11**).
6. Clear glass GC-sample vials (2.0 mL) and polypropylene screw tops with PTFE liners (Agilent Technologies, Santa Clara, California, United States).
7. Pulled point conical glass inserts (250 μL) for GC-sample vials (Agilent Technologies, Santa Clara, California, United States).
8. Solvent evaporator (nitrogen) with heating block (Bonna-Agela Technologies Inc., Wilmington, USA) (*see* **Note 12**).
9. Glass volumetric flasks (100 mL and 10 mL).
10. Glass Pasteur pipettes and plunger (rubber teat for pipette).

11. Glass measuring cylinder (100 mL).
12. Tube rack for 2.0 μL microcentrifuge tubes.
13. GC-sample vial tray for GC vials.
14. Stainless steel tweezers.
15. Duran® Schott glass bottle (100 mL).

3 Methods

Carry out all procedures at room temperature. A total of 24 samples can be extracted simultaneously (dependent on the number of sample vial positions and nitrogen gas nozzles on the drying block). If the sample cohort exceeds this number, we suggest that samples are analyzed in randomly assigned batches and that these batches are extracted on consecutive days, in order to prevent a time dependent batch effect. Additionally, it is essential that a quality control (QC) sample is made up by using a mixture of equally small amounts of all the collected bacterial cell samples. This stock sample is then aliquoted into several smaller, but identical QC samples, and one of which is extracted and analyzed in each sample batch. The data from these QC samples is subsequently used to monitor the repeatability of the extraction method and analysis of the extracts, and corrected for any analytical drift or batch effects, if necessary.

3.1 Washing the Mycobacterial Cells with PBS

Take note that these washing steps are only necessary if the cultured cells were not washed with PBS directly after harvesting.

1. Remove samples from −80 °C storage and allow to thaw in a sample tray at room temperature.
2. Pellet cells by placing the thawed sample vials into the appropriately sized wells in the centrifuge rotor (balance out the samples and the mass across the rotor) and centrifuge at 17,530 × *g* (14,000 rpm) for 30 min (*see* **Note 8**).
3. Remove the sample from the centrifuge and carefully place into a sample tray so as not to disturb the pellet. Uncap the sample vials and decant and discard any fluids from the wet cell mass pellet using a glass Pasteur pipette and teat (*see* **Note 13**). If lyophilized cells are used, as in the case of this experiment, **step 3** can be ignored.
4. Add 500 μL of PBS solution, (*see* **Note 2**) to the pelleted cells and resuspend the cells by gently swirling each Eppendorf tube by hand.
5. Centrifuge the samples at 17,530 × *g* (14,000 rpm) for 30 min as described in **step 2** (*see* **Note 8**).

6. Remove supernatant again using a glass Pasteur pipette and discard as described in **step 3** (*see* **Note 13**).
7. Repeat **steps 4–6**.
8. If the samples were not aliquoted into the appropriate sample sizes at the time of isolation, cells can be weighed at this stage (*see* **Note 1**) and these aliquots can be extracted immediately or stored at −80 °C until the day of analysis.

3.2 Extraction Methodology

A schematic presentation of the extraction method is shown in Fig. 1.

1. On the day of analysis, remove sample aliquots from the −80 °C storage freezer and allow them to thaw at room temperature in a sample tray.
2. Uncap all sample microcentrifuge tubes and add 50 μL of the internal standard to each using a 100 μL glass syringe (*see* **Note 3**).
3. Add 1 mL of the extraction solvent mixture (*see* **Note 4**).

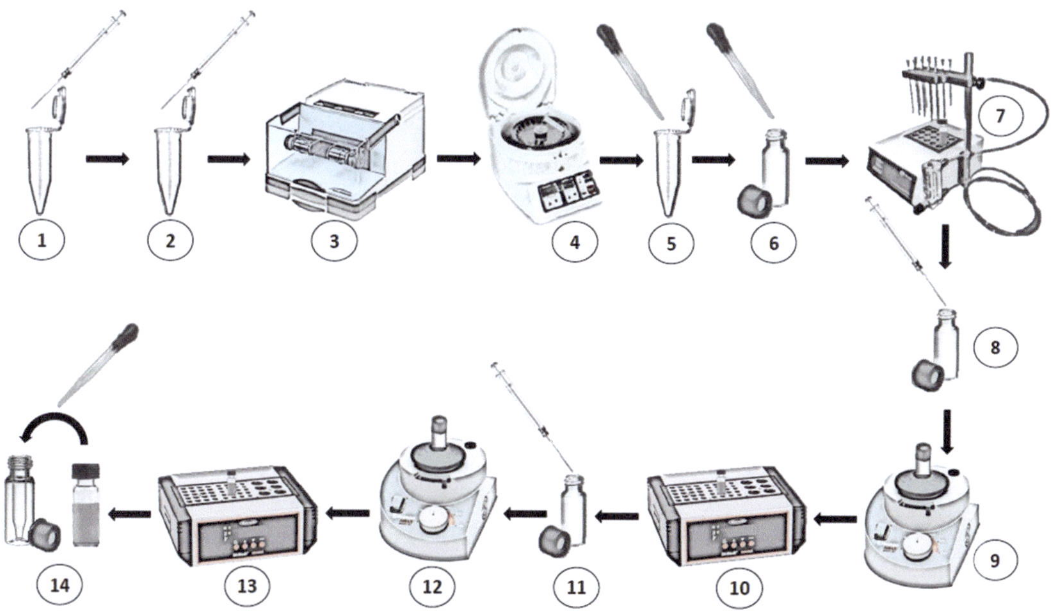

Fig. 1 Whole metabolome extraction method. (1) Add 50 μL 3-phenyl butyric acid, (2) add 1.0 mL of the extraction solvent mixture, (3) shake in vibration mill at 30 Hz for 10 min, (4) centrifuge the sample mixture at 17,530 × *g* (14,000 rpm) for 5 min, (5) remove supernatant using a glass Pasteur pipette (6) transfer supernatant to glass GC-vial, (7) place GC vials containing supernatant in the heating block and dry solute under nitrogen at 30 °C until solvent completely evaporated, (8) add 50 μL methoxyamine hydrochloride to dried sample, (9) vortex sample for 1 min, (10) place sample in heating block and heat for 90 min at 50 °C, (11) add 40 μL BSTFA with 1% TMCS, (12) vortex sample for 1 min, (13) transfer sample to heating block and heat for 60 min at 60 °C, (14) transfer sample solution to a clean glass GC-vial containing a glass vial insert using a glass Pasteur pipette

4. Cap the tubes securely and place into the sample vial adaptor wells in the vibration mill and shake at 30 Hz for 5 min on each side (*see* **Note 9**).
5. Centrifuge the microcentrifuge tubes containing the sample mixture at 17,530 × *g* (14,000 rpm) for 5 min as described in **step 2** (*see* **Note 8**).
6. Remove microcentrifuge tubes from the centrifuge and place into the appropriate sample tray. Open all of the centrifuge tubes and transfer the supernatant (pellet should remain behind), using a glass Pasteur pipette and teat, into a glass GC-sample vial (*see* **Note 13**).
7. Place uncapped GC vials containing the supernatant into the heating block of the solvent evaporator set at 30 °C and evaporate all the solvent from the mixture under a gentle nitrogen gas stream, until completely dry (*see* **Note 12**).
8. Add 50 μL of methoxyamine hydrochloride solution to each GC-sample vial using a glass microliter syringe, cap securely and heat for 90 min at 50 °C on a preheated heating block.
9. Remove all GC-sample vials from the heating block and place into the appropriately sized well of the sample tray. Uncap the GC-sample vials and add 80 μL of BSTFA with 1% TMCS using a glass microliter syringe, recap and heat for 60 min at 60 °C on the preheated heating block (*see* **Note 12**).
10. Remove the samples from the heating block and place in the sample tray, consecutively uncap each sample and transfer the derivatized sample extract into a glass vial insert in a new GC-sample vial. Place the capped GC-sample vials into the sample tray of the GC-MS and analyze (*see* **Note 14**).

4 Notes

1. This extraction method requires 30 mg (equivalent to approximately 1×10^8 bacteria) of wet cell mass or 0.3 mg of lyophilized cell mass, for optimal results, although far smaller cell mass volumes have been successfully used (250 CFU). The cells required for such an experiment can be obtained by conventional mycobacterial culturing methods and/or cell isolation techniques from patient collected samples. Although these procedures fall beyond the scope of the methodology described here, it should be kept in mind that for metabolomics analyses, all metabolic activities should be stopped/quenched at the time of sampling, while at the same time also ensuring no or minimal cell leakage. Various quenching methods, such as rapid freezing and acidic treatments, have been developed and described for metabolomics applications [13, 14], and should

be adhered to when harvesting these bacteria. Additionally, all samples should be harvested at the exact same growth phase (preferably the mid-log phase), and the collected samples should be stored at −80 °C immediately after harvesting and quenching. It is highly recommended that all samples are stored in smaller aliquots (approximately the biomass required for metabolomics analyses) instead of depositing each culture as a whole, considering that repeated freeze–thaw cycles will influence the metabolic profiles. These aliquots will also allow for repeat analyses of specific samples if necessary, or for preparing the QC samples used for monitoring the efficiency of the experiment. Ideally, the collected cells should be washed with PBS directly after harvesting, before the aliquots are frozen. If this is not possible, the PBS washing procedure (Subheading 3.1) should be followed before the extraction commences in order to remove any trace substances (sugars and phosphates) commonly found in the culture media, or patient sample matrix, which may still be present in the sample. If cultures are supplied by a different institution to that where the metabolomics analyses are performed, the collected cells should be transported frozen (using dry ice or a portable liquid nitrogen Dewar). All culturing steps and reagents used should be well documented in order to allow the correct interpretation of the identified metabolite biomarkers in the context of the growth substrates used.

2. Phosphate-buffered saline (PBS) is a water-based salt solution commonly used in biological research experiments. The simplest way to prepare a PBS solution is to use PBS buffer tablets or pouches and the final solution can be stored at room temperature or in a refrigerator. Since the osmolality and ion concentrations of the final solution match those of the human body (isotonic), it is considered nontoxic to most cells. Transfer one PBS tablet (yields 0.01 M phosphate buffer, 0.0027 M potassium chloride and 0.137 M sodium chloride buffer, pH 7.4) from Sigma-Aldrich (St. Louis, MO, USA) to a glass measuring cylinder and add 150 mL of purified water (autoclaved double distilled water). Add a magnetic stirrer to the PBS mixture and stir on a stirring plate until the tablet is dissolved. Remove the magnetic stirrer and fill the cylinder to 200 mL with purified water.

3. The internal standard is prepared by weighing 13.125 mg of 3-phenylbutyric acid using an analytical balance and weighing boat, and then transferring this to a 100 mL glass volumetric flask. Dissolve this weighed powder by swirling in 50 mL of water and then fill to volume with water again. Calculate the total volume of internal standard solution needed (considering the amount needed and the total number of samples and QC's

which need to be extracted) prior to the analyses. We recommended using the same internal standard batch for extraction of the entire sample cohort, in order to reduce the chance for any additional variation that may exist when a new stock solution is prepared.

4. The required solvent volumes can be measured using a glass measuring cylinder or a calibrated bottle top dispenser (Boeckel and Co (GmbH and Co), Hamburg, Germany), such as the Boeco® brand for instance (used together with a glass Duran® Schott bottle). Combine 20 mL of chloroform, 60 mL of methanol and 20 mL of water in a 100 mL Duran® Schott bottle, cap the bottle and shake gently for 30 s. Bear in mind that when mixing these three solvents, an exothermic (heat releasing) reaction will occur. It is therefore recommended that the flask is briefly uncapped after the initial mixing step to release the built-up pressure, where after it can be recapped and mixed again for an additional 30 s. This solution should be stored in a capped glass bottle to prevent evaporation of the solvents and to maintain the original solvent ratios.
5. Add 150 mg of methoxyamine hydrochloride to a glass volumetric flask and fill the flask to 10 mL with pyridine. Shake until completely dissolved. Extreme caution should be taken when working with the pyridine or other highly toxic volatile compounds, and these should under no circumstances be inhaled.
6. BSTFA with 1% TMCS can be purchased as a premixed solution from Sigma Aldrich (St. Louis, MO, USA). The solution can be capped and stored at room temperature and used as is.
7. The vibration mill requires the appropriate adaptors to house 2.0 mL Eppendorf tubes and a vibrating function at 30 Hz. Adapters containing sample tubes should be flipped vertically after the first 5 min in the vibration mill, in order to reposition the tungsten beads to the opposite side of the microcentrifuge tube, and then be allowed to mill for an additional 5 min at 30 Hz, thereby maximizing the milling efficiency (hence, total milling time is 10 min). For safety reasons, ensure that the vibration mill is firmly grounded and far away from the edges of the bench before commencing with the pulverization step. It is crucial to make sure that the microcentrifuge tubes are properly sealed before placing them into the vibration mill. In the case of accidental spillage (which can be determined by the presence of moisture in the milling cup after shaking), all microcentrifuge tubes should be inspected to identify which samples have leaked. Discard these samples (which should then be reextracted with the next sample batch) and spray the milling cup and milling chamber with an antituberculocidal (such as the Q/T tuberculocidal spray disinfectant cleaner from

Canberra Corporation, Toledo, OH, USA) to kill the infectious agents which might still be present.

8. The microcentrifuge should be equipped with a sample chamber cover which is able to contain an accidental spillage. If a spillage should happen, the chamber and cover should be sprayed with an antituberculocidal (such as the Q/T tuberculocidal spray disinfectant cleaner from Canberra Corporation, Toledo, OH, USA) and wiped clean.
9. After use, beads can be washed with a foam free soap such as Sterigel™ from Quadex International, rinsed with water and stored in a beaker filled with the extraction solvent mixture. The tweezers can be washed by rinsing it with the same extraction solvent mixture.
10. We recommend using a 100 μL glass syringe with a stainless steel plunger. Conventional pipettes with polypropylene or polyethylene plastic tips should not be used for this extraction since they are only conditionally resistant to organic solvents such as chloroform, and the dissolved plasticides will cause interfering peaks in the sample metabolite profiles. Rinse the glass syringe three times by pumping to waste with hexane before every use. While in use, should the needle of the syringe accidentally touch a sample tube, glove, fume hood, or any other surface other than that of the liquid that is being transferred, the needle should be washed and rinsed with hexane immediately in order to prevent contamination of subsequent samples. We find it best to dedicate individual syringes to each of the liquids (internal standard extraction solvent mixture, and methoxyamine hydrochloride) in order to minimize contamination and corrosion of the syringes.
11. Similar tubes can be used as long as they are resistant to acids and organic solvents.
12. Place the samples in the heating block after it has been set to the correct temperature for the derivatization step. Open the nitrogen gas and regulate flow such that the stream of nitrogen barely breaks the surface tension of the sample liquid.
13. Use a new glass Pasteur pipette for each sample. Be careful not to disturb the pelleted cell debris. Similarly, a new clean glass Pasteur pipette should be used for transferring the derivatized sample to the glass vial insert in the GC-sample vial before GC-MS analysis of the samples.
14. Recommended GC-MS settings: An Rxi-5Sil MS (low polarity, from Restek Corporation, Bellefonte, USA) capillary column (30 m × 0.25 mm i.d., 0.25 μm film thickness) as the primary column. In the case of a GC×GC system, a Rxi-17 (medium polarity, from Restek Corporation, Bellefonte, USA) capillary column (1 m × 0.25 mm i.d., 0.25 μm film thickness) is the

recommended secondary column. Sample extracts (1 μL) can be randomly injected at a split ratio of 1:3, injector temperature of 270 °C for the total run and helium as carrier gas at a constant flow of 1 mL/min. The temperature program (primary oven) used should be: an initial GC oven temperature of 70 °C for 2 min followed by an initial increase in oven temperature of 4 °C/min to a final temperature of 300 °C, at which the temperature should be maintained for 2 min. The recommended temperature program for the secondary oven in the case of GC×GC, is identical to that of the primary column, only with an offset of +15 °C. Cryomodulation and a hot pulse of nitrogen gas of 0.7 s, every 3 s can be used to control the effluent emerging from the primary column onto the secondary column for GC×GC analyses. With regard to the MS settings, we recommend that no mass spectra are recorded for the first 450 s of each run, which serves as a solvent delay. The transfer line is kept at a constant 270 °C and the ion source temperature at 200 °C, for the entire run. For time-of-flight (TOF)—MS detectors, the recommended detector voltage is 1500 V and the filament bias −70 eV. Mass spectra can be collected at an acquisition rate of 200 spectra/s from 50 to 800 *m/z*. Although these parameters can be a good starting point, the optimum GC-MS conditions should, however, be determined before each study, and should be optimized according to the specific samples and instrument used. If a quadrupole MS detector is used the mass spectra can be collected using a scan of 60–800 *m/z* at 120 scans/min.

Acknowledgments

The financial assistance of the National Research Foundation (NRF) of South Africa for this research is gratefully acknowledged (UID: 95269). The opinions expressed and conclusions derived are those of the authors and are not necessarily those of the NRF.

References

1. Wishart DS, Tzur D, Knox C, Eisner R, Guo AC, Young N, Cheng D, Jewell K, Arndt D, Sawhney S, Fung C, Nikolai L, Lewis M, Coutouly M, Forsythe I, Tang P, Shrivastava S, Jeroncic K, Stothard P, Amegbey G, Block D, Hau DD, Wagner J, Miniaci J, Clements M, Gebremedhin M, Guo N, Zhang Y, Duggan E, MacInnis GD, Weljie AM, Dowlatabadi R, Bamforth F, Clive D, Greiner R, Li L, Marrie T, Sykes BD, Vogel HJ, Querengesser L (2007) HMDB: the human Metabolome database. Nucleic Acids Res 35:D521–D526
2. Dunn WB, Bailey NJC, Johnson HE (2005) Measuring the metabolome: current analytical technologies. Analyst 30:606–625
3. Olivier I, Loots DT (2012) A comparison of two extraction methods for differentiating and characterising various *Mycobacterium* species and *Pseudomonas aeruginosa* using GC-MS metabolomics. Afr J Microbiol Res 6:3159–3172

4. Du Preez I, Loots DT (2013) Detection limit for differentiating between various Mycobacterium species and Pseudomonas aeruginosa using gas chromatography-mass spectrometry (GC-MS) metabolomics: a comparison of two extraction methods. Afr J Microbiol Res 7:797–801
5. Loots DT, Meissner-Roloff RJ, Newton-Foot M, van Pittius G, Nicolaas C (2013) A metabolomics approach exploring the function of the ESX-3 type VII secretion system of *M. smegmatis*. Metabolomics 9:631–641
6. Loots DT, Swanepoel CC, Newton-Foot M, Gey van Pittius NC (2016) A metabolomics investigation of the function of the ESX-1 gene cluster in mycobacteria. Microb Pathog 100:268–275
7. Meissner-Rolloff RJ, Koekemoer G, Warren RM, Loots DT (2012) A metabolomics investigation of a hyper- and hypo-virulent phenotype of Beijing lineage *M. tuberculosis*. Metabolomics 8:1194–1203
8. Loots DT (2014) An altered *Mycobacterium tuberculosis* metabolome induced by katG mutations resulting in isoniazid resistance. J Antimicrob Chemother 58(4):2144–2149
9. Loots DT (2016) New insights into the survival mechanisms of rifampicin-resistant *Mycobacterium tuberculosis*. J Antimicrob Chemother 71(3):655–660
10. Du Preez I, Loots DT (2013) New sputum metabolite markers implicating adaptations of the host to *Mycobacterium tuberculosis*, and vice versa. Tuberculosis 93(3):330–337
11. Luies L, Loots D (2016) Tuberculosis metabolomics reveals adaptations of man and microbe in order to outcompete and survive. Metabolomics 2(3):1–9
12. World Health Organization (2012) Tuberculosis laboratory biosafety manual. http://www.who.int/tb/publications/2012/tb_biosafety/en/
13. Brown M, Dunn WB, Ellis DI, Goodacre R, Handl J, Knowles JD et al (2005) A metabolome pipeline: from concept to data to knowledge. Metabolomics 1(1):39–51
14. Mashego MR, Rumbold K, De Mey M, Vandamme E, Soetaert W, Heijnen JJ (2007) Microbial metabolomics: past, present and future methodologies. Biotechnol Lett 29 (1):1–16

Chapter 7

High-Throughput Solid-Phase Microextraction–Liquid Chromatography–Mass Spectrometry for Microbial Untargeted Metabolomics

Fatemeh Mousavi, Barbara Bojko, and Janusz Pawliszyn

Abstract

Nowadays, metabolomics data, when combined with other "omics" data, can provide important information regarding systems biology. Acquiring a comprehensive untargeted metabolome snapshot of complex sample matrices requires proper sample preparation, and access to sophisticated analytical instrumentation such as mass spectrometry. In metabolomics, sample preparation has substantial influence on the quality of the obtained metabolome profile. To achieve a real snapshot of the metabolome, the analysis method must be capable of inhibiting metabolite interconversion by immediately quenching all metabolome activity. Application of solid-phase microextraction (SPME), particularly in its in vivo set up, when undertaken in conjunction with a conscious selection of coating type based on the chosen sample matrix and the physicochemical properties of the analytes under study, is capable of providing extraction of representative metabolomes for many biological matrices. Metabolomes identified by SPME include low-abundance species and short-lived or unstable metabolites hardly captured by traditional extraction techniques. SPME coupled to liquid chromatography–high-resolution mass spectrometry has recently been introduced as an innovative alternative technique that integrates sampling, sample preparation, and extraction for metabolic profiling and isolation of candidate biomarkers. This chapter presents a detailed protocol for microbial metabolome analysis of *Escherichia coli* as a model organism, applying the high-throughput SPME-LC-MS workflow.

Key words High-throughput analysis, Metabolomics, LC-MS, SPME, Sample preparation, Automation, *E. coli*

1 Introduction

Metabolomics is an influential technique for the study of biological samples [1]. In metabolomics studies, liquid chromatography coupled with mass spectrometry (LC-MS) is extensively implemented for metabolic characterization, including in targeted and untargeted approaches [2]. Metabolomics comprises the comprehensive snapshot of cells' metabolome and their fluctuation in response to drugs, environment, diet, lifestyle, stimuli, and genetic inflection

Edward E.K. Baidoo (ed.), *Microbial Metabolomics: Methods and Protocols*, Methods in Molecular Biology, vol. 1859, https://doi.org/10.1007/978-1-4939-8757-3_7, © Springer Science+Business Media, LLC, part of Springer Nature 2019

[3]. The isolation and extraction of metabolites from biological samples, as the first step of a typical metabolomics workflow, can affect the quality of the obtained results [4].

Solid-phase microextraction (SPME), as a non-exhaustive, environmentally friendly sample preparation technique, combines the steps of sampling, sample preparation, and analyte extraction, which have often been attributed as the bottleneck steps of the entire analytical procedure [5]. Application of SPME provides multiple advantages, including simplification of the analysis procedure in comparison to conventional sample preparation methods, which typically include multistep, labor-intensive processes, in addition to generally requiring the use of large amounts of organic solvents [5].

The SPME sampling procedure generally includes exposure of an SPME coating (extraction phase immobilized on solid support) directly to the sample matrix or to the headspace of that for a predetermined amount of time [6]. The biocompatibility of the selected coating toward the analytes of interest aids in the prevention of fouling effects caused by protein adsorption during extraction from complex matrices generally targeted in food, biological fluids, and tissue [7]. During exposure of the coating to the matrix, analytes diffuse from the sample and into the extraction phase through a boundary layer, with maximum sensitivity being reached under equilibrium conditions [8]. In most cases, a washing step is added to the workflow after the extraction step so as to remove particulates, salts, and other contaminants loosely attached on the coating surface by non-specific interactions, as the presence of such contaminants could potentially cause matrix effects and instrument contamination [9]. Once the extraction and washing steps are completed, the extracted metabolites can then be desorbed to the optimized solvent (most often an organic/aqueous mixture), and submitted to instrumental analysis [5].

High-throughput sample preparation can be accomplished with the use of the Concept 96-autosampler, which facilitates simultaneous preparation of up to 96 samples, thus significantly shortening overall time of analysis [7]. In addition to enabling fast analysis, since the entire procedure is automated, good precision can be obtained even when working at preequilibrium conditions [10].

This chapter illustrates the potential of high-throughput SPME coupled to the liquid chromatography–high-resolution mass spectrometry (LC-HRMS) platform in comprehensive studies of bacterial metabolome, using *Escherichia coli* (*E. coli*) as a model organism exposed to environmental stress or drug treatment. The use of SPME coupled to different chromatographic platforms can assist in the identification of new potential targets, metabolic pathway screening, biomarker discovery, and drug development [10–12]. The presented protocol can also be adapted for analysis of intracellular and extracellular metabolites from diverse microbial cells.

The protocol enables extraction of a wide range of metabolites, from water-soluble to lipophilic species, with the use of a single SPME coating. Extraction is followed by separation of compounds on three chromatographic platforms, RPLC (reversed phase liquid chromatography) and HILIC (hydrophilic interaction liquid chromatography), and their detection in both positive and negative electrospray ionization modes so as to ensure effective analysis.

Different extractive phases and different LC-MS methods were evaluated with the purpose of ensuring the most wide-spanning, effective coverage of metabolites. While certain coatings demonstrated superior applicability toward extraction of compounds belonging to specific classes of metabolites, PS-DVB-WAX/HLB 50:50 [w/w] was identified as the most effective extractive phase, providing extraction of metabolites with a wide range of polarities ($-7 < \log P < 22$) and the highest peak intensity, including amino acids, peptides, nucleotides, carbohydrates, polycarboxilic acids, vitamins, phosphorylated compounds, as well as lipids such as hydrophilic phospholipids, prenol lipids, and fatty acids. For selection of coating chemistries offering coverage of analytes beyond the range of PS-DVB-WAX/HLB 50:50 [w/w], the reader is encouraged to refer to the original research article on the topic [13].

For analyses with a purpose other than the metabolic profiling described herein, the adapted method should be evaluated through selection of coating type, optimization of the coating procedure, investigation of coating preconditioning requirements, evaluation of carryover, type of desorption solvent, extraction and desorption time, evaluation of well plate cross-contamination, validation of identical agitation in all wells, agitation speed, and position of blades inside each well.

2 Materials

Prepare all solutions using LC-MS grade water and analytical grade reagents. Prepare and store mobile phase reagents at room temperature (unless indicated otherwise). Diligently follow all waste disposal regulations when disposing waste materials.

2.1 Sample Preparation Procedure Chemicals

Chemicals used:

1. LC-MS grade solvents and LC-MS grade formic acid (1 mL glass ampules).
2. Ammonium acetate, peptone, yeast extract, and sodium chloride.
3. Phosphate-buffered saline (PBS) solution pH 7.4 is prepared by dissolving 8.0 g of sodium chloride, 0.2 g of potassium

chloride, 0.2 g of potassium phosphate, and 1.44 g of sodium phosphate in 1 L of purified water and adjusting the pH to 7.4.

4. Prepare preconditioning solution composed of methanol--water 50:50 (v/v); wash solution composed of water + 0.1% formic acid; and acetonitrile–water 50:50 (v/v) as desorption solvent.

Cell culture:

1. Prepare standard Luria–Bertani (LB) agar media composed of 10 g tryptone, 5 g yeast extract, 5 g NaCl, and 15 g agar per L, for cultivation of *E. coli*.
2. Plate the strain on LB agar plates.
3. From a single agar plate colony, make a preinoculum culture, and inoculate culture flasks by that until the culture reaches an OD_{600} value of 0.5–1.
4. Prepare serial dilutions of inoculated culture to achieve proper OD the next day. 10^7 CFU/mL provides enough sensitivity for the present study. It is important to note that culture size influences the quantity and type of metabolites produced; as such, the number of cells should always be kept constant within the same set of experiments.
5. For investigations related to drug effects on cell metabolome, cell cultures should be supplemented with subminimal inhibitory concentrations of drug to assess antibiotic induction of target metabolites. The minimum inhibitory concentration (MIC) of the drug under study must be investigated for any individual case study (*see* **Note 1**).
6. Perform the metabolism quenching step by soaking the cells in liquid nitrogen followed by storing samples in a −80 °C freezer.

2.2 Sample Preparation Procedure Equipment

1. Liquid nitrogen.
2. −80 °C freezer.
3. Centrifuge.
4. Polypropylenes deep 96-well plates.
5. Robotic Concept 96 autosampler (Fig. 1) from Professional Analytical Systems (PAS) Technology (Magdala, Germany) and 96-SPME-blade unit (Fig. 2) from Professional Analytical Systems (PAS) Technology (Magdala, Germany) or Kolbut Sp. z o.o. (Bytom, Poland) are applied for sample preparation.
6. Alternatively, the 96-blade unit can be prepared in the lab according to protocols described elsewhere [7, 13, 14].

Fig. 1 Automated Concept 96-autosampler

7. The automated and software-controlled Concept 96 SPME system (Fig. 1) can perform all steps of SPME, including preconditioning, extraction, wash, and desorption [7].
8. The device consists of four agitators, accommodating commercial 96-well plates used in subsequent steps of the analytical procedure with a brush-like device that holds 96 blades of SPME, and is mounted to a moving arm [7].
9. The orbital agitators are assigned to shake the 96-well plates at a speed predetermined by the operating software [15].
10. The system also contains an evaporation unit applied for optional evaporation, reconstitution of extracts, or preconcentration of analytes [15]. Alternatively, to the automated Concept 96 system, studies can be performed using a manual 96-Concept system (Fig. 2), which consists of one agitation- and temperature-controlled unit. The unit does not support evaporation and requires manual operation i.e. immersion and retraction SPME devices to/from the wells (no software control).
11. Manual operation, such as manual immersion and retraction of SPME devices to/from the wells (with no software control).
12. The 96-SPME blade device consists of eight rows of blade sets made from 1.4310 grade stainless steel (Fig. 3). Each blade is

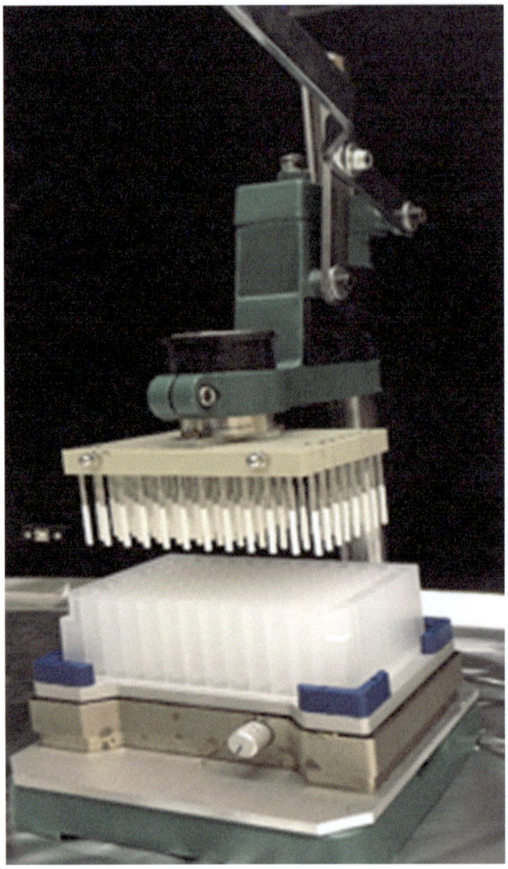

Fig. 2 Manual Concept 96-autosampler

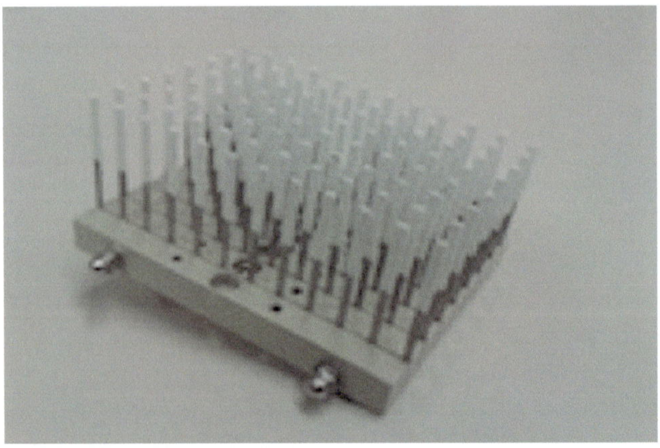

Fig. 3 96-Blade SPME brush

composed of 12 thin film pins (length, 50 mm; width, 2.5 mm, depth, 0.7 mm). The eight rows of the blade sets are kept together by nine interblade holders, forming the 96-blade SPME brush that is compatible with commercial 96-well plates [7, 13, 14].

2.3 Analytical Methodology Equipment

1. An LC-MS system equipped with a Thermo Accela autosampler, pumps, and an Exactive benchtop Orbitrap system (Thermo, San Jose, California, USA) for LC-MS analyses.
2. Kinetex PFP column [100 × 2.1 mm, 1.7 μm] (Phenomenex, Torrance, CA, USA) with a guard filter (SecurityGuard ULTRA Cartridges UHPLC PFP for 2.1 mm).
3. Luna HILIC Si, 100 × 2 mm, 3 μm particle size (Phenomenex, Torrance, CA, USA).
4. XBridge C18 (150 × 4.6 mm, 3.5 μm) (Waters, MA, USA).
5. Column temperature in all cases is 25 °C.
6. LC gradient information for Kinetex PFP: flow rate of 300 μL/min; mobile phase A is composed of water–formic acid 99.9:0.1%, and mobile phase B is composed of acetonitrile--formic acid 99.9:0.1%. The starting mobile phase conditions for Kinetex PFP were 90% A from 0 to 1.0 min, followed by a linear gradient to 10% A from 1.0 to 9.0 min, and an isocratic hold at 10% A until 12.0 min. The total run time is 16 min per sample, including a 4 min column reequilibration period [13] (*see* **Note 2**).
7. LC gradient information for Luna HILIC Si: flow of 400 μL/min; mobile phase A is composed of acetonitrile–ammonium acetate buffer (9:1, v/v, effective salt concentration 20 mM), and mobile phase B is composed of acetonitrile–ammonium acetate buffer (1:1, v/v, effective salt concentration 20 mM). The starting mobile phase composition is 100% A held for 3.0 min, with ramp to 100% B in 5.0 min. 100% B is held for 4.0 min. The column reequilibration for 8 min prior to the next injection, thus making the total run time 20 min per sample [13] (*see* **Note 3**).
8. LC gradient information for XBridge C18: mobile phase A is composed of acetonitrile–water 90:10 (v/v) with 15 mM effective concentration of ammonium acetate, and mobile phase B is composed of 0.5% acetic acid in isopropanol, with a flow rate of 200 μL/min. The starting mobile phase conditions for the XBridge C18 column: 90% A from 0 to 6.0 min, followed by a linear gradient to 40% A from 6.0 to 8.0 min, then another linear gradient to 30% from 8.0 to 18.0 min, and finally a linear gradient to 90% from 18 to 18:30 min. An isocratic hold is set at 90% A until 22.0 min. The total run time is 22 min per sample [13] (*see* **Note 4**).

9. The sample injection volume for all methods is set to 10 μL.
10. Sample tray temperature is 4 °C.
11. Trays for 96-well plate are used.
12. LC wash solution is acetonitrile–water 70:30 (v/v).
13. Analyses are performed in both positive and negative ESI modes for all three LC methods.
14. MS acquisitions are performed using AGC = balanced (1,000,000 ions), 100,000 resolution at m/z of 200, with injection times into the C-trap of 100 and 250 ms for the two RP and HILIC methods, respectively. Sheath gas (arbitrary units), auxiliary gas (arbitrary units), ESI voltage (kV), capillary voltage (V), capillary temperature (°C), and tube lens voltage (V) were (a) 40, 25, 4.0, 27.5, 275, and 100, respectively, for positive ESI RP LC-MS; (b) 50, 25, −2.7, −67.5, 325, and −85, respectively, for negative ESI RP LC-MS; (c) 60, 30, 2.8, 90, 325, and 155, respectively, for positive ESI HILIC LC-MS; and (iv) 50, 25, −2.5, −55, 275, and 100, respectively, for negative ESI HILIC LC-MS (*see* **Notes 5** and **6**).
15. LC-MS acquisition software is Xcalibur software version 2.1 (Thermo) by isolating the extracted ion chromatograms (XIC), using a 5 ppm window around the accurate mass of a given compound.
16. The data processing of raw data is done by XCMS Online arranged in three phases: data upload, parameter assortment, and result analysis.

3 Methods

3.1 Sample Preparation for Extracellular Metabolite Analysis

1. At the selected growth curve time point, centrifuge the cultures at 5000 × g for 20 min.
2. Transfer 1 mL supernatant in each well of a sterile deep well plate for storage at −80 °C, maintaining supernatant stored until extraction.
3. Noninoculated culture must be preserved to be used as blank media.

3.2 Sample Preparation for Intracellular Metabolite Analysis

1. Spin cell suspension at 5000 × g for 20 min after cell growth is finished.
2. Either discard supernatant, or keep it for extracellular metabolite study.
3. Wash the cell pellet with cold phosphate buffered saline (PBS) twice to scrape them off the flask surface.

4. Spin the cell suspension at 5000 × *g* for 20 min, and discard supernatant.
5. Flash-freeze cell pellets in liquid nitrogen.
6. Store cell pellets at −80 °C until further analysis.
7. At the time of analysis, thaw bacterial pellets at room temperature, and resuspend in PBS. Gently spin until complete suspension is obtained.
8. 1 mL of sample is exposed to high-throughput extraction.

3.3 High-Throughput SPME Methodology

1. Choose the type of coating to be used, based on the mode of study, polarity of metabolites under study, and sample matrix (*see* **Notes 7–13**).
2. Prepare four sterile 96-deep well plates.
3. Pipet 1.5 mL of ethanol–water 7:3 (v/v) mixture to the first 96-deep well plate.
4. Check pH of your sample and adjust as needed (*see* **Note 12**).
5. Pipet 1 mL of 5.0 log CFU/mL *E. coli* (initial concentration) in sterile LB media to the second 96-deep well plate.
6. Pipet 1 mL of distilled water with 0.1% formic acid to the third 96-deep well plate.
7. Pipet 1 mL of acetonitrile–water 1:1 (v/v) to the fourth 96-deep well plate.

When using the automated version of Concept 96:

1. Run the start-up script in the Concept software in order to prepare the autosampler.
2. The X and Y positions of all 96 blades must be positioned in the middle of the wells of 96-well plate in all four agitators, without touching the bottom of the wells. (Coatings should be kept at a 2–3 mm distance from the bottom of the wells.)
3. Insert values for preconditioning, extraction, wash, and desorption time with corresponding agitation speed values for all four steps (120 min at 850 rpm, 60 min at 10 × *g*, 30 s at 850 rpm, and 60 min at 25 × *g*, respectively).

When using the manual version of Concept 96:

1. Place the first 96-deep well plate, containing the ethanol–water mixture, on the agitator. Immerse coatings in the solution.
2. Start Concept 96. After 120 min have elapsed, stop the device and lift the SPME brush (*see* **Notes 14–16**).
3. Replace the first plate, containing the preconditioning solution, with the plate containing samples to be analyzed.

4. Ensure this is done in an expeditious manner so as to not allow coatings to dry between plate changes.
5. Set agitation speed at 1000 rpm.
6. Immerse coatings in the solution.
7. Start Concept 96. After 60 min have elapsed, stop the device and lift the SPME brush (*see* **Note 17**).
8. Replace the plate containing samples with the plate that contains water.
9. Set agitation speed at 850 rpm. Immerse coatings in the solution. Start Concept 96.
10. After 30 s have elapsed, stop the device and lift the SPME brush (*see* **Note 18**).
11. Replace the plate that contains water with the plate holding the desorption solvent (*see* **Note 19**).
12. Set agitation speed at 1500 rpm. Immerse coatings in the solution.
13. Start Concept 96. After 60 min have elapsed, stop the device and lift the SPME brush (*see* **Note 20**).
14. Place the plate now containing your extracts in the autosampler of your LC-MS system (*see* **Note 21**).

Optional steps:

1. Perform a second desorption by repeating the last step with a new plate, containing 1 mL of acetonitrile–water 1:1 (v/v) under the same conditions (60 min, agitation speed at 1500 rpm) (*see* **Note 22**).
2. Evaporate your extract to dryness using a 96-nitrogen blow system (*see* **Note 23**).
3. Reconstitute the evaporated extract with the desired solvent by pipetting the solvent to the previously evaporated extracts in order to make the final sample extract compatible with the mobile phase (*see* **Note 23**).

3.4 LC-MS Acquisition

1. Submit the obtained extracts to analysis by three different LC columns so as to ensure that all extracted analytes are retained and detected. It is important to note that the LC-MS technique must always be optimized in view of the goal of analysis, the targeted metabolites interest, and sensitivity of the method. The method must provide stable retention time, constant signal intensity, and mass accuracy (*see* **Notes 24–30**).
2. Load batch method sequence or work list via appropriate software.
3. Run batch method sequence or work list.

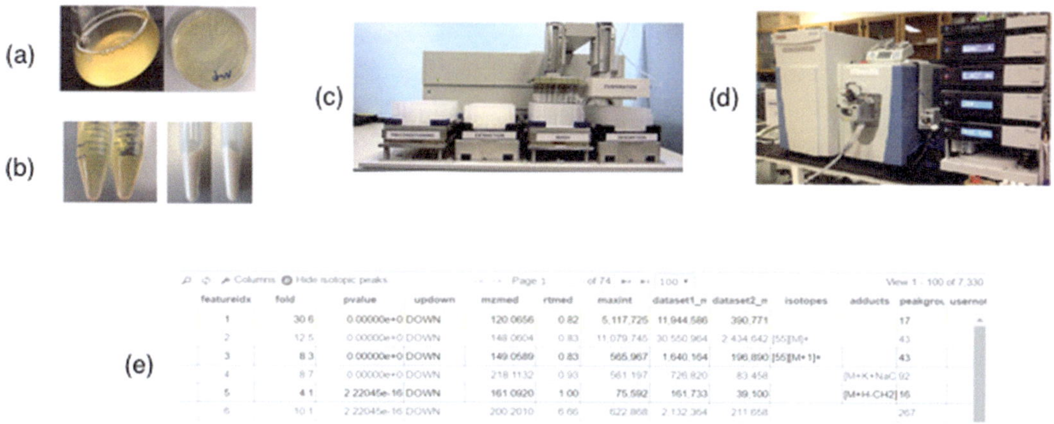

Fig. 4 Workflow for high-throughput SPME-LC-MS analysis: (**a**) cell culture (apply drug if needed for study); (**b**) separate intracellular and extracellular metabolites for sampling; (**c**) Concept 96 SPME system; (**d**) LC-MS Exactive orbitrap for metabolomics analysis; (**e**) data analysis and interpretation using XCMS online

4. Wash injection needle two times with needle wash solvent (refer to Subheading 2).
5. Load sample and inject 10 μL of sample.
6. Run LC-MS method using the aforementioned LC gradients and flow rates.
7. Acquire mass spectrometric data via the positive/negative ion mode and appropriate MS or MS/MS parameters. The SPME-LC-MS analysis workflow is summarized in Fig. 4.

3.5 Data Processing and Data Analysis

1. Convert raw data to mzXML using MS Convert software.
2. Compress your data if necessary.
3. Upload data to the XCMS Online.
4. The samples for each group that are to be compared are uploaded through a specific Java applet, which permits users to only drag and drop their files into the upload part of the program.
5. Characterize the parameter settings to change the signal–noise threshold, to adjust the mass tolerance for the identification step, or to change the methods that are used for feature detection, retention time correction, alignment, and annotation [16–18].
6. XCMS Online applies nonlinear methods to compensate for retention time drifts between samples [17].
7. The feature table (Fig. 4e) presents comprehensive material for each single feature including *m/z* value, retention time, and fold change, an extracted ion chromatogram, spectrum details, mass difference (ppm), multiple adducts, and putative

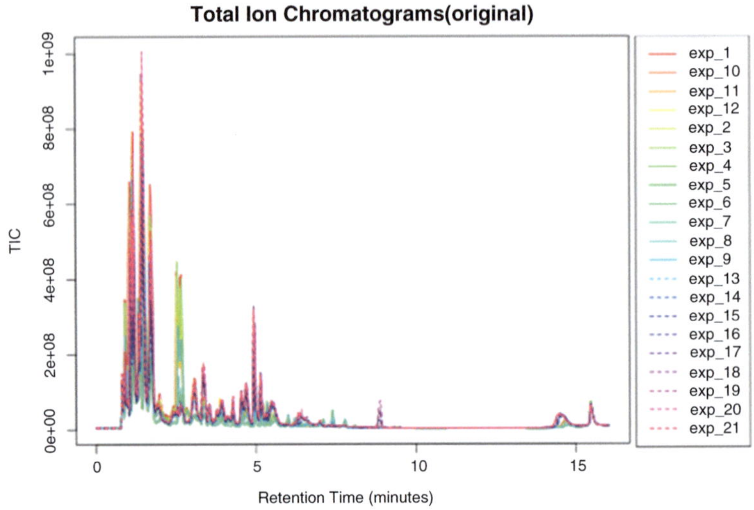

Fig. 5 Total ion chromatogram

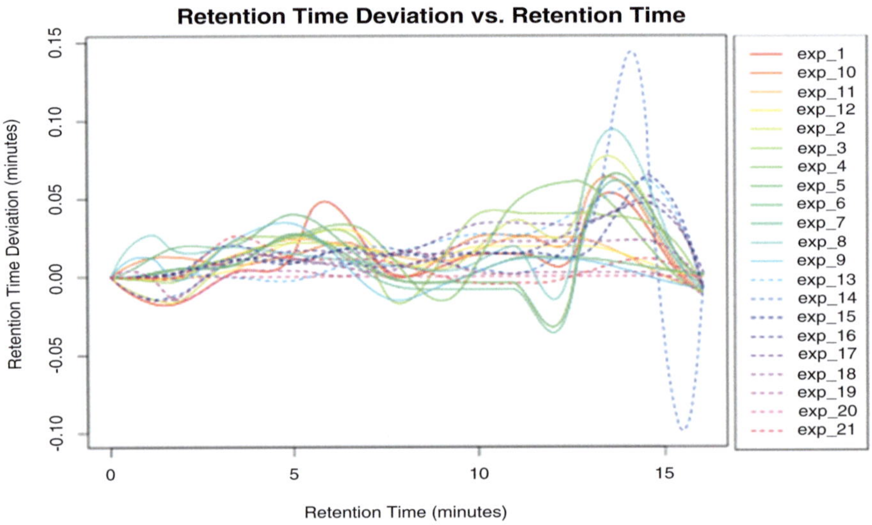

Fig. 6 Retention time deviation vs. retention time

METLIN assignments. Filtering can be done based on any of the investigated parameters. METLIN metabolite database is directly linked to the XCMS online, illustrating metabolite detail and MS/MS data (if available) [16].

8. In order to provide visualization and for quality control of this correction procedure, an overlay of all total ion chromatograms (TICs) assimilated is demonstrated before and after retention time correction, in addition to the retention time correction curves (Figs. 5, 6, and 7).

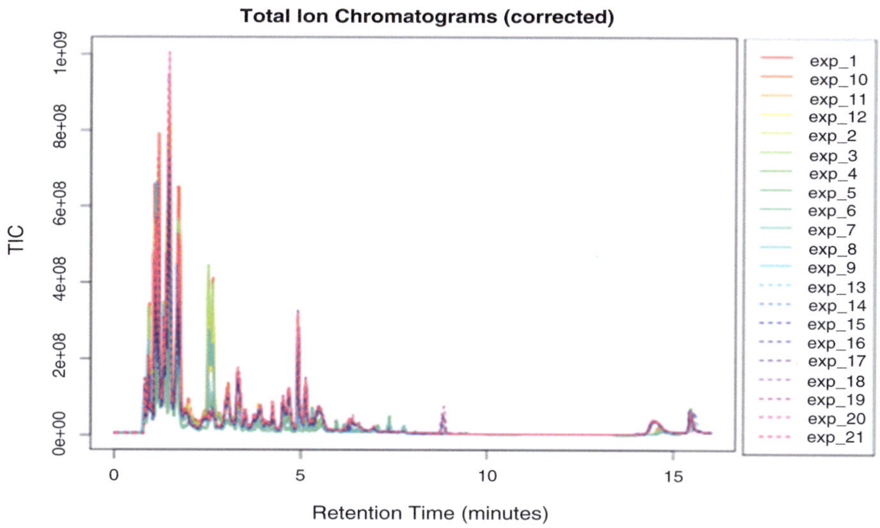

Fig. 7 Total ion chromatogram (corrected)

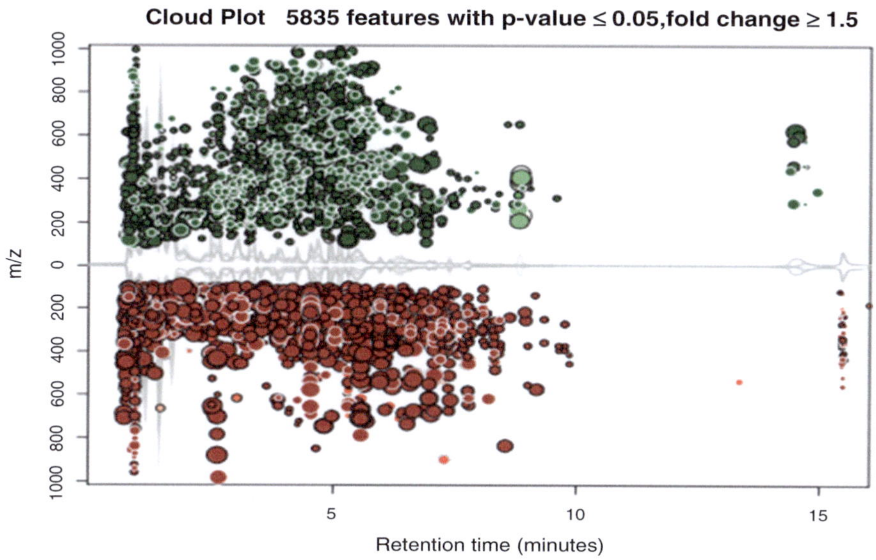

Fig. 8 Cloud plot

9. All TICs are in alignment after retention time correction process. Theoretically tricky samples with high deviations can be documented from the correction curve plot and may be eliminated from the data set (Fig. 6).
10. A cloud plot showing dysregulated features (Fig. 8) is used to denote ions with different intensities changed between sample groups according to statistical thresholds set by the user (e.g., *P*-value ≤ 0.001, fold change ≥ 2).

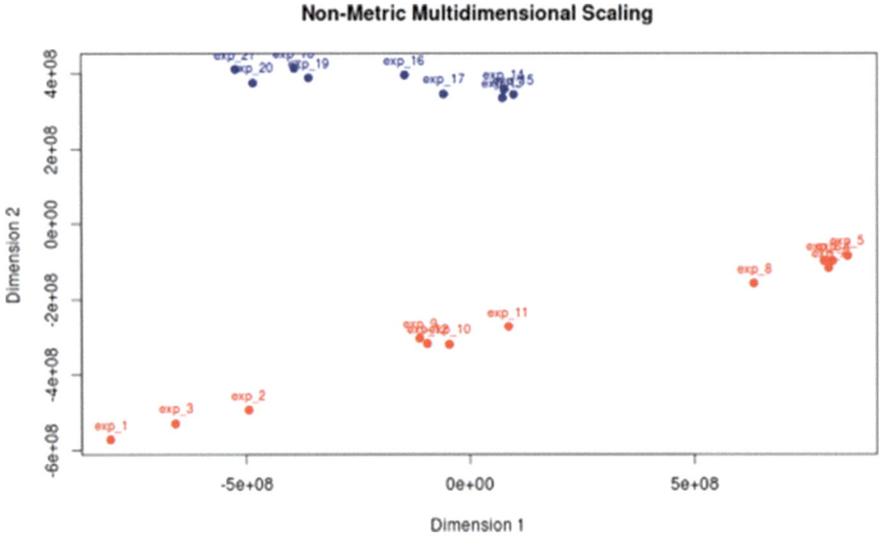

Fig. 9 Nonmetric multidimensional scaling plot

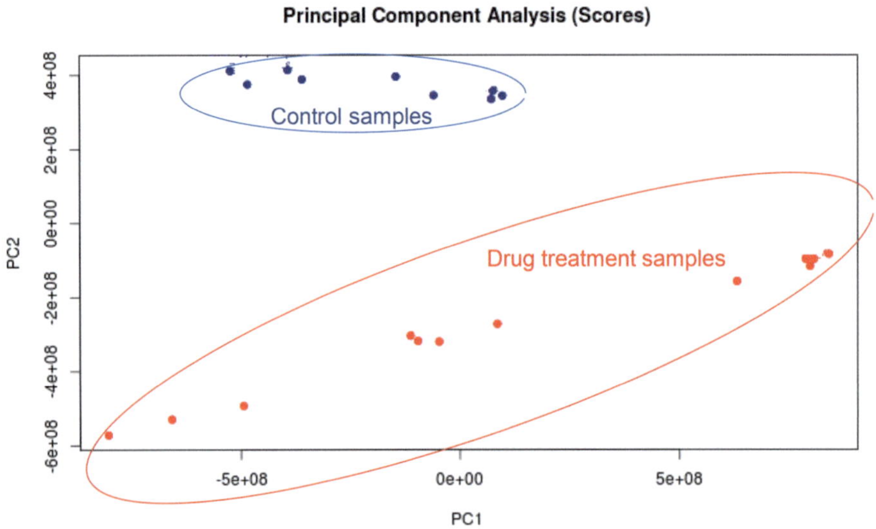

Fig. 10 PCA plot

11. Down-regulated features signified as circles on the bottom of the plot, while upregulated features are symbolized as circles on the top. Fold change of the features is characterized by size of each circle. The bright of color is used to denote *P*-value.
12. There are multidimensional scaling (Fig. 9) and principal component analysis (PCA) plots created by the software performed on the centered and scaled data in order to visualizing high dimensional data sets (Fig. 10). These plots are based on the

intensities of all aligned features and can illustrate relationships between the samples or assist in classifying probable outliers [19] (*see* **Notes 31** and **32**).

4 Notes

1. For investigations of other biological aspects, analysis can be carried out by collecting samples at different time points of the bacteria growth cycle.
2. The kinetex PFP column separates the highest number of features, identifying metabolites in the range of $-7 < \log P < 15$. Metabolites may include carboxylic acids and derivatives, amino acids and derivatives, peptides, prenol lipids, glycosyl compounds, monosaccharide/disaccharide/trisaccharide, glycerophospholipids, fatty acids, and conjugates [13].
3. The identified metabolites separated by HILIC are in the range of $-8.0 < \log P < 9.0$, with the majority of the separated metabolites being related to glycosyl compounds, peptides, phenols and derivatives, amino acids and derivatives, nucleotides, and monosaccharides and disaccharides [13].
4. The separated metabolites identified by XBridge C18 column have the polarity range of $-3 < \log P < 22$, indicating better usage of this column for nonpolar extracted metabolites.
5. Ion source and spray needle must be cleaned frequently, as they may be contaminated by complex biological samples.
6. The high-resolution mass spectrometer must be calibrated daily before analyzing samples, including standard masses that cover the mass range of the method (100–1000 Da). External instrument mass calibrations were performed every 24 h, and found to be within 2 ppm for all ions.
7. Based on increasing peak intensities for the various coating chemistries studied, the degree of simultaneous extraction of both polar and nonpolar metabolites is ordered as follows: PS-DVB-WAX/HLB 50:50 [w/w] > PS-DVB-WAX/HLB 80:20 [w/w] > PS-DVB-WAX/SiIL 80:20 [w/w] > PS-DVB-WAX > HLB > Si-IL > Si-RP > PBA [13].
8. PBA, Si-IL, and Si-RP-WCX coatings have similar extraction behaviors, as all three coatings showed approximately the same extraction recoveries for the same classes of metabolites, with particular affinity toward polar metabolites ($-4 < \log P < 4$) such as amino acids, peptides, carboxylic acids, monosaccharides, and sugar acids [13].
9. PS-DVB-WAX provides high extraction efficiency for both polar and nonpolar metabolites ($-7 < \log P < 7$). The HLB

coating is shown to exhibit mostly nonpolar characteristics ($-4 < \log P < 22$).

10. PS-DVB-WAX: HLB 50:50 (w:w), as the best candidate for untargeted analysis using the 96-blade system, is able to extract a wide variety of metabolites in terms of polarities characterized by $-7 < \log P < 15$, and $100 < MW < 1000$, including metabolites related to central carbon metabolism (glycolytic, and pentose phosphate pathway including glucose, ribulose-5-phosphate, glucose-6-phosphate, and intermediates of citric acid cycle such as aconitic acid, succinic acid, and malic acid), biosynthetic metabolites (amino acids such as S-adenosyl-L-methionine, asparagine, cystathionine, betaine, and homoserine, and nucleotides including cytidine 3′-triphosphate, nicotinic acid mononucleotide, adenosine diphosphate ribose, adenosine 3′,5′-bisphosphate, and adenosine triphosphate) classified as polar metabolites, and hydrophobic metabolites such as lipids linolenic acid, hexacosanoic acid, and stearic acid, PG(22:0/19:0), PG(16:0/16:0), PC(P-20:0/12:0), which are widely distributed within the *E. coli* membrane. These metabolites can be separated by employing three different LC columns [13]. For selection of coating chemistries offering coverage of analytes beyond the range of PS-DVB-WAX/HLB 50:50 [w/w], the reader is encouraged to refer to the original research article on the topic [13].
11. Keep track of SPME coating saturation while doing extraction of new sample matrix. Saturation of the coating surface results in nonlinear adsorption isotherms. Analytes can be competitively displaced from the coatings by other analytes or other interferences present in the sample. This phenomenon depends on the magnitude of distribution constant of the interferences between the coating and sample matrix and their concentration in the matrix.
12. The pH and ionic strength of the sample under study should be checked prior to analysis. Metabolites have different physicochemical properties that affect their interaction by different stationary phases. As SPME extracts neutral forms of metabolites, the pH of the sample should be adjusted to molecular form. For untargeted studies, it is recommending that a neutral pH be kept for samples. For extraction of acidic metabolites, pH should be 2 units more than pKa, while extraction of basic metabolites of interest should be carried out under a pH that is 2 units lower than the p*K*a of the metabolites of interest. Irrespective of the selected pH value of the sample being studied, it is important that such a value is kept constant over the entire set of experiments. Adjustments can be done by addition of buffer solutions such as carbonate, citrate, or phosphate buffers, thus providing better buffer capacity and

influencing coating tolerance rather than through direct addition of acids and bases.

13. During subsequent steps, fluid volume must be enough to immerse the whole length of the coating, but not as large as to prevent any cross-contamination between adjacent wells.
14. Precondition coatings by overnight exposure to methanol-- water (1:1 (v/v)) solution at 850 rpm before the first application.
15. Based on the type of coatings applied for the study, the preconditioning time may vary from 30 min to 3 h. For example, HILIC coatings need more time in this step than hydrophobic types of stationary phases. For this step, it is important to ensure that no analyte residue remains on the surface of the coatings from previous extraction analysis.
16. Preconditioning in an ethanol–water 7:3 (v/v) mixture combines activation of the sorbent and its sterilization. Another sterilization procedure, namely autoclaving, can also be used. However, in such a case, it is recommended that any possible effects related to such a change in protocol are examined in regard to coating performance. This can be carried out by evaluating the extraction efficiency of the coating, and by performing sterilization prior to the preconditioning step.
17. The time needed to achieve equilibrium extraction is different for each analyte, and cannot be determined for untargeted analysis. Extraction can be performed either at preequilibrium or equilibrium extraction conditions. Reproducible results are obtainable as long as precise control of extraction time is ensured. Longer extractions provide better sensitivity, but may cause evaporation, oxidation, or degradation of analytes, which in turn can impact the result of analysis.
18. Between the extraction and desorption steps, a fast wash step is required to eliminate particulates or cell debris from the coating surface. Optimally, a 30-s immersion of coatings in nanopure water should be implemented. However, highly water-soluble compounds can be lost in the washing process; therefore, if those are of particular interest in a given analysis, the contents of the rinsing solution, as well as time and agitation should be optimized for this step. Analyte loss in the wash step can be monitored by direct injection of a wash solution into the LC-MS.
19. The desorption step should be optimized in terms of desorption solvent and time. The selected solvent needs to be compatible with the mobile phase applied for LC-MS analysis so as to obtain an acceptable chromatographic separation. Methanol–water (1:1 (v/v)), acetonitrile–water (1:1 (v/v)), or methanol–acetonitrile–water (x:y:z (v/v/v)) are the most

common compositions used as mobile phase for LC-MS analysis. The desorption solvent may have an acidic or basic composition to boost the extraction recovery, based on the type of analytes being targeted. Aqueous solvents decrease the chance of metabolite evaporation, extraction efficiency reduction, and reproducibility variation, following lack of evaporation/reconstitution requirement. However, for analysis of highly nonpolar metabolites, selection of high organic compositions such as dichloromethane or isopropanol may be required. In such cases, the evaporation of desorption solvent and subsequent reconstitution of the sample with mobile phase would be done. Desorption time length could span from 30 min to 2 h; factors influencing desorption time include the type of coating selected, coating thickness, type of metabolites under study, and the complexity of matrix.

20. After method completion, the 96-well plate is ready for LC-MS analysis. Plates must be covered by multiwell plate sealing films and refrigerated at 4 °C so as to avoid solvent evaporation or metabolite degradation.
21. For on-site analysis, coatings must be stored at −80 °C and covered with aluminum foil after sampling so as to protect the extracted metabolites on the coating surface from evaporation or degradation.
22. Second desorption must be evaluated, particularly when coatings are used for subsequent extractions. This additional step can improve the result of the extraction recovery as well as eliminate any residue maintained on the coating as carryover, aiding in the preparation of the coating for its next use.
23. The last step of the high-throughput procedure is the evaporation/reconstitution step. This step can be applied to preconcentrate analytes in the extract by dissolving extracts in smaller volumes or by replacing the desorption solvent, which is not compatible with the mobile phase, with the compatible one.
24. When carrying out quantitative metabolomics studies, standard analytes and samples must be prepared and analyzed using a proper LC-MS/MS method.
25. Do not exceed the linear dynamic range of the instrument.
26. Investigate blank injection and blank coating extraction before running samples.
27. The preparation of QC for LC-MS analysis is done through pool 10 μL aliquots of each sample within entire sample set.
28. QC samples should be injected 5–10 times at the start of a run to evaluate column preconditioning, evaluation of peak retention times and intensities as well as throughout the run to assess instrument stability and data quality.

29. Samples were run in a randomized order, and a QC sample was run periodically to verify instrument performance.
30. It is important to validate the LC-MS system on a daily basis. Quality control (QC) samples with a given concentration of representative standards should be analyzed within the sample batches (every 10–12 samples).
31. QC between sample runs assists in checking system suitability and stability, assessing system contamination, identifying instrument breakdown issues, as well as in detecting decreases in sensitivity, shifts in retention time, and mass accuracy deviations. When performing PCA, if QC samples are not tightly clustered, SPME coating performance must be investigated in addition to instrument conditions.
32. SPME provides clear cluster separation among different treatment samples and time points of the *E. coli* growth cycle, demonstrating the method's strength for evaluating the effect of different drugs and drug-dosing regimens, stress conditions, or growth stages [11, 12].

Acknowledgments

The authors thank the Natural Sciences and Engineering Research Council (NSERC) of Canada (IRCPJ 184412-10 050165, IRCPJ 184412-10 050165) for financial support.

References

1. Vernocchi P, Chierico FD, Putignani L (2016) Gut microbiota profiling: metabolomics based approach to unravel compounds affecting human health. Front Microbiol 7(1144):1–21
2. Zhou B, Xiao JF, Tuli L, Ressom HW (2012) LC-MS-based metabolomics. Mol BioSyst 8 (2):470–481
3. Bruce SJ, Tavazzi I, Parisod V, Rezzi S, Kochhar S, Guy P (2009) Investigation of human blood plasma sample preparation for performing metabolomics using ultrahigh performance liquid chromatography/mass spectrometry. Anal Chem 81:3285–3296
4. Dettmer K, Aronov PA, Hammock BD (2007) Mass spectrometry-based metabolomics. Mass Spectrom Rev 26:51–78
5. Pawliszyn J (2009) Handbook of solid phase microextraction. Chemical Industry Press, Beijing
6. Risticevic S, Lord H, Górecki T, Arthur CL, Pawliszyn J (2010) Protocol for solid-phase microextraction method development. Nat Protoc 5:122–139
7. Mirnaghi FS, Chen Y, Sidisky LM, Pawliszyn J (2011) Optimization of the coating procedure for a high-throughput 96-blade solid phase microextraction system coupled with LC–MS/MS for analysis of complex samples. Anal Chem 83:6018–6025
8. Risticevic S, Chen Y, Kudlejova L, Vatinno R, Baltensperger B, Stuff JR, Hein D, Pawliszyn J (2010) Protocol for the development of automated high-throughput SPME–GC methods for the analysis of volatile and semivolatile constituents in wine samples. Nat Protoc 5:162–176
9. Mirnaghi FS, Mousavi F, Rocha SM, Pawliszyn J (2013) Automated determination of phenolic compounds in wine, berry, and grape samples using 96-blade solid phase microextraction system coupled with liquid chromatography–tandem mass spectrometry. J Chromatogr A 1276:12–19

10. Bojko B, Reyes-Garces N, Bessonneau V, Gorynski K, Mousavi F, Souza Silva EA, Pawliszyn J (2014) Solid-phase microextraction in metabolomics. Trends Anal Chem 61:168–180
11. Mousavi F, Bojko B, Bessonneau V, Pawliszyn J (2016) Cinnamaldehyde characterization as an antibacterial agent toward E. coli metabolic profile using 96-blade solid-phase microextraction coupled to liquid chromatography–mass spectrometry. J Proteome Res 15(3):963–975
12. Mousavi F, Gionfriddo E, Carasek E, Souza-Silva EA, Pawliszyn J (2016) Coupling solid phase microextraction to complementary separation platforms for metabotyping of E. colimetabolome in response to natural antibacterial agents. Metabolomics 12 (11):169–181
13. Mousavi F, Bojko B, Pawliszyn J (2015) Development of high throughput 96-blade solid phase microextraction-liquid chromatrography-mass spectrometry protocol for metabolomics. Anal Chim Acta 892:95–104
14. Mousavi F, Pawliszyn J (2013) Silica-based ionic liquid coating for 96-blade system for extraction of aminoacids from complex matrixes. Anal Chim Acta 803:66–74
15. Boyaci E, Gorynski K, Rodriguez-Lafuente A, Bojko B, Pawliszyn J (2013) Introduction of solid-phase microextraction as a high-throughput sample preparation tool in laboratory analysis of prohibited substances. Anal Chim Acta 809:69–81
16. Tautenhahn R, Bottcher C, Neumann S (2008) Highly sensitive feature detection for high resolution LC-MS. BMC Bioinformatics 9(504):1–16
17. Prince JT, Marcotte EM (2006) Chromatographic alignment of ESI-LC-MS proteomics data sets by ordered bijective interpolated warping. Anal Chem 78(17):6140–6152
18. Kuhl C, Tautenhahn R, Böttcher C, Larson TR, Neumann S (2012) CAMERA: an integrated strategy for compound spectra extraction and annotation of liquid chromatography/mass spectrometry data sets. Anal Chem 84:283–289
19. Stacklies W, Redestig H, Scholz M, Walther D, Selbig J (2007) pcaMethods—a bioconductor package providing PCA methods for incomplete data. Bioinformatics 23(9):1164–1167

Part III

Current Analytical Techniques

Chapter 8

Targeted Metabolomics of Xylose-Fermenting Yeasts Based on Mass Spectrometry

Christiane Gonçalves Campos, José Antônio de Aquino Ribeiro, João Ricardo Moreira de Almeida, Betania Ferraz Quirino, and Patrícia Verardi Abdelnur

Abstract

Mass spectrometry is a sensitive and selective analytical technique that enables detection and quantitation of low abundance compounds in a complex sample matrix. Targeted metabolomics allows for quantitative analysis of metabolites, providing kinetic information of production and consumption rates, an essential step to investigate microbial metabolism. Here, we describe a targeted metabolomics protocol for yeast samples, from sample preparation to mass spectrometry analysis, which enables the identification of metabolic fluxes after xylose consumption. Sample preparation methods were optimized for quenching of yeast metabolism followed by intracellular metabolite extraction, using cold methanol and boiling ethanol protocols. Ultrahigh performance liquid chromatography tandem mass spectrometry (UHPLC-MS/MS) methods using ion pair chromatography (IPC) and hydrophilic interaction liquid chromatography (HILIC) allowed for the quantitation of 18 metabolites involved in central carbon metabolism (glycolysis, pentose phosphate pathway, and tricarboxylic acid cycle). The protocol here described was successfully applied to quantify metabolites in *Scheffersomyces stipitis*, *Spathaspora passalidarum*, *Spathaspora arborariae*, and *Candida tenuis* samples after xylose consumption.

Key words Microbial metabolomics, Biotechnology, Yeast, UHPLC-MS/MS, LC-MS

1 Introduction

Metabolomics allows for qualitative and quantitative analysis of metabolites in a biological system [1]. There are two main approaches in metabolomics: untargeted and targeted [2]. Untargeted metabolomics aims to "get a picture" of all (or almost all) metabolites in a system, through metabolic fingerprinting or metabolic profiling analysis [3]. On the other hand, targeted metabolomics enables quantitative analysis of key metabolites from specific pathways to answer specific biological questions [4].

A general workflow for metabolomics involves the following steps: (a) choosing the sample, (b) sample preparation, (c) analytical

Edward E.K. Baidoo (ed.), *Microbial Metabolomics: Methods and Protocols*, Methods in Molecular Biology, vol. 1859, https://doi.org/10.1007/978-1-4939-8757-3_8,

methods, (d) data processing, and (e) data analysis and interpretation [5]. Sample preparation and analytical methods are crucial steps in metabolomics, since they affect the integrity of extracted metabolites and identification of compounds [6, 7]. Mass spectrometry has been widely used in metabolomics because it is a sensitive and selective technique, enabling identification of low-abundance metabolites in a complex matrix sample [8–10].

This chapter describes a targeted metabolomics protocol for yeast samples, giving details from sample preparation to mass spectrometry analysis. Cold methanol (−40 °C) was used for quenching of yeast metabolism, followed by metabolite extraction using boiling ethanol [11]. UHPLC-MS/MS using IPC [12, 13] or HILIC [14, 15] were used to quantify, respectively, 13 and 5 metabolites from central carbon metabolism (glycolysis, pentose phosphate pathway, and tricarboxylic acid cycle). Specifically, these quantified metabolites were acetyl coenzyme A (ACCOA), alpha ketoglutarate (AKG), D-malate (L-MAL), D-(+)-glucose, xylose, glucose-6-phosphate (G6P), fructose-6-phosphate (F6P), dihydroxyacetone phosphate (DHAP), glyceraldehyde-3-phosphate (GAP), glycerol-3-phosphate (GLY3P), ribose-5-phosphate (R5P), ribulose-5-phosphate (RU5P), xylulose (Xylu), phospho(enol)pyruvate (PEP), glycerol, pyruvate (PYR), sedoheptulose-7-phosphate (S7P), and xylitol. Quantitative analysis of these metabolites allows for prediction of metabolic flux after xylose consumption, thus making it possible to identify pathway bottlenecks in xylose-fermenting yeasts.

The targeted metabolomics protocol has proved to be robust since it was applied to four xylose-fermenting yeasts: *Scheffersomyces stipitis*, *Spathaspora passalidarum*, *S. arborariae*, and *Candida tenuis*.

2 Materials

All standards (ACCOA, AKG, L-MAL, D-(+)-glucose, xylose, G6P, F6P, DHAP, GAP, GLY3P, R5P, RU5P, Xylu, PEP, glycerol, PYR, S7P, and xylitol), tributylamine, ammonium hydroxide, acetic acid, and ammonium acetate should be acquired with purity higher than 95%. Prepare all solutions using reagents of LC-MS grade and deionized water with an electric resistance of 18.2 MΩ at 25 °C. Diligently follow all appropriate waste disposal regulations when disposing of waste materials.

2.1 Strain and Cultivation Conditions

1. Yeasts: *Scheffersomyces (Pichia) stipitis* NRRL Y-7124, *Spathaspora passalidarum* NRRL Y-27907, *Spathaspora arborariae* NRRL Y-48658, *and Candida tenuis* NRRL Y-1498.
2. YPD (10 g/L yeast extract, 20 g/L peptone, 20 g/L glucose) plate.
3. 500 mL Erlenmeyer flask.

4. Bioreactor (Multifors 2, Infors HT) (working volume of 500 mL).
5. Defined mineral medium: 2.5 times concentrated and supplemented with 40 g/L of xylose as the carbon source. This medium is based on Verduyn et al. [16], and should be prepared as follows: Salt solution: $(NH_4)_2SO_4$ 12.5 g/L, $MgSO_4{\cdot}7H_2O$ 1.25 g/L, KH_2PO_4, 7.5 g/L. Trace elements solution: dissolve 1.5 g EDTA (disodium) in 50 mL H_2O, add 0.45 g $ZnSO_4{\cdot}7H_2O$. Adjust pH to 6.0 and add following components (after each addition, adjust pH to 6.0): $MnCl_2{\cdot}2H_2O$ 0.1 g, $CoCl_2{\cdot}6H_2O$ 0.03 g, $CuSO_4{\cdot}5H_2O$ 0.03 g, $Na_2MoO_4{\cdot}H_2O$ 0.04 g, $CaCl_2{\cdot}2H_2O$ 0.45 g, $FeSO_4{\cdot}7H_2O$ 0.3 g, H_3BO_3 0.1 g, KI 0.01 g. Finally, adjust pH to 4.0 and the volume to 100 mL. The solution should be filter-sterilized and stored in the refrigerator. A volume of 2.5 mL of trace element solution should be added to 1 L of medium. Vitamin solution: dissolve 0.005 g biotin in 1 mL 0.1 M NaOH and add it to 80 mL H_2O. Adjust pH to 6.5 with 1 M HCl/1 M NaOH. The following vitamins should be dissolved into the biotin solution, one at a time. The pH needs to be adjusted to 6.5 after each addition. Ca-pantothenate 0.1 g, nicotinic acid 0.1 g, myo-inositol 2.5 g, thiamine–HCl 0.1 g, pyridoxine–HCl 0.1 g, p-aminobenzoic acid, 0.02 g, riboflavin 0.02 g, folic acid 0.0002 g. Adjust the volume to 100 mL with H_2O and set the pH to 6.5. A total volume of 2.5 mL of this vitamin solution should be added to 1 L of medium. The solution should be filter-sterilized and stored in the refrigerator. Ergosterol–Tween 80: Dissolve 1.6 g ergosterol in 64 mL pure boiling ethanol, add 67.2 g Tween 80 and adjust volume to 200 mL with pure ethanol. Filter-sterilize. A volume of 3.125 mL of this solution should be added to 1 L of final medium.

2.2 Quenching of Cellular Metabolites

1. 15 mL Falcon® tubes. These tubes are used to collect 2 mL of each sample.
2. Liquid nitrogen.
3. Small polystyrene box.
4. Thermostatic bath set to −40 °C.
5. Refrigerated centrifuge set to −9 °C with rotors compatible with 15 mL Falcon® tubes.
6. Buffered 60% methanol (v/v): methanol, 10 mM ammonium acetate, water; the pH should be approximately 7.5. Weigh 0.77 g ammonium acetate in a glass beaker and dissolve the salt with 400 mL of water, then mix it. Transfer the solution to a 1 L glass bottle and add 600 mL of methanol. Store at room temperature.

2.3 Metabolite Extraction

1. 2 mL microcentrifuge tubes.
2. Thermometer.
3. 250 mL Erlenmeyer flask.
4. Heating and magnetic stirring plate.
5. 10 mL pipette.
6. Thermomixer set to 85 °C and 800 rpm.
7. Centrifugal vacuum concentrators.
8. Thermostatic bath set to −40 °C.
9. Vortex mixer.
10. Refrigerated centrifuge set to −9 °C with rotor for 2 mL microcentrifuge tubes.
11. Buffered 75% ethanol (v/v): ethanol, 10 mM ammonium acetate, water; the pH should be approximately 7.5. Weigh 0.77 g ammonium acetate in a glass beaker and dissolve the salt with 250 mL of water gradually. Transfer the solution to a 1 L glass bottle and make up to 750 mL with ethanol, store at room temperature.

2.4 UHPLC-MS

1. UHPLC (Acquity™ UPLC system, Waters) coupled to a triple quadruple mass spectrometer (Xevo TQD, Waters) equipped with an electrospray ionization source.
2. Wash system solution: acetonitrile/water (50:50, v/v).
3. Glass LC vials with screw caps.
4. Ultrasonic bath.
5. Minicentrifuge.
6. IPC mobile phases: Mobile phase A is composed of 5 mM tributylamine, 10 mM acetic acid, and 5% (v/v) methanol, the pH should be approximately 4.8. Add to graduated cylinder 50 mL of methanol, 1.19 mL of tributylamine, and 560 μL of acetic acid (*see* **Note 1**). Transfer the solution to a glass bottle containing 950 mL of water (*see* **Note 2**). Store at room temperature. Mobile phase B is methanol.
7. A Waters reversed phase HSS-T3 (2.1 × 150 mm × 1.8 μm) column is used for the IPC method.
8. HILIC mobile phases: Mobile phase A is composed of 0.1% (v/v) ammonium hydroxide aqueous solution. Add 500 μL of ammonium hydroxide to 500 mL of ultrapure water. Mobile phase B is composed of acetonitrile with 0.1% (v/v) ammonium hydroxide (*see* **Note 2**). Add 500 μL of ammonium hydroxide in 500 mL of acetonitrile. Store at room temperature.
9. A Waters BEH amide (2.1 × 150 mm × 1.7 μm) column is used for the HILIC method.

2.5 Metabolite Quantitation Analysis

1. Stock solution: Weigh each standard in a 5 mL volumetric flask and dilute in water to a concentration of 1.0 mg/mL. Transfer the stock solution to a glass flask and store at −20 °C.
2. Intermediate stock solution: Dilute the stock solution in water to a concentration of 100 μg/mL for each metabolite (except ACOOA) in a 5 mL volumetric flask. Transfer the intermediate solution to a glass flask and store at −20 °C.
3. Standard mixture solutions for the IPC method: Prepare six levels of standard mixture from the stock solutions of 1.0 mg/mL (S1–S4) and 100 μg/mL (S5–S6) (Table 1) in a 5 mL volumetric flask. Pipette ACCOA from stock solutions of 1.0 mg/mL (S1–S6). Transfer the standard mixture solutions to 1 mL microcentrifuge tubes and store in a freezer (−80 °C). Table 1 shows the final concentration of each metabolite in the S1–S6 solutions used in the IPC separation method.
4. Standard mixture solutions for the HILIC method: Prepare six levels of standard mixture from the stock solutions of 5.0 mg/mL (S1–S3) and 50 μg/mL (S4–S6) (Table 2) in a 2 mL volumetric flask. Transfer the standard mixture solutions to 1 mL microcentrifuge tubes and store in a freezer (−80 °C). Table 2 shows the final concentration of each metabolite in the S1–S6 solutions used in the HILIC separation method.

Table 1
Final concentration (μg/mL) of each metabolite in standard mixture solutions (S1–S6) used for calibration curves in the IPC method

Metabolites	S1 (μg/mL)	S2 (μg/mL)	S3 (μg/mL)	S4 (μg/mL)	S5 (μg/mL)	S6 (μg/mL)
ACCOA	80	60	40	25	15	7.4
AKG	50	40	20	15	7.5	3.5
DHAP	50	40	20	15	5.0	2.0
F6P	50	40	20	15	7.5	3.5
G6P	50	40	20	10	5.0	1.5
GAP	50	40	20	10	7.5	3.5
GLY3P	50	40	20	10	5.0	0.5
L-MAL	50	25	10	5.0	1.0	0.2
PEP	50	40	20	10	5.0	2.5
PYR	50	30	15	10	5.0	0.5
R5P	50	40	20	10	5.0	2.0
RU5P	50	40	20	10	5.0	2.5
S7P	50	30	15	10	5.0	1.4

Table 2
Final concentration (μg/mL) of each metabolite in the standard mixture solutions (S1–S6) used for calibration curves in the HILIC method

Metabolites	S1 (μg/mL)	S2 (μg/mL)	S3 (μg/mL)	S4 (μg/mL)	S5 (μg/mL)	S6 (μg/mL)
Glucose	50	25	10	5.0	2.5	0.5
Glycerol	50	25	10	5.0	1.0	0.5
Xylitol	75	25	10	5.0	1.0	0.5
Xylose	75	25	10	5.0	2.5	0.5
Xylulose	50	25	10	5.0	2.5	0.5

Quality control (QC) solution: Prepare the QC solution by mixing all the standards in water to a final concentration of 25 μg/mL.

3 Methods

3.1 Yeast Cultivation

1. Grow the xylose-fermenting yeast (*S. stipitis*, *S. passalidarum*, *S. arborariae*, or *C. tenuis*) in an YPD plate overnight.
2. Use a single colony to start growth in 50 mL of defined mineral medium in a 500 mL Erlenmeyer flask overnight at 28 °C.
3. Use the bioreactor for cell cultivation under controlled conditions at 28 °C, stirring at 400 rpm, minimum air flow at 0.05 L/min, pH 5.5 (by addition of KOH 3 M) and initial O_2 at 100%. The starter culture is inoculated into the bioreactor containing the same medium to an OD_{600nm} of 0.5.
4. For a microaerobic condition, synthetic air (20% pure oxygen and 80% pure nitrogen) is injected into the reactor at 0.05 L/min, whereas for an aerobic condition a flow of 0.5 L/min is used.
5. Collect samples in triplicate at three specific time points during exponential growth.

3.2 Quenching of Cellular Metabolites

Figure 1a illustrates the steps for quenching of yeast cellular metabolites.

1. Prepare the quenching solution as described in the Subheading 2.2.
2. Turn on the thermostatic bath to −40 °C for at least 40 min before collecting samples so that it is stable at the correct temperature.
3. Add 8 mL of methanol buffer solution to 15 mL Falcon® tubes (*see* **Note 3**). Place the tubes in a polystyrene support and transfer them to a thermostatic bath at −40 °C for at least

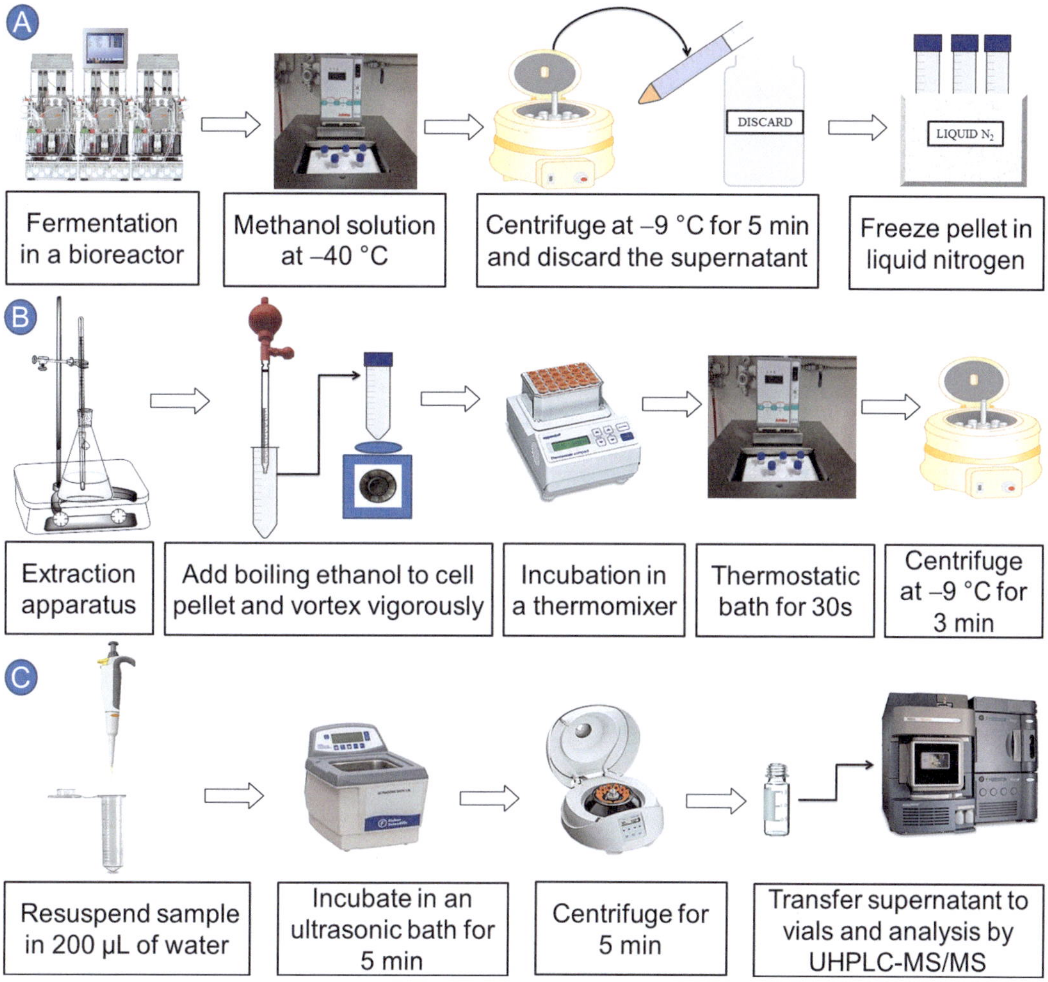

Fig. 1 Workflow of sample preparation steps: (**a**) quenching of cellular metabolites; (**b**) extraction of intracellular metabolites; and (**c**) metabolite analysis by UHPLC-MS/MS

10 min before sample collection to stabilize the methanol solution temperature.

4. Add 2 mL of cell culture to the 15 mL Falcon® tube containing the methanol buffer solution at −40 °C. After 30 s, centrifuge the sample at −9 °C and 3589 × *g* for 5 min.
5. Discard the supernatant and freeze the resulting pellet immediately in liquid nitrogen. Store the sample at −80 °C until metabolite extraction (*see* **Note 4**).

3.3 Metabolite Extraction

Figure 1b illustrates the steps for the extraction of yeast intracellular metabolites.

1. Attach a condenser grip to a universal plate stand and place the heating and magnetic stirring plate on it. Hold the thermometer with the condenser grip.

2. Place a Teflon magnetic stirring bar in a 250 mL Erlenmeyer flask and add the buffered 75% ethanol solution prepared as described in the Subheading 2.2 (*see* **Note 5**).
3. Place the Erlenmeyer flask on the heating and magnetic stirring plate and adjust the thermometer height to monitor the temperature of the extraction solution.
4. Turn on the heating and magnetic stirring plate and stir the extraction solution until it reaches 85 °C.
5. Withdraw the samples from the −80 °C freezer and place them in the thermostatic bath at −40 °C for 5 min.
6. Take out the samples from thermostatic bath and add 2 mL of boiling ethanol to the pellets (*see* **Note 6**). Then, homogenize the samples thoroughly using a vortex mixer.
7. Transfer samples from Falcon® tube to 2 mL microcentrifuge tubes and incubate samples for 3 min at 85 °C in a thermomixer with constant shaking at 800 rpm (Fig. 1b).
8. Cool the microcentrifuge tubes containing the samples to −40 °C in the thermostatic bath for 30 s and centrifuge at 3080 × *g* for 3 min at −9 °C (Fig. 1b).
9. Transfer the supernatant to a new 2 mL microcentrifuge tube and discard the cell debris. Then, evaporate the solvent in a vacuum concentrator.
10. Store the samples at −80 °C until UHPLC-MS/MS analysis (*see* **Note 7**).

3.4 Metabolite Analysis

Figure 1c illustrates the steps for sample preparation for UHPLC-MS/MS analysis.

1. Remove the samples from the −80 °C freezer and resuspend them in 200 μL of water.
2. Mix the samples vigorously in a vortex mixer for 30 s and then further homogenize the samples by incubating them in an ultrasonic bath for 5 min at room temperature.
3. Centrifuge the homogenized samples at 12,100 × *g* for 5 min at room temperature.
4. Transfer the clear supernatant to glass vials for analysis by UHPLC-MS/MS.
5. For the HILIC method, samples should be diluted 100 times (*see* **Note 8**). Pipette 10 μL of sample into a clean glass vial, add 990 μL of water and mix thoroughly. Store both sets of samples (diluted and undiluted) in a refrigerator until UHPLC-MS/MS analysis.

Table 3
Mass spectrometer (MS) parameters for the electrospray ionization (ESI) source: capillary voltage, desolvation and source temperature, cone and solvation gas flow

MS parameters	
Capillary voltage	3500 V
Desolvation temperature	450 °C
Source temperature	130 °C
Cone gas flow	20 L/h
Desolvation gas flow	700 L/h

3.5 Mass Spectrometry (MS)

1. Prepare each metabolite standard to a concentration of 50 μg/mL.
2. Set the ionization source temperature at 130 °C and capillary voltage at 3.5 kV.
3. Set the cone gas flow at 20 L/h, and desolvation gas flow at 700 L/h.
4. Operate the source in the negative electrospray ionization mode, ESI(−)-MS.
5. Infuse each metabolite standard in the mass spectrometer using mode 'combined'.
6. Use 0.1% ammonium hydroxide/acetonitrile (95.5, v/v) at 0.4 mL/min, and adjust the flow rate of the infusion syringe to 20.0 μL/min.
7. First, optimize the full scan MS experiment by adjusting the cone voltage parameter for each metabolite to obtain the most intense signal.
8. Next, optimize the MS/MS experiment by adjusting the collision energy value parameter for each metabolite to choose the *m/z* values that yield the most intense and/or selective fragment ion.

Table 3 shows the mass spectrometer parameters used in the full scan MS experiment. Table 4 shows the optimized MS/MS parameters used in the multiple reaction monitoring (MRM) experiments (Figs. 2 and 3) (*see* **Notes 9** and **10**).

3.6 UHPLC-IPC

1. Prepare the mobile phase as described in Subheading 2.4. Parameters used for IPC separation are shown in Table 5.

Table 4
Multiple reaction monitoring (MRM) parameters for the analysis of eighteen metabolites using ESI (−)-MS/MS

Metabolite	Cone voltage (V)	Q1 (*m/z*)	Q3 (*m/z*)	Collision energy (eV)
ACCOA	60	808.1	408.0	50
AKG	18	144.6	56.8	15
DHAP	25	169.0	97.0	18
F6P	32	258.6	96.8	18
G6P	32	258.7	96.8	20
GAP	20	168.8	96.8	18
Glucose	15	179.1	58.9	18
GLY3P	30	170.5	78.7	25
Glycerol	18	91.0	59.1	18
L-MAL	18	132.6	114.8	10
PEP	20	166.6	78.8	10
PYR	20	86.8	42.8	8
R5P	30	228.8	96.8	20
RU5P	30	228.8	79.0	20
S7P	32	288.4	138.9	25
Xylitol	25	151.0	58.9	20
Xylose	10	149.0	59.0	14
Xylu	10	149.0	59.0	14

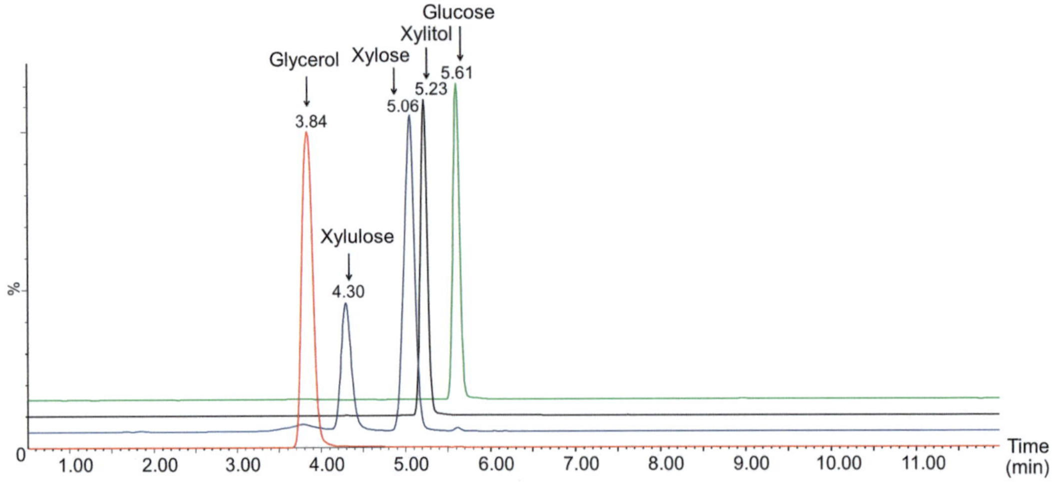

Fig. 2 Multiple reaction monitoring (MRM) chromatograms of five metabolites quantified by HILIC method: glycerol, xylulose, xylose, xylitol, and glucose

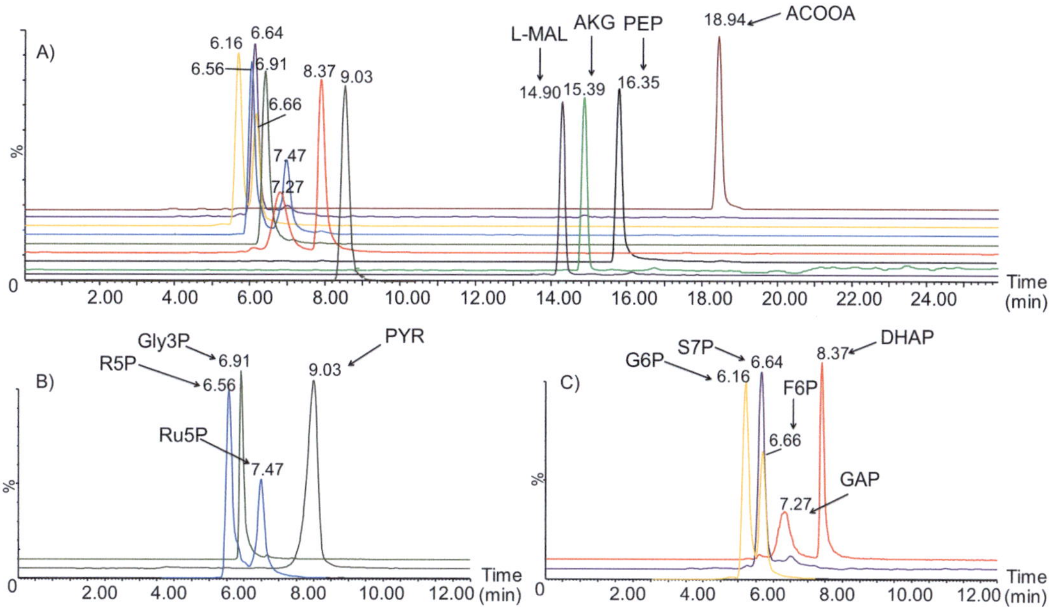

Fig. 3 (**a**) Multiple reaction monitoring (MRM) chromatograms of thirteen metabolites quantified by IPC method. (**b** and **c**) Expanded MRM chromatograms showing metabolites in detail

Table 5
Ion pair chromatography (IPC) separation parameters

IPC separation				
Injection volume	2.0 μL			
Sample temperature	10 °C			
Column temperature	45 °C			
Column	HSS-T3 reversed phase (2.1 × 150 mm × 1.8 μm)			
Mobile phase	A: 5 mM tributylamine, 10 mM acetic acid, and 5% (*v*/*v*) methanol and B: methanol			
Gradient	Time (min)	Flow (mL/min)	Eluent A (vol. %)	Eluent B (vol. %)
	0.0	0.4	100	0
	10.0	0.4	89.5	10.5
	18.0	0.4	47.4	52.6
	19.0	0.4	47.4	52.6
	20.0	0.4	100	0
	26.0	0.4	100	0

2. Before starting the run, flow rate and column temperature are set to 0.4 mL/min (100% A) and 45 °C, respectively, for 20 min to equilibrate the system.
3. The elution gradient begins with 100% A at a flow rate of 0.4 mL/min.

Table 6
Parameters used for HILIC separation

HILIC separation				
Injection volume	2.0 μL			
Sample temperature	10 °C			
Column temperature	50 °C			
Column	BEH amide (2.1 × 150 mm × 1.7 μm)			
Mobile phase	A: 0.1% ammonium hydroxide and B: acetonitrile with 0.1% ammonium hydroxide			
Gradient	Time (min)	Flow (mL/min)	Eluent A (vol. %)	Eluent B (vol. %)
	0.0	0.2	15	85
	6.5	0.2	50	50
	7.5	0.4	50	50
	8.0	0.2	15	85
	12.0	0.2	15	85

4. Increase eluent B linearly in the next 10 min until 10.5%, and then to 52.6% in the next 8.0 min and hold for 1 min.
5. Then return to the starting condition in 1 min and hold for 6 min to reequilibrate the column before the next sample injection.

3.7 UHPLC-HILIC

1. Prepare the mobile phase as described in Subheading 2.4. Parameters used for HILIC separation are shown in Table 6.
2. Before starting the run, set the column temperature to 50 °C, and equilibrate the system with 15% A using a flow rate of 0.4 mL/min for 20 min.
3. The elution gradient begins with 15% A at flow rate of 0.2 mL/min.
4. Increase eluent A linearly in the next 6.5 min until 50%, and hold for 1 min.
5. Return to the starting conditions in 0.5 min and hold for 4 min to reequilibrate the column before the next sample injection.

3.8 Quantitation of Metabolites in Yeast Samples

1. For the IPC separation method, prepare a calibration curve using standard mixture solutions (S1–S6) (Subheading 2.5, Table 1) of thirteen standards: ACCOA, AKG, L-MAL, G6P, F6P, DHAP, GAP, GLY3P, R5P, RU5P, PEP, PYR, and S7P (*see* **Note 11**).
2. For the HILIC separation method, prepare a calibration curve using standard mixture solutions (S1–S6) (Subheading 2.5, Table 2) of five metabolites: glucose, glycerol, xylitol, xylose, and xylulose (*see* **Note 12**).

3. Add each standard mixture solution into a 1 mL vial and then prepare the sample table in the software for each method of separation (IPC and HILIC).
4. Start the batch analysis with the blank (water), followed by six levels of concentrations from a standard mixture (S1–S6) and then inject the yeast samples in "random" mode. Standard mixtures and yeast samples should be analyzed in triplicate. After every set of nine yeast samples, inject the QC solution.
5. Check the retention time, MS response, and peak shape of the standards in the QC solution to monitor the stability of the UHPLC-MS system.
6. Data processing should be performed according to manufacturers′ software instructions (*see* **Note 13**).

4 Notes

1. Tributylamine (TBA) is more soluble in methanol than in water. Therefore, first dissolve TBA in methanol to prepare the mobile phase.
2. Ideally, the mobile phase solution should be prepared every day before UHPLC analysis.
3. A total of 8 mL of methanol buffer is used for 2 mL of cell culture medium. However, the methanol volume should be adjusted according to the volume of collected culture medium, using a 1:4 ratio (v/v) (culture medium–methanol).
4. The extraction procedure should be performed within 48 h, or as soon as possible.
5. The Erlenmeyer flask volume depends on the number of samples. It is important to add an extra volume of buffer solution and keep the Erlenmeyer flask closed (with aluminum foil) to avoid ethanol evaporation.
6. A total of 2 mL of ethanol buffer is used for 2 mL of cell culture medium. However, the ethanol volume should be adjusted according to the volume of collected culture medium, using a 1:1 ratio (v/v) (culture medium–ethanol).
7. Ideally, samples should be analyzed by UHPLC-MS/MS immediately after the extraction procedure.
8. HILIC methods require diluted samples since xylose and xylitol are present in high concentration in samples.
9. Ion source cleaning and MS calibration should follow the manufacturer's instructions.
10. Tuning of MS parameters may be required for optimal results.

11. For the IPC calibration curve, the concentration of the standard metabolites was varied between 0.1 and 50 μg/mL; however, these values need to be adjusted for each instrument.
12. For the HILIC calibration curve, the concentration of the standard metabolites was varied between 0.5 and 75 μg/mL; however, these values need to be adjusted for each instrument.
13. A graph of peak area versus concentration should be plotted for each metabolite. The polynomial equation is then used to calculate the final concentration of each metabolite in the yeast sample.

Acknowledgment

The authors are grateful to the Brazilian Agricultural Research Corporation (EMBRAPA) for the financial support.

References

1. Ricci-Silva EM, Vaz BG, Vasconcelos AG, Romão W, Aricetti JA, Caldana C, Abdelnur PV (2016) Mass spectrometry for metabolomics and biomass composition analyses. In: Vaz S (ed) Analytical techniques and methods for biomass. Springer, Cham, pp 115–141
2. Vargas LHG, Neto JCR, de Aquino Ribeiro JA et al (2016) Metabolomics analysis of oil palm (Elaeis guineensis) leaf: evaluation of sample preparation steps using UHPLC–MS/MS. Metabolomics. https://doi.org/10.1007/s11306-016-1100-z
3. Krastanov A (2016) Metabolomics—the state of art metabolomics – the state of art. doi: https://doi.org/10.2478/V10133-010-0001-Y
4. Mathew AK, Padmanaban VC (2013) Metabolomics: the apogee of the omics trilogy. Int J Pharm Pharm Sci 5:45–48. https://doi.org/10.1038/nrm3314
5. Abdelnur PV, Caldana C, Martins MCM (2014) Metabolomics applied in bioenergy. doi:https://doi.org/10.1186/s40538-014-0022-0
6. Zakhartsev M, Vielhauer O, Horn T, et al (2015) Fast sampling for quantitative microbial metabolomics: new aspects on cold methanol quenching: metabolite co-precipitation. 286–301. doi:https://doi.org/10.1007/s11306-014-0700-8
7. Putri SP, Yamamoto S, Tsugawa H, Fukusaki E (2013) Current metabolomics: technological advances. J Biosci Bioeng 116:9–16. https://doi.org/10.1016/j.jbiosc.2013.01.004
8. Pyke JS, Callahan DL, Kanojia K et al (2015) A tandem liquid chromatography–mass spectrometry (LC–MS) method for profiling small molecules in complex samples. Metabolomics 11:1552–1562. https://doi.org/10.1007/s11306-015-0806-7
9. Wamelink MMC, Struys EA, Huck JHJ et al (2005) Quantification of sugar phosphate intermediates of the pentose phosphate pathway by LC–MS/MS: application to two new inherited defects of metabolism. J Chromatogr B Anal Technol Biomed Life Sci 823:18–25. https://doi.org/10.1016/j.jchromb.2005.01.001
10. Calderón-Santiago M, Fernández-Peralbo M, Priego-Capote F, Luque de Castro MD (2016) MSCombine: a tool for merging untargeted metabolomic data from high-resolution mass spectrometry in the positive and negative ionization modes. Metabolomics 12:43. https://doi.org/10.1007/s11306-016-0970-4
11. Smedsgaard J, Nielsen J (2005) Global metabolite analysis of yeast: evaluation of sample preparation methods. 1155–1169. doi: https://doi.org/10.1002/yea.1308
12. Buescher JM, Moco S, Sauer U, Zamboni N (2010) Ultrahigh performance liquid chromatography–tandem mass spectrometry method for fast and robust quantification of anionic and aromatic metabolites. Anal Chem 82:4403–4412
13. Seifar RM, Ras C, Deshmukh AT et al (2013) Quantitative analysis of intracellular coenzymes in Saccharomyces cerevisiae using ion pair

reversed phase ultra high performance liquid chromatography tandem mass spectrometry. J Chromatogr A 1311:115–120. https://doi.org/10.1016/j.chroma.2013.08.076

14. Nováková L, Havlíková L, Vlcková H (2014) Hydrophilic interaction chromatography of polar and ionizable compounds by UHPLC. Trends Anal Chem 63:55–64. https://doi.org/10.1016/j.trac.2014.08.004
15. Kawachi Y, Ikegami T, Takubo H et al (2011) Chromatographic characterization of hydrophilic interaction liquid chromatography stationary phases: hydrophilicity, charge effects, structural selectivity, and separation efficiency. J Chromatogr A 1218:5903–5919. https://doi.org/10.1016/j.chroma.2011.06.048
16. Verduyn C, Van Kleef R, Frank J et al (1985) Properties of the NAD(P)H-dependent xylose reductase from the xylose-fermenting yeast Pichia stipitis. Biochem J 226:669–677. https://doi.org/10.1042/bj2260669

Chapter 9

Exploiting High-Resolution Mass Spectrometry for Targeted Metabolite Quantification and ^{13}C-Labeling Metabolism Analysis

Zhucui Li, Yujing Li, Yinjie J. Tang, and Wenqing Shui

Abstract

Quantification of targeted metabolites, especially trace metabolites and structural isomers, in complex biological materials is an ongoing challenge for metabolomics. In this chapter, we summarize high-resolution mass spectrometry-based approaches mainly used for targeted metabolite and metabolomics analysis, and then introduce an MS1/MS2-combined PRM workflow for quantification of central carbon metabolism intermediates, amino acids, and shikimate pathway-related metabolites. Major steps in the workflow, including cell culture, metabolite extraction, LC-MS analysis and data processing, are described. Furthermore, we adapt this new approach to a dynamic ^{13}C-labeling experiment and demonstrate its unique advantage in capturing and correcting isotopomer labeling curves to facilitate nonstationary ^{13}C-labeling metabolism analysis.

Key words High-resolution MS, Metabolite quantification, Parallel reaction monitoring (PRM), ^{13}C-labeling metabolism analysis

1 Introduction

Mass spectrometry (MS) coupled to different separation techniques such as liquid chromatography (LC), gas chromatography (GC), and capillary electrophoresis has become the leading platform for large-scale quantification of metabolites in various biological systems [1–3]. MS platforms widely used for targeted analysis of metabolites and metabolomes include both hybrid low-resolution instruments such as triple quadrupole (QQQ) [4, 5] and quadrupole-ion trap (QIT) [6, 7], and high-resolution instruments such as Orbitrap [8, 9], time-of-flight (TOF) [10, 11] and Fourier transform ion cyclotron resonance mass spectrometry (FT-ICR-MS) mass spectrometer (FTICR-MS) [12–14]. Owing to the fast scan speed, high sensitivity and relatively low cost, low-resolution MS instruments have been traditionally employed for quantification of specific pharmaceuticals and metabolites mainly using the selected

Edward E.K. Baidoo (ed.), *Microbial Metabolomics: Methods and Protocols*, Methods in Molecular Biology, vol. 1859, https://doi.org/10.1007/978-1-4939-8757-3_9,

reaction monitoring (SRM) or multiple reaction monitoring (MRM) approach [15–17]. In fact, SRM or MRM performed on triple quadrupole instruments is recognized as the "gold standard" MS technique for targeted bioanalyte quantification due to its high sensitivity and robustness [18–20]. The MRM technique has been recently adopted in metabolomics studies focusing on a defined subset of metabolites representative of key pathways [21–24]. This type of targeted metabolomics is a valuable tool for assessing changes of metabolic pathways or networks resulting from genetic mutation, altered gene expression, protein dysfunction or environmental perturbation. For example, Wei and colleagues developed an LC/MS/MRM platform to monitor 205 endogenous metabolites and applied it to targeted metabolite analysis in plasma samples [20]. Nevertheless, SRM or MRM assays performed on low-resolution mass spectrometers sometimes are limited by ambiguous metabolite identifications as a result of interfering peaks from complex matrix that resemble the compound of interest under a specific Q1/Q3 transition [25].

Technological improvements in high-resolution MS have led to new acquisition methods of remarkable quantification performance which have challenged the dominance of MRM for targeted analysis as well as enabled high-throughput nontargeted analysis for metabolomics [26]. Nontargeted MS-based metabolomics predominantly rely on high-resolution instruments such as Orbitrap and TOF to profile hundreds of endogenous and drug metabolites (Fig. 1) [27–29]. In these studies, metabolite identification is typically afforded by MS/MS analysis of isolated precursors using a data-dependent acquisition (DDA) strategy, while metabolite quantification is derived from high-resolution MS1 scan data [30, 31]. In this regard, relative quantification of individual metabolites in different biological samples is based on the MS1 responses of extracted ion chromatograms for specific metabolites. However, it has been recognized that metabolite quantification relying on MS1 scan data could suffer from low reproducibility due to inadequate sensitivity of MS1 scans for metabolites that are less abundant, prone to matrix interference or have isomeric compounds [3].

Targeted metabolite quantification aims at the determination or relative comparison of the concentrations of a limited number of preselected metabolites [32–34]. A robust method of high accuracy, high reproducibility and wide linear range is the prerequisite for reliable quantification results. Both MS1 and MS2-based approaches have been developed for targeted metabolite or metabolome analysis. Selected ion monitoring (SIM) that records the MS1 signal of targeted metabolites in a narrow mass range is a classical quantification approach that can be employed on low-resolution or high-resolution MS instruments [6, 7]. However, SIM often suffers from low specificity in metabolite identification

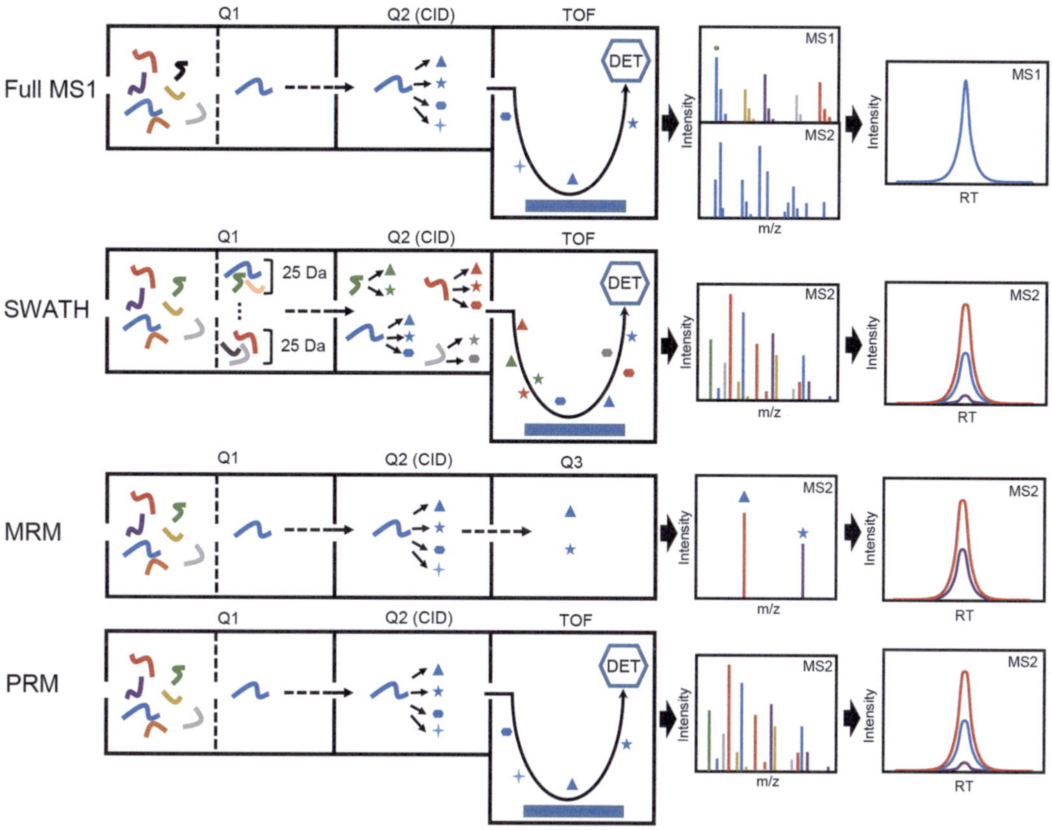

Fig. 1 Overview of major MS techniques for targeted metabolite quantification. Full MS1 indicates quantification of various metabolites based on the MS intensity of precursor ions from high-resolution MS1 scan data. Metabolite quantification by SWATH is derived from fragmentation spectra for all precursor ions eluting at a given time point. The fragmentation spectra are acquired over a mass range in sequential small windows of a defined size (commonly 20–25 Da). MRM quantification of metabolites relies on specific precursor-fragment transitions that are detected on low-resolution QQQ instruments. PRM quantification of metabolites relies on multiple fragment peaks from a specific precursor detected on high-resolution TOF or Orbitrap instruments

due to lack of MS/MS information [35]. SIM has become obsolete since MS2-based quantification techniques were established and is now primarily implemented in bioanalysis of drugs and relevant metabolites [18–20].

MS2-based quantification approaches currently used in metabolite analysis or metabolomics include SRM/MRM, parallel reaction monitoring (PRM) [36, 37] and sequential window acquisition of all theoretical fragment ion spectra (SWATH) [38–41]. PRM implemented on high-resolution MS platforms has emerged as a competent alternative to MRM for targeted metabolite analysis. In a PRM assay, the precursor ion is isolated in Q1 and fragmented in Q2, and all generated fragment ions are monitored in MS2 scans on a high resolution, accurate mass spectrometer [42]. Then fragment ion intensities from MS2 spectra are used for specific precursor quantification (Fig. 1). However, in an MRM assay, quantification of metabolites relies on specific precursor–fragment transitions that are typically monitored with low-resolution QQQ instruments (Fig. 1). The PRM approach has been successfully implemented in a panel of targeted proteomic studies and demonstrated superior specificity, accuracy and multiplexing capability compared to the MRM assay [43–46]. In contrast, there has been rare reports of PRM application to targeted metabolomic analysis except the very latest research from Zhou et al. [47] and Qiu et al. [48]. Both groups developed PRM assays on a hydrid Q-Exactive instrument (Thermo) for measurement of targeted metabolites. Zhou et al. reported monitoring 237 endogenous metabolites for which structural identification solely relied on MS/MS spectra available in public databases [47]. Qiu et al. constructed PRM assays using chemical standards for quantification of 25 intracellular primary metabolites [48].

Very recently, a data-independent acquisition (DIA) workflow using the SWATH MS technique initially developed for proteomic quantification has been adapted to global metabolomics to allow for sensitive, reproducible, and high-throughput metabolite quantification based on fragment ion intensities in MS2 scans correlated to wide isolation windows (Fig. 1) [38–41]. This new type of MS2 quantification technique is subject to ongoing debate given that the link between precursors and their fragment ions is dissociated in the DIA process, which could compromise structural assignment of metabolite targets [37]. Furthermore, software for processing SWATH MS data is under developed, with MS-DIAL being the only option so far [37].

In this chapter we summarized four types of MS techniques for targeted metabolite quantification in Fig. 1. It should be noted that full MS1, SWATH and PRM need to be employed on high-resolution MS platforms and full MS1 can be also applied to non-targeted metabolomics analysis.

^{13}C-metabolic flux analysis (MFA) is a well-established technology which uses isotopic tracers to determine *in vivo* reaction rates (fluxes) in the metabolic network of cells [49]. Dynamic ^{13}C-MFA based on transient labeling of intracellular metabolites has shown great potential in control and optimization of biotechnological processes not accessible by conventional MFA [50]. Dynamic ^{13}C-MFA relies on time-resolved ^{13}C-labeling metabolism analysis which refers to the measurement of isotopic labeling enrichment in metabolite pools with a maximal information gain [51]. A big challenge for dynamic ^{13}C-labeling metabolism analysis is precise measurement of the isotopomer labeling fractions for a large number of intracellular metabolites after the pulse of ^{13}C substrates. Because the complexity of metabolite isotopomer peaks is substantially expanded in ^{13}C-labeled metabolomic pools, the conventional quantification method of extracting high-resolution

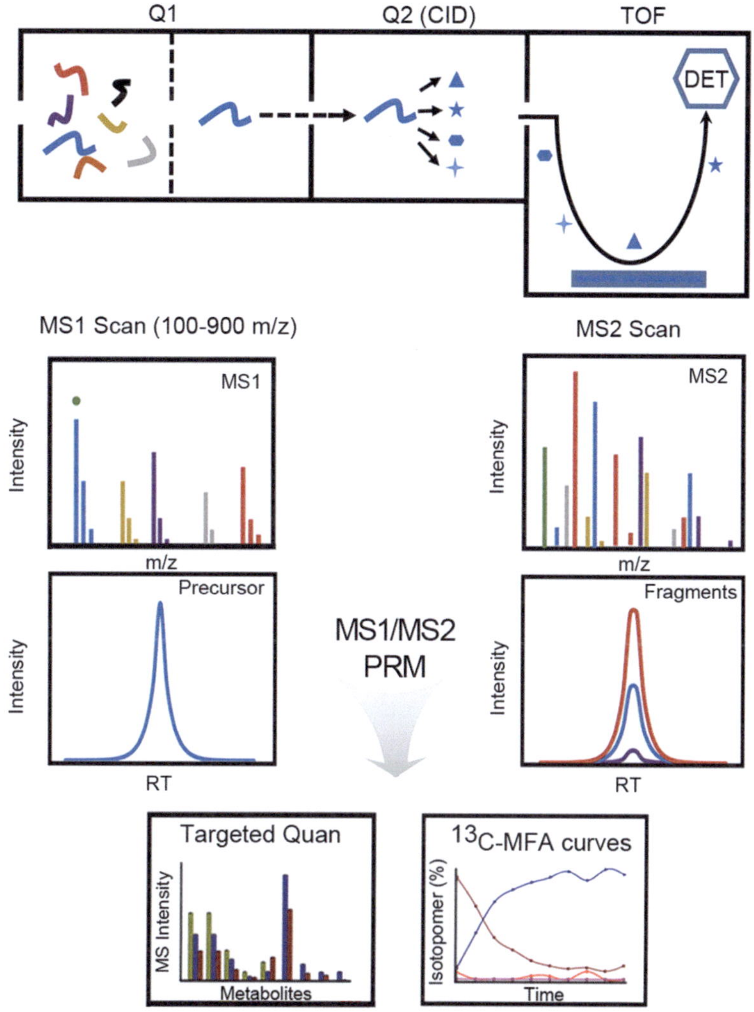

Fig. 2 The workflow of MS1/MS2 combined PRM analysis. The precursor ion is isolated in Q1 and fragmented in Q2, and all generated fragment ions are monitored in MS2 scans on a high-resolution mass spectrometer. Then the precursor ion intensity and fragment ion intensities are integrated for quantification of metabolites. This approach can be applied to both targeted metabolite analysis and ^{13}C-labeling metabolism analysis

MS1 data for different isotopomers of the same metabolite species would be severely limited due to frequent interferences.

In our recent study, we developed a new MS1/MS2-combined PRM workflow for targeted metabolite quantification as well as ^{13}C-labeling metabolism analysis on an orthogonal QqTOF instrument (TripleTOF 6600) (Fig. 2) [52]. Sixty-five targeted metabolites include central carbon metabolism (CCM) and shikimate pathway intermediates, and free amino acids. Skyline software was used to automatically extract MS2 ion chromatograms after we defined specific transitions for each metabolite precursor. As MS1 full scan was also performed in each cycle of the PRM assay, quantitative data were extracted for a given precursor ion using the same "Transition List" module in Skyline. Therefore, we were able to extract both MS1 and MS2 transitions in a simultaneous data-processing workflow with Skyline for targeted metabolite quantification. Concentration curve assessment implicated that exploiting both MS1 and MS2 scans in PRM analysis afforded higher sensitivity, wider dynamic ranges and better reproducibility than relying on either scan mode for quantification. Furthermore, Skyline allowed us to efficiently process these MS1/MS2 ion chromatograms and readily eliminate noisy signals and transitions with interferences [52].

In this protocol, we described the integrated MS1/MS2 PRM approach, which was applied to targeted metabolite quantification in engineered *E. coli* strains for understanding metabolic pathway modulation. Additionally, this approach was implemented in a dynamic ^{13}C-labeling experiment to achieve accurate measurement of diverse isotopomers for tracing cell metabolism.

2 Materials

Diligently follow all appropriate waste disposal regulations when disposing of waste materials.

2.1 Cell Culture and Metabolite Extraction

1. Cell culture media: NBS minimal complete media with a glucose concentration optimal for the strain to be studied. Store the media at 4 °C before use.
2. Quenching solution: Prepare saline in a plastic bottle and then prechill at −80 °C. Saline: a solution of 0.9% w/v of NaCl.
3. Washing solution: Prepare and store PBS at 4 °C. PBS: 8 g/L NaCl, 0.2 g/L KCl, 1.42 g/L Na_2HPO_4, and 0.24 g/L KH_2PO_4.
4. Extraction solvent: a mixture of methanol, acetonitrile, and water (4:4:2 (v/v) with 0.1% FA). Add the internal standard (d4-succinate) to the extraction solvent that is prechilled at −20 °C before extraction. The final concentration of internal standard is 250 ng/mL. Use this solvent to extract polar primary metabolites with high efficiency.
5. ^{13}C-labeled glucose: a stock solution of the ^{13}C-labeled glucose at 20 mg/mL and store it at 4 °C.

2.2 Metabolite analysis by HILIC-MS

1. HILIC column: ZIC®-HILIC 100 mm × 2.1 mm, 3.5 μm, Merck, Germany (Item NO. 1.50441.0001).
2. Mobile phases for HILIC: (A) 10 mM ammonium acetate in water; (B) acetonitrile. Store at 4 °C and degas before use.
3. Metabolite standards: metabolite stock solutions in 50% acetonitrile–50% water stored at −80 °C. These stock solutions are

used to prepare a mixture of 20–30 metabolite standards ~10 ng/μL for direct infusion into the mass spectrometer to acquire reference MS/MS spectra.

4. LC-MS instrument: Metabolite standards or total cell extracts are analyzed on a Nexera HPLC system (SHIMADZU, Japan) which is connected to TripleTOFTM 6600 mass spectrometer (AB SCIEX, USA). Other types of Q-TOF or hybrid Orbitrap systems can be used as well.
5. Software: PeakView (AB SCIEX, USA), Skyline (v3.1); R; ProteoWizard; SIMCA-P (Umetrics, Umea, Sweden); MeV (v4.8.1).

3 Methods

3.1 Cell Culture

1. For seed culture preparation, grow wild-type *E. coli* strain AB2834 (WT) and two engineered strains (TIB01 and TIB02) with improved production of 3-dehydroshikimic acid in LB media aerobically at 30 °C and 250 rpm to produce the seed cultures.
2. For cell culture preparation, transfer the seed culture into a 250 mL flask containing the NBS minimal complete media and grow at 37 °C and 250 rpm until OD_{600} reaches 2.0 (exponential phase *see* **Note 1**).
3. To prepare cell culture for dynamic ^{13}C-labeling metabolism analysis, transfer the seed culture into a 250 mL flask containing the NBS minimal complete media (*see* **Note 2**) and grow at 37 °C and 250 rpm until OD_{600} reaches 2.5. Isotope labeling is initialized during the logarithmic phase ($OD_{600} = 2.5$) of the engineered strain TIB01 by fast addition of ^{13}C-labeled glucose into NBS culture media at 20 mg/mL. Divide the cell suspension into eleven aliquots, which are cultured for a given time (0, 10, 30, 60, 120, 240, 450, 600, 900, 1800, or 7200 s) for dynamic labeling. Then collect 10 mL of bacterial cell culture, for each aliquot, at each labeling time point.

3.2 Metabolite Extraction

1. Quench bacterial cell cultures (5–10 mL) by mixing with the same volume of frozen saline buffer (*see* **Note 3**) quickly. Gently shake the mixture to protect cell membrane integrity and quench enzyme activities. Then harvest the cell pellet via centrifugation at 3500 × *g* for 3 min at 4 °C.
2. Wash the cell pellet three times with PBS quickly, and pellet the bacteria cells by centrifugation at 3500 × *g* for 3 min at 4 °C.
3. Extract metabolites from the cell pellet using a 1.5 mL prechilled solvent mixture of methanol–acetonitrile–water (4:4:2, v/v/v) with 0.1% FA, which contains d_4-succinic acid as the

internal standard (*see* **Note 4**). Snap-freeze metabolite extracts with liquid nitrogen, then thaw and sonicate them in a water bath (*see* **Note 5**). Repeat the freeze–thaw–sonication cycle twice. Incubate the samples at −20 °C for 2 h. Centrifuge the resulting sample at 16,000 × *g* for 20 min at 4 °C and harvest the supernatant. Dry the supernatant by speed vacuum and store the dried extract at −80 °C. Reconstitute metabolite extracts in acetonitrile–water (1:1; v/v) before HILIC-MS analysis (*see* **Note 6**).

4. For the dynamic ^{13}C-labeling experiment, bacterial cells are washed with the same volume of prechilled saline buffer and then quenched by placing the culture tubes in liquid nitrogen (this step only takes a few seconds, *see* **Note 7**). Metabolite extraction from total cells is performed as described hereinbefore.

3.3 Standard Curves of Targeted Metabolites in Simple Matrix

Eleven standards of targeted metabolites listed are used to generate standard concentration curves in simple matrix. These standards are spiked into a mixture of 60 other metabolites at nine different concentrations (0.25, 0.5, 1, 5, 10, 50, 100, 1000, 10,000 ng/mL). The internal standard (d_4-succinic acid) is present at 250 ng/mL. Standard curves are acquired on a HILIC column coupled to a TripleTOF 6600 mass spectrometer operating in the PRM mode ($N = 3$).

3.4 Concentration Curves of Targeted Metabolites in Serially Diluted E. coli Extracts

Reconstitute metabolite extracts from wild-type E. coli cells in an acetonitrile–water (1:1; *v/v*) solvent mixture at high concentration and then dilute at ten different factors (1/2, 1/4, 1/8, 1/16, 1/32, 1/64, 1/128, 1/256, 1/512, and 1/1024) in the extraction buffer. Concentration curves of all targeted metabolites are acquired on a HILIC column coupled to a TripleTOF 6600 mass spectrometer operating in the PRM mode ($N = 3$).

3.5 HILIC-MS Analysis of Targeted Metabolites in PRM Mode

1. Carry out LC separations on a ZIC®-HILIC column with the mobile phases specified in Subheading 2.2. Set the flow rate and sample injection volume to 0.2 mL/min and 5 μL, respectively. Run the following linear gradient: 0–3 min, 90% B; 3–18 min, 90–60% B; 18–25 min, 60–55% B; 25–27 min, 55–50% B; 27–30 min, 50% B; and the column is reequilibrated for 10 min.
2. Acquire mass spectra in the negative ion ESI mode within a mass range of 100–1000 *m/z*. Set the ion source parameters: ion spray voltage, −4500 V; curtain gas, 35 psi; nebulizer gas, 55 psi; heater gas, 55 psi; and source temperature, 550 °C. Directly infuse each metabolite standard into the mass spectrometer to acquire reference MS/MS spectra. Assess rolling CE and fixed CE using standards to acquire optimal CEs for individual metabolite targets. MS data acquisition for PRM

analysis is programmed in the product ion mode, and consists of one MS1 scan (150 ms) with the mass range from 50 to 900 *m/z*, followed by targeted MS2 scans (50 ms) under the optimal CE condition across the mass range from 30 to *X* (*X* = precursor mass + 10 Da) *m/z*. The maximal cycle time is 3.25 s for a total of 65 targets.

3.6 Data Analysis

1. Targeted metabolite identification using the MS1/MS2 PRM approach. In the analysis of both metabolite standards and total cell extracts, each metabolite target is first identified with Peakview software (v1.2.0.3, AB SCIEX) according to three criteria: (a) mass error < 15 ppm, (b) RT shift relative to the pure standard <1 min, (c) PRM MS2 spectrum matching the reference MS2 spectrum.
2. Targeted metabolite quantification using the MS1/MS2 PRM approach. Quantitative data analysis is conducted using the Transition List module in the Skyline software (v3.1) in a manner similar to peptide quantification by PRM [46], except that in the transition settings "precursor charges" and "ion charges" are set to 1. One to six fragment ions of relatively strong intensity and structural specificity in the reference MS2 spectra are selected to construct the transition list for each metabolite precursor (*see* **Note 8**). After the raw data is imported into Skyline, all extracted ion chromatograms (XICs) for precursor and selected fragment ions are manually inspected to ensure proper peak picking and integration. The retention times of targeted metabolites in *E. coli* cell extracts are reconfirmed by comparison with the standard mixture. Both precursor ion (MS1) and fragment ion (MS2) peak areas are exported from Skyline (Fig. 3). The MS1 peak area of the precursor ion and sum of MS2 peak areas of selected fragment ions provided independent quantitative measure of the targeted metabolite at MS1 and MS2 levels (*see* **Note 9**).

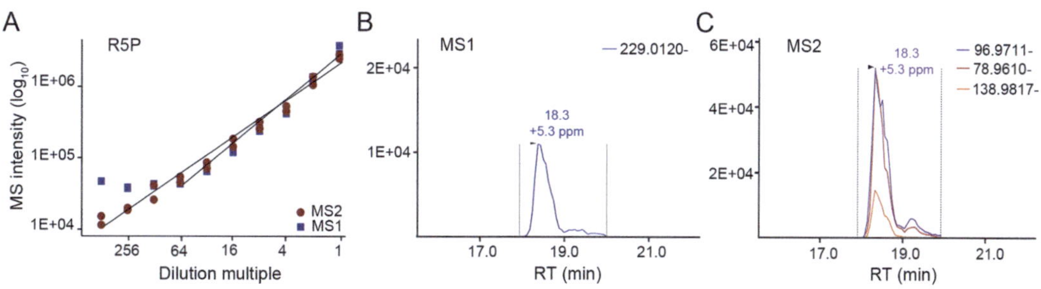

Fig. 3 MS1 and MS2-based quantification of metabolites in the PRM assay. (**a**) Concentration curves of R5P by MS1 or MS2 quantification. The single MS1 ion chromatogram (**b**) and several MS2 ion chromatograms (**c**) of R5P in E. coli cell extracts diluted by 512-fold

3. Dynamic ^{13}C-labeling metabolism data analysis. The PRM raw data is first converted to a text file using ProteoWizard. An in-house perl script is used to extract isotopomer peak intensities for each metabolite from its MS1 spectrum with constraints on accurate mass (<15 ppm mass error) and RT compared to the standard (<1 min RT shift). Isotopomer fractions of a targeted metabolite are calculated from the MS1 peak intensities of all detected isotopomers. Any labeled isotopomer showing a targeted MS2 spectrum discordant with that of the unlabeled M0 peak is excluded, as shown for metabolite deoxy-5-phosphoric acid (d-R5P) (Fig. 4). The dynamic labeling curves are then prepared based on the isotopomer fraction data obtained at different time points for a specific strain.

4 Notes

1. Cell culture at the exponential phase is usually preferred for metabolomics studies if only one time point is needed. For cell culture with OD_{600} around 2, the metabolite extract from 1 mL culture is sufficient for LC-MS analysis using the state-of-the-art platform. We often collect ~10 mL cell culture in one experiment to perform multiple processing replicates and injection replicates for reproducibility assessment of the technical workflow and the instrument platform.
2. Minimal cell media used here should contain no other carbon source except glucose. Rich media is not suitable for ^{13}C-labeling metabolism analysis or metabolic flux experiments.
3. Prechill the centrifuge (4 °C), centrifuge tubes (−80 °C), saline solution (−80 °C), PBS (4 °C) and extraction solvent (−20 °C). Saline in a plastic bottle can be smashed and transferred into a 50 mL centrifuge tube.
4. Metabolites with ^{13}C isotope labels and having identical or similar chemical structures of the metabolite targets can be considered as internal standards. Add the internal standard(s) into the extraction solvent before doing metabolite extraction.
5. The prechilled extraction solvent is added to each cell pellet, vortexed for 30 s, and incubated in liquid nitrogen for 1 min. The samples then are allowed to thaw at room temperature, taking ~5 min, after which they are sonicated in a water bath with ice for 10 min.
6. Biological duplicates of each strain are prepared and, therefore, metabolite extraction is processed in duplicate for each biological replicate.

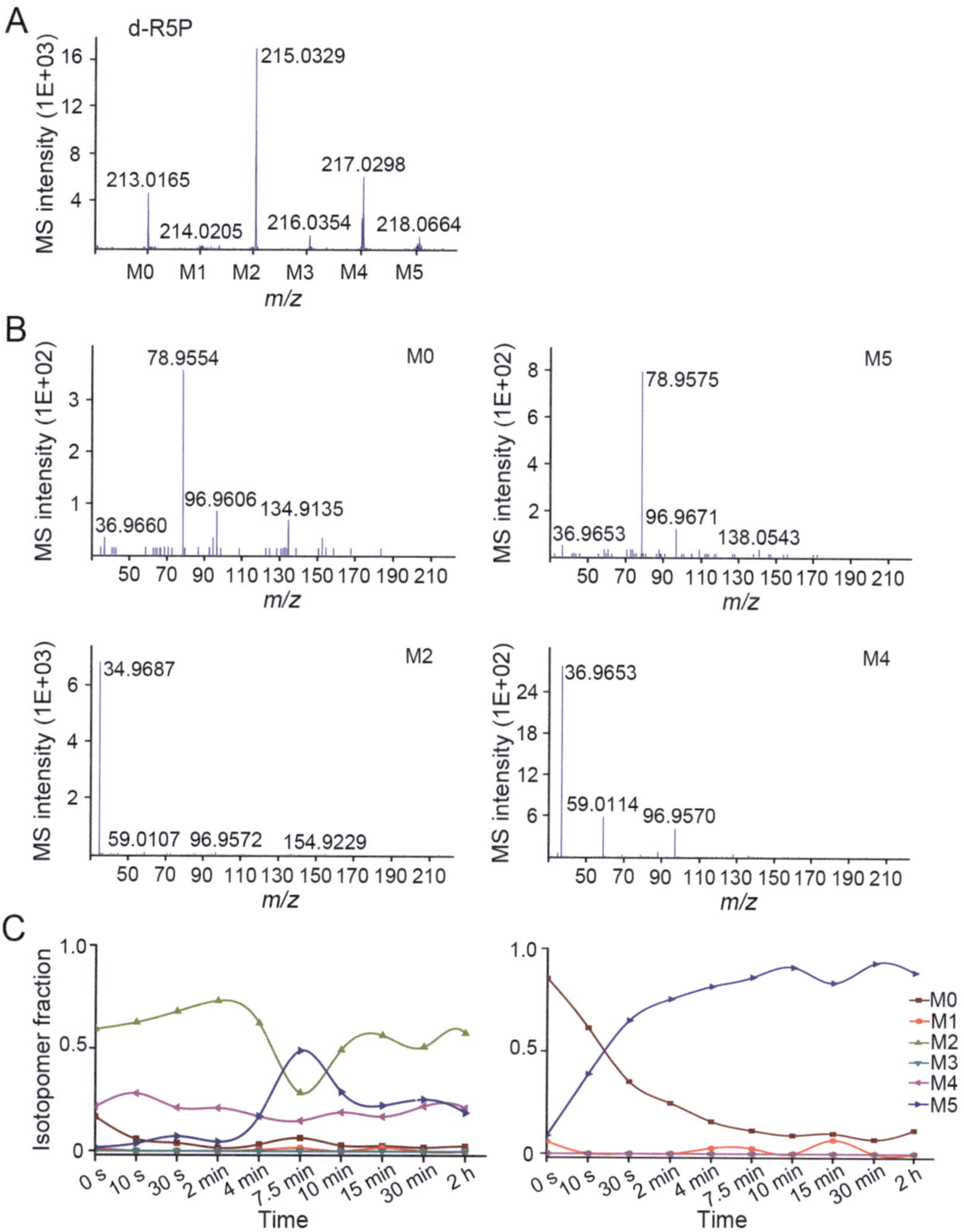

Fig. 4 Distinguishing interferences of d-R5P isotopomers in dynamic ^{13}C-labeling metabolism analysis using PRM MS2 scans. (**a**) MS1 spectrum of d-R5P in the unlabeled T0 sample. The *m/z* positions of M0–M5 isotopomers of d-R5P are annotated. (**b**) MS/MS spectra of precursors at specific isotopomer positions acquired on the T0 sample. (**c**) Isotopomer labeling curves derived from the original MS1 ion intensity data (left) or the corrected MS1 ion intensity data after removal of interfering peaks at the M2 and M4 positions (right)

7. Using a digital thermometer as the stirring apparatus will ensure that the temperature of the cell suspension is dropped to 0 °C.
8. When selecting fragments appropriate for metabolite quantification by PRM analysis, we take into account not only the sensitivity of fragment peaks in MS/MS data but also the specificity of fragments that can be used to distinguish the metabolite target from other structurally similar metabolites.
9. To compare the quantitative performance by MS1- and MS2-based analysis of targeted metabolites, we generated concentration curves in both simple and complex matrices. Depending on the sensitivity, precision, and linear range of metabolite quantification in different matrices, we classified 50 metabolite targets to be MS2 quantifiable and 11 to be MS1 quantifiable [52].

References

1. Patti GJ, Yanes O, Siuzdak G (2012) Innovation: metabolomics: the apogee of the omics trilogy. Nat Rev Mol Cell Biol 13(4):263–269
2. Theodoridis G, Gika HG, Wilson ID (2011) Mass spectrometry-based holistic analytical approaches for metabolite profiling in systems biology studies. Mass Spectrom Rev 30 (5):884–906
3. Junot C, Fenaille F, Colsch B, Becher F (2014) High resolution mass spectrometry based techniques at the crossroads of metabolic pathways. Mass Spectrom Rev 33(6):471–500
4. Chen S et al (2013) Pseudotargeted metabolomics method and its application in serum biomarker discovery for hepatocellular carcinoma based on ultra high-performance liquid chromatography/triple quadrupole mass spectrometry. Anal Chem 85(17):8326–8333
5. Klockmann S, Reiner E, Cain N, Fischer M (2017) Food targeting: geographical origin determination of hazelnuts (Corylus avellana) by LC-QqQ-MS/MS based targeted metabolomics application. J Agric Food Chem 65 (7):1456–1465
6. Wang Y et al (2016) An improved pseudotargeted metabolomics approach using multiple ion monitoring with time-staggered ion lists based on ultra-high performance liquid chromatography/quadrupole time-of-flight mass spectrometry. Anal Chim Acta 927:82–88
7. Rosenfelder N, Van Zee NJ, Mueller JF, Gaus C, Vetter W (2010) Gas chromatography/electron ionization-mass spectrometry-selected ion monitoring screening method for a thorough investigation of polyhalogenated compounds in passive sampler extracts with quadrupole systems. Anal Chem 82 (23):9835–9842
8. Barnaba C, Nardin T, Pierotti A, Malacarne M, Larcher R (2017) Targeted and untargeted characterisation of free and glycosylated simple phenols in cocoa beans using high resolution-tandem mass spectrometry (Q-Orbitrap). J Chromatogr A 1480:41–49
9. Del Mar Gomez-Ramos M, Rajski L, Heinzen H, Fernandez-Alba AR (2015) Liquid chromatography Orbitrap mass spectrometry with simultaneous full scan and tandem MS/MS for highly selective pesticide residue analysis. Anal Bioanal Chem 407 (21):6317–6326
10. Zhu X, Chen Y, Subramanian R (2014) Comparison of information-dependent acquisition, SWATH, and MS(All) techniques in metabolite identification study employing ultrahigh-performance liquid chromatography-quadrupole time-of-flight mass spectrometry. Anal Chem 86(2):1202–1209
11. Ferrer I, Thurman EM (2010) Identification of a new antidepressant and its glucuronide metabolite in water samples using liquid chromatography/quadrupole time-of-flight mass spectrometry. Anal Chem 82(19):8161–8168
12. Yilmaz A, Rudolph HL, Hurst JJ, Wood TD (2016) High-throughput metabolic profiling of soybean leaves by fourier transform ion cyclotron resonance mass spectrometry. Anal Chem 88(2):1188–1194
13. Pollier J, Goossens A (2013) Metabolite profiling of plant tissues by liquid

chromatography Fourier transform ion cyclotron resonance mass spectrometry. Methods Mol Biol 1011:277–286

14. Baidoo EE et al (2008) Capillary electrophoresis-fourier transform ion cyclotron resonance mass spectrometry for the identification of cationic metabolites via a pH-mediated stacking-transient isotachophoretic method. Anal Chem 80(9):3112–3122
15. Yan Z, Yan R (2015) Increase the accessibility and scale of targeted metabolomics: construction of a human urinary metabolome-wide multiple reaction monitoring library using directly-coupled reversed-phase and hydrophilic interaction chromatography. Anal Chim Acta 894:65–75
16. Luo P et al (2015) Multiple reaction monitoring-ion pair finder: a systematic approach to transform nontargeted mode to pseudotargeted mode for metabolomics study based on liquid chromatography-mass spectrometry. Anal Chem 87(10):5050–5055
17. Shao Y et al (2015) Development of urinary pseudotargeted LC-MS-based metabolomics method and its application in hepatocellular carcinoma biomarker discovery. J Proteome Res 14(2):906–916
18. Zhang X, Xiu ZM, Li XW (2011) Preparation and characterization of (HAp/SiO_2)/Ti biocomposites. Adv Mater Res 217-218:88–92
19. Kitteringham NR, Jenkins RE, Lane CS, Elliott VL, Park BK (2009) Multiple reaction monitoring for quantitative biomarker analysis in proteomics and metabolomics. J Chromatogr B Analyt Technol Biomed Life Sci 877 (13):1229–1239
20. Ciccimaro E, Blair IA (2010) Stable-isotope dilution LC-MS for quantitative biomarker analysis. Bioanalysis 2(2):311–341
21. Yang M et al (2016) A targeted metabolomics MRM-MS study on identifying potential hypertension biomarkers in human plasma and evaluating acupuncture effects. Sci Rep 6:25871
22. Chen W et al (2013) A novel integrated method for large-scale detection, identification, and quantification of widely targeted metabolites: application in the study of rice metabolomics. Mol Plant 6(6):1769–1780
23. Yuan M, Breitkopf SB, Yang X, Asara JM (2012) A positive/negative ion-switching, targeted mass spectrometry-based metabolomics platform for bodily fluids, cells, and fresh and fixed tissue. Nat Protoc 7(5):872–881
24. Wei R, Li G, Seymour AB (2010) High-throughput and multiplexed LC/MS/MRM method for targeted metabolomics. Anal Chem 82(13):5527–5533
25. Jemal M, Ouyang Z, Xia YQ (2010) Systematic LC-MS/MS bioanalytical method development that incorporates plasma phospholipids risk avoidance, usage of incurred sample and well thought-out chromatography. Biomed Chromatogr 24(1):2–19
26. Cajka T, Fiehn O (2016) Toward merging untargeted and targeted methods in mass spectrometry-based metabolomics and lipidomics. Anal Chem 88(1):524–545
27. Denery JR, Nunes AA, Dickerson TJ (2011) Characterization of differences between blood sample matrices in untargeted metabolomics. Anal Chem 83(3):1040–1047
28. Liu X, Ser Z, Locasale JW (2014) Development and quantitative evaluation of a high-resolution metabolomics technology. Anal Chem 86(4):2175–2184
29. Wang X et al (2012) Urine metabolomics analysis for biomarker discovery and detection of jaundice syndrome in patients with liver disease. Mol Cell Proteomics 11(8):370–380
30. Oberacher H, Schubert B, Libiseller K, Schweissgut A (2013) Detection and identification of drugs and toxicants in human body fluids by liquid chromatography-tandem mass spectrometry under data-dependent acquisition control and automated database search. Anal Chim Acta 770:121–131
31. Broecker S, Herre S, Wust B, Zweigenbaum J, Pragst F (2011) Development and practical application of a library of CID accurate mass spectra of more than 2,500 toxic compounds for systematic toxicological analysis by LC-QTOF-MS with data-dependent acquisition. Anal Bioanal Chem 400(1):101–117
32. Michopoulos F et al (2014) Targeted profiling of polar intracellular metabolites using ion-pair-high performance liquid chromatography and -ultra high performance liquid chromatography coupled to tandem mass spectrometry: applications to serum, urine and tissue extracts. J Chromatogr A 1349:60–68
33. Li H, Cai Y, Guo Y, Chen F, Zhu ZJ (2016) MetDIA: targeted metabolite extraction of multiplexed MS/MS spectra generated by data-independent acquisition. Anal Chem 88 (17):8757–8764
34. Gu H, Zhang P, Zhu J, Raftery D (2015) Globally optimized targeted mass spectrometry: reliable metabolomics analysis with broad coverage. Anal Chem 87(24):12355–12362
35. Zhou J, Yin Y (2016) Strategies for large-scale targeted metabolomics quantification by liquid

chromatography-mass spectrometry. Analyst 141(23):6362–6373

36. Gertsman I, Gangoiti JA, Barshop BA (2014) Validation of a dual LC-HRMS platform for clinical metabolic diagnosis in serum, bridging quantitative analysis and untargeted metabolomics. Metabolomics 10(2):312–323
37. Tsugawa H et al (2015) MS-DIAL: data-independent MS/MS deconvolution for comprehensive metabolome analysis. Nat Methods 12(6):523–526
38. Gao Y et al (2015) Targeted data-independent acquisition and mining strategy for trace drug metabolite identification using liquid chromatography coupled with tandem mass spectrometry. Anal Chem 87(15):7535–7539
39. Prasain JK, Wilson L, Hoang HD, Moore R, Miller MA (2015) Comparative lipidomics of Caenorhabditis elegans metabolic disease models by SWATH non-targeted tandem. Mass Spectr Metabol 5(4):677–696
40. Scheidweiler KB, Jarvis MJ, Huestis MA (2015) Nontargeted SWATH acquisition for identifying 47 synthetic cannabinoid metabolites in human urine by liquid chromatography-high-resolution tandem mass spectrometry. Anal Bioanal Chem 407(3):883–897
41. Bonner R, Hopfgartner G (2016) SWATH acquisition mode for drug metabolism and metabolomics investigations. Bioanalysis 8 (16):1735–1750
42. Peterson AC, Russell JD, Bailey DJ, Westphall MS, Coon JJ (2012) Parallel reaction monitoring for high resolution and high mass accuracy quantitative, targeted proteomics. Mol Cell Proteomics 11(11):1475–1488
43. Gallien S et al (2012) Targeted proteomic quantification on quadrupole-orbitrap mass spectrometer. Mol Cell Proteomics 11 (12):1709–1723
44. Tang H et al (2014) Multiplexed parallel reaction monitoring targeting histone modifications on the QExactive mass spectrometer. Anal Chem 86(11):5526–5534
45. Gallien S, Bourmaud A, Kim SY, Domon B (2014) Technical considerations for large-scale parallel reaction monitoring analysis. J Proteome 100:147–159
46. Schilling B et al (2015) Multiplexed, scheduled, high-resolution parallel reaction monitoring on a full scan QqTOF instrument with integrated data-dependent and targeted mass spectrometric workflows. Anal Chem 87 (20):10222–10229
47. Zhou J et al (2016) Development and evaluation of a parallel reaction monitoring strategy for large-scale targeted metabolomics quantification. Anal Chem 88(8):4478–4486
48. Qiu N et al (2016) Simultaneous determination of multiple intracellular primary metabolites by ultrahigh performance liquid chromatography coupled with a Q exactive HF mass spectrometer. Anal Chem 88 (19):9647–9653
49. Zamboni N (2011) 13C metabolic flux analysis in complex systems. Curr Opin Biotechnol 22 (1):103–108
50. Noh K, Wiechert W (2011) The benefits of being transient: isotope-based metabolic flux analysis at the short time scale. Appl Microbiol Biotechnol 91(5):1247–1265
51. Wiechert W, Noh K (2013) Isotopically non-stationary metabolic flux analysis: complex yet highly informative. Curr Opin Biotechnol 24(6):979–986
52. Li Z et al (2016) Integrating MS1 and MS2 scans in high-resolution parallel reaction monitoring assays for targeted metabolite quantification and dynamic 13C-labeling metabolism analysis. Anal Chem 89(1):877–885

Chapter 10

Quantitative Profiling of Endogenous Metabolites Using Hydrophilic Interaction Liquid Chromatography–Tandem Mass Spectrometry (HILIC-MS/MS)

Attila Teleki and Ralf Takors

Abstract

Dynamic modeling of metabolic reaction networks requires absolute quantification of intracellular and extracellular metabolite concentrations with high precision and accuracy. This chapter presents a robust HILIC-ESI-MS/MS procedure for targeted quantitative profiling of more than 50 polar key metabolites in multicomponent endogenous extracts. Without using ion-pairing-agents or prior derivatization protocols, organic acids, amino acids, sugar phosphates, coenzymes, and nucleotides are measured on a triple quadrupole platform in positive and negative electrospray ionization modes with preoptimized MRM transitions. Robust polymer-based zwitterionic stationary phases (ZIC®-pHILIC) support alkaline mobile phase conditions (pH 9.2) for enhancing retention and chromatographic performance of polar analytes in bicratic elution mode without unfavourable column bleed. The quality of the method was extensively validated and demonstrated by absolute metabolite quantification in endogenous *Escherichia coli* extracts by comparative use of standard-based external calibration, isotope dilution, and standard addition as quantification strategies. In sum, alkaline ZIC®-pHILIC chromatography emerged as an efficient approach providing high selectivity and sensitivity for comprehensive metabolic studies.

Key words Metabolomics, Quantitative metabolic profiling, Intracellular metabolite pools, Hydrophilic interaction liquid chromatography (HILIC), Electrospray ionization (ESI), Liquid chromatography–tandem mass spectrometry (LC-MS/MS)

1 Introduction

Metabolomics aims at the systematic quantification of low-molecular-weight compounds acting as intermediates (metabolites) in complex biochemical reaction networks in defined cellular states. Compared to genome, transcriptome and proteome, the metabolome directly reflects all upstream cellular regulatory mechanisms and thus represents, most closely, the actual physiological phenotype [1, 2]. In this context, the analysis of metabolic dynamics after stimulus–response experiments enables the investigation of enzyme kinetics under in vivo conditions and gives direct access to unravel

Edward E.K. Baidoo (ed.), *Microbial Metabolomics: Methods and Protocols*, Methods in Molecular Biology, vol. 1859,
https://doi.org/10.1007/978-1-4939-8757-3_10,

regulatory structures of corresponding cellular reaction networks [3, 4]. Whereas non-targeted high-throughput approaches like *metabolic fingerprinting* aim for qualitative or at best semiquantitative analysis of overall metabolic patterns, targeted approaches like *quantitative metabolic profiling* are focused on a non-biased absolute quantification of distinct metabolites in selected pathways [1, 5, 6]. Considering the tremendous number of endogenous metabolites, the multitude of biological matrices with high chemical diversity and concentration ranges across nine orders of magnitude (pM–mM), quantitative approaches still impose great analytical challenge on the precision requirements [6–8]. Modern analytical methods comprise different approaches such as nuclear magnetic resonance spectroscopy (NMR) or mass spectrometry coupled to liquid chromatography (LC-MS), gas chromatography (GC-MS) or capillary electrophoresis (CE-MS) [9–12]. In this regard, LC-MS platforms offer significantly higher sensitivity compared to NMR. Usually they do not require sophisticated derivatization of non-volatile polar compounds like GC, and are able, depending on column material, to separate a broad analytical spectrum unlike CE [2, 13, 14]. Consequently, LC-MS approaches are becoming commonly accepted as the most universal analytical technology for absolute metabolite quantification at present [2, 15]. However, for comprehensive quantification studies covering a wide range of metabolic compounds, a laborious and time-consuming combination of multiple analytical platforms with varying precision and accuracy is often unavoidable. In this context, hydrophilic interaction liquid chromatography (HILIC) is increasingly preferred, as it enables a direct separation of frequently occurring charged and polar endometabolites (e.g., central metabolism) that are poorly retained on reversed stationary phases [16–18]. The partition-driven retention mechanism is primarily based on an analyte distribution between a water-enriched layer, partially immobilized on the stationary phase, and a predominantly polar-organic mobile phase, which exhibits excellent compatibility with ESI-MS detection [16, 19].

Here we describe an HILIC-ESI-MS/MS procedure for comprehensive and simultaneous quantitative metabolic profiling of a wide range of key metabolites of the cellular metabolism, comprising organic acids, amino acids, sugar phosphates, coenzymes, and nucleotides [6]. Without use of ion-pairing-agents or prior derivatization protocols, direct separation is achieved on polymer-based zwitterionic HILIC stationary phases (ZIC®-pHILIC) under alkaline mobile phase conditions (pH 9.2) and a bicratic elution mode. Metabolites are detected by tandem mass spectrometry (MS/MS) on a triple quadrupole platform (Agilent 6410B) in the *multiple reaction monitoring* (MRM) mode with enhanced selectivity and sensitivity. The method has been validated for a representative cross-section of 59 common intracellular metabolites with preoptimized MRM transitions and particular focus on corresponding chromatographic performances, linear dynamic ranges, detection limits, retention time stabilities, and reproducibilities. The applicability of the approach was demonstrated for the determination of absolute metabolite concentrations in intracellular extracts of *Escherichia coli* biomasses. Different quantification strategies, namely standard-based external calibration, isotope dilution and standard addition with internal standards were compared. Following the instructions, the method is applicable for absolute quantification of a broad, and depending on the focus, extendable range of polar metabolites in various organisms or biological samples, enabling simultaneous quantitation of dozens of metabolites with high precision in a single LC-MS/MS run.

2 Materials

Diligently follow all appropriate waste disposal regulations when disposing of waste materials.

2.1 Sampling and Metabolite Extraction

1. Safe-lock microcentrifuge tubes (1.5 or 2 mL).
2. Refrigerated benchtop centrifuge (e.g., Eppendorf 5430 R, Hamburg, D).
3. Liquid nitrogen (−196 °C).
4. Temperature-adjustable water bath (e.g., Lauda 006T, Lauda-Königshofen, D).
5. Vortex mixer.
6. Isotonic 0.9% (wv^{-1}) sodium chloride solution (HPLC-or MS-grade).
7. HPLC- or MS-grade water.
8. Internal standard solution (e.g., 150 μM L-norvaline, HPLC- or MS-grade).
9. Syringe filters, PTFE (polytetrafluoroethylene) membrane, non-sterile, 0.45 μm, 25 mm (e.g., Carl Roth, D).

2.2 Hydrophilic Interaction Liquid Chromatography (HILIC)

2.2.1 HPLC Columns

1. Sequant® ZIC®-pHILIC® polymeric HPLC column, 150 by 2.1 mm, 5 μm (Merck Millipore, Darmstadt, D).
2. Sequant® ZIC®-pHILIC® polymeric guard column, 20 by 2.1 mm, 5 μm (Merck Millipore, Darmstadt, D).

2.2.2 Mobile Phases

1. MS-grade water and acetonitrile.
2. Ammonium acetate stock solution (e.g., 1 M, HPLC- or MS-grade) adjusted to pH 9.2 with 25% (vv^{-1}) ammonium hydroxide (MS-grade) and stored at 4 °C.

3. Eluent A (non-polar): 90% acetonitrile–10% water (vv^{-1}), 10 mM ammonium acetate, pH 9.2.
4. Eluent B (polar): 10% acetonitrile–90% water (vv^{-1}), 10 mM ammonium acetate, pH 9.2.

2.2.3 Calibration Strategies

1. Metabolite standard stock solutions (e.g., 1 mM, analytical-grade) in LC-MS water and stored at −70 °C.
2. U-^{13}C-labeled lyophilized algal cells (>99 atom % ^{13}C, e.g., Sigma-Aldrich, D).

2.3 LC-MS/MS Instrumentation

1. Agilent 1200 or 1260 HPLC system consisting of a binary pump, a degasser, a thermostated column compartment and a bioinert (*see* **Note 1**) thermostated autosampler (Agilent, Waldbronn, D).
2. Agilent 6410B triple quadrupole mass spectrometer (MS-QQQ) with an electrospray ion source (ESI) (Agilent, Waldbronn, D).
3. Nitrogen generator: compressed air supply (≥7 bar).
4. Collision gas for *collision-induced dissociation* (CID): Nitrogen compressed (UN 1066, ≥99.999%).
5. Calibration solution: ESI-L Low Concentration Tuning Mix (Agilent, Waldbronn, D).
6. Acquisition and data analysis: MassHunter software package (Agilent, Waldbronn, D).

3 Methods

3.1 Sampling and Metabolite Extraction

Reliable quantifications of intracellular metabolites in defined cellular states requires suitable sampling strategies. Sequential methods typically include fast sampling, an instant arrest of metabolic activity (*quenching*) and a quantitative metabolite extraction step. The following sample processing procedure is based on a *fast centrifugation treatment* (FCT) [20] and a subsequent *hot water extraction* (HWE) of intracellular metabolites [21].

1. Manually transfer 2 mL of *E. coli* culture (e.g., 2 g_{CDW}/L) during the exponential growth phase in 2 mL microcentrifuge tubes (*see* **Note 2**).
2. Centrifuge defined biomass amounts (e.g., 4 mg) at 20,000 × *g* and 4 °C for 20 s.
3. Remove the cultivation supernatant quantitatively (*see* **Note 3**) (optional analog extracellular sample) (*see* **Note 4**).
4. Optionally wash the cells with 2 mL isotonic 0.9% (wv^{-1}) sodium chloride solution (*see* **Note 5**).
5. Centrifuge washed cell pellets at 20,000 × *g* and 4 °C for 20 s.

6. Remove the wash supernatant quantitatively (*see* **Note 3**) and quench the cells immediately by liquid nitrogen (−196 °C).
7. Store intracellular (and optional extracellular) samples temporarily at −70 °C.
8. Add defined amounts (50–200 μL) of 150 μM internal standard solution (e.g., L-norvaline) to the cell pellets to obtain an extraction biomass concentration $c_x$of for example 15 g_{CDW}/L (*see* **Note 6**).
9. Immediately preincubate intracellular samples in a water bath for 1 min at 100 °C for an irreversible enzymatic inactivation.
10. Homogenize the resulting suspensions by short-time vortexing for 20 s.
11. Incubate the sample again in a water bath for 5 min at 100 °C for exhaustive metabolite extraction and, afterward, chill the sample on ice–water (*see* **Note 7**).
12. Separate extracts from cell debris by centrifugation at 20,000 × *g* and 4 °C for 10 min.
13. Store metabolome extracts temporarily at −70 °C.

3.2 Preparation of $U^{13}C$-Labeled Standard

Preparations of uniformly labeled $(U)^{13}C$ metabolite solutions follow the aforementioned hot water extraction (HWE) procedure:

1. Transfer defined amounts of commercially available $U^{13}C$-labeled lyophilized algal cells (*see* **Note 8**) to 2 mL microcentrifuge tubes.
2. Add preheated water (100 °C) for an extraction concentration of 90 g/L.
3. Incubate samples in a water bath for 2 min at 100 °C for metabolite extraction.
4. Homogenize resulting suspensions by short-time vortexing for 20 s.
5. Repeat the procedure two more times.
6. Chill samples on ice-water.
7. Separate extracts from cell debris by centrifugation at 20,000 × *g* and 4 °C for 10 min and, afterward, filter with PTFE (polytetrafluoroethylene) membranes.
8. Store $U^{13}C$-metabolome extracts at −70 °C.

3.3 HILIC Sample Preparation

In regard to the hydrophobic mobile phase starting condition of HILIC chromatography (here 90% (vv^{-1}) acetonitrile), a sample predilution with organic solvent is strongly recommended (e.g., >50% (vv^{-1}) acetonitrile). A direct injection of aqueous samples provokes phase separation with reduced metabolite retention and sensitivity [22]. Sample and standard preparations are performed manually in conical 1.5 mL microcentrifuge tubes in following order:

1. Add 20 μL of intracellular extracts (or standard solution) in the desired aqueous solution.
2. Add 4 μL of a suitable measuring internal standard (e.g., 1 mM α-aminoisobutyrate or γ-aminobutyrate).
3. Add 15 μL of LC-MS-water (optionally $U^{13}C$ algal extract (*see* Subheading 3.6) or ^{12}C metabolite standards, *see* **Note 9**).
4. Add 1 μL of 1 M ammonium acetate stock solution (pH 9.2) and homogenize by fast vortexing for 2 s.
5. Add 60 μL of acetonitrile (60%, vv^{-1}) and homogenize by fast vortexing for 2 s.
6. Chill resulting formulations (100 μL, 60% (vv^{-1}) acetonitrile) on ice for 10 min.
7. Centrifuge the formulations at 20,000 × *g* and 4 °C for 10 min.
8. Transfer 90 μL of residue-free preparations (*see* **Note 10**) to screw top HPLC vials (1.5 mL) with suitable microinserts (0.1 mL).

3.4 LC-MS/MS Analysis

HILIC-ESI-MS/MS analyses are performed under alkaline mobile phase conditions (*see* **Note 11**) on a tandem quadrupole mass spectrometer (Agilent 6410B) in MRM mode with a metal-free bioinert HPLC system (*see* **Note 1**). The standard-based optimization of bicratic chromatographic conditions and source parameters was based on a selection of highly polar central metabolic intermediates (G6P, F6P, 6PG, PEP, Mal, Suc and Cit, *see* Table 1 for full metabolite names) [6]. The comprehensive method should be easily transferable to comparable platforms from different manufacturers and operation modes (e.g., *selected ion monitoring*, SIM).

1. Before starting extensive measurement series, the mass spectrometer performance should be checked as specified by the manufacturer:
   (a) Check and optionally adjust global MS parameters (vacuum level, compressed air pressure, collision gas flow, etc.).
   (b) Clean nebulizer and ESI source components thoroughly (*see* **Note 12**).
   (c) Check and optionally recalibrate MS quadrupole and detector parameters (mass accuracy, selectivity, and responsivity) by direct infusion (column bypassing) of commercial calibration solution.
2. Set the ESI source parameters as follows:
   (a) Adjust nitrogen gas flow and temperature to 10 L/min and 350 °C, respectively.

Table 1
MS/MS parameters (Agilent 6410B) for 60 common intracellular metabolites including MRM transitions for non- and full-labeled analogues (parts 1 and 2)

Metabolite		Detection parameters						
			$U^{12}C$-MRM		MS/MS		$U^{13}C$-MRM	
Abbreviation	Metabolite	Pol.	Precursor ion (*m/z*)	Ion (*m/z*)	CE [V]	*F* [V]	Precursor ion (*m/z*)	Ion (*m/z*)
(a) *Part 1*								
aKB	α-Ketobutyrate	(−)	145	101	4	70	150	105
Fum	Fumarate	(−)	115	71	3	70	119	74
Suc	Succinate	(−)	117	73	9	82	121	76
Mal	Malate	(−)	133	115	7	80	137	119
aKG	α-Ketoglutarate	(−)	145	101	4	70	150	105
Cit	Citrate	(−)	191	111	10	90	197	116
IsoCit	Isocitrate	(−)	191	173	6	90	197	179
cis-Aco	*cis*-Aconitate	(−)	173	85	10	60	179	89
Xu5P	Xylulose-5-phosphate	(−)	229	97	4	70	234	97
6PG	6-Phosphogluconate	(−)	275	97	12	100	281	97
E4P	Erythrose-4-phosphate	(−)	199	97	4	70	203	97
R5P	Ribose-5-phosphate	(−)	229	97	6	90	234	97
Ru5P	Ribulose-5-phosphate	(−)	229	97	6	90	234	97
F6P	Fructose-6-phosphate	(−)	259	97	10	100	265	97
G6P	Glucose-6-phosphate	(−)	259	97	12	100	265	97
FbP	Fructose-1,6-bisphosphate	(−)	339	97	15	100	345	97
DHAP	Dihydroxyacetonphosphate	(−)	169	97	4	60	172	97
PEP	Phosphoenolpyruvate	(−)	167	79	14	70	170	79
2PG	2-Phosphoglycerate	(−)	185	79	38	90	188	79
3PG	3-Phosphoglycerate	(−)	185	97	14	80	188	97
NAD	Nicotinamide adenine dinucleotide	(+)	666	649	65	135	687	670
NADP	Nicotinamide adenine dinucleotide phosphate	(+)	744	136	58	135	765	141
AMP	Adenosine monophosphate	(−)	346	79	60	135	356	79
ADP	Adenosine diphosphate	(−)	426	79	98	135	436	79
ATP	Adenosine triphosphate	(−)	506	79	108	135	516	79

(continued)

Table 1
(continued)

Metabolite		Detection parameters						
			$U^{12}C$-MRM		MS/MS		$U^{13}C$-MRM	
Abbreviation	Metabolite	Pol.	Precursor ion (*m/z*)	Ion (*m/z*)	CE [V]	*F* [V]	Precursor ion (*m/z*)	Ion (*m/z*)
GMP	Guanosine monophosphate	(−)	362	79	60	135	372	79
GDP	Guanosine diphosphate	(−)	442	79	108	135	452	79
GTP	Guanosine triphosphate	(−)	522	79	108	135	532	79
CMP	Cytosine monophosphate	(−)	322	79	60	135	331	79
(b) *Part 2*								
CDP	Cytosine diphosphate	(−)	402	79	98	135	411	79
CTP	Cytosine triphosphate	(−)	482	79	108	135	491	79
UMP	Uridine monophosphate	(−)	323	79	60	135	332	79
UDP	Uridine diphosphate	(−)	403	79	98	135	412	79
UTP	Uridine triphosphate	(−)	483	79	108	135	492	79
cAMP	cyclic Adenosine monophosphate	(−)	328	134	22	135	338	139
AIBA	α-Aminoisobutyrate	(+)	104	58	10	75	–	–
GABA	γ-Aminobutyrate	(+)	104	87	10	75	–	–
HSer	L-Homoserine	(+)	120	74	10	75	124	77
*O*AcSer	*O*-Acetyl-L-serine	(+)	148	60	20	60	153	62
*O*SHS	*O*-Succinyl-L-homoserine	(+)	220	102	8	70	228	106
Cysth	LL-Cystathionine	(+)	223	88	27	85	230	91
Gly	Glycine	(+)	76	30	8	70	78	31
Ala	L-Alanine	(+)	90	44	10	70	93	46
Ser	L-Serine	(+)	106	60	10	75	109	62
Pro	L-Proline	(+)	116	70	17	75	121	74
Val	L-Valine	(+)	118	72	10	75	123	76
Thr	L-Threonine	(+)	120	74	9	75	124	77
Leu	L-Leucine	(+)	132	44	20	75	138	47
Ile	L-Isoleucine	(+)	132	44	20	75	138	47
Asn	L-Asparagine	(+)	133	74	15	70	137	76
Asp	L-Aspartate	(+)	134	74	13	75	138	76

(continued)

Table 1
(continued)

Metabolite			Detection parameters						
			$U^{12}C$-MRM		MS/MS		$U^{13}C$-MRM		
Abbreviation	Metabolite	Pol.	Precursor ion (*m/z*)	Ion (*m/z*)	CE [V]	*F* [V]	Precursor ion (*m/z*)	Ion (*m/z*)
Gln	L-Glutamine	(+)	147	84	18	70	152	88
Lys	L-Lysine	(+)	147	84	17	75	153	89
Glu	L-Glutamate	(+)	148	84	17	75	153	88
Met	L-Methionine	(+)	150	56	17	75	155	59
Phe	L-Phenylalanine	(+)	166	120	12	75	175	128
Arg	L-Arginine	(+)	175	70	30	75	181	74
Trp	L-Tryptophan	(+)	205	188	4	75	216	199
His	L-Histidine	(+)	156	110	13	85	162	115
Nva	L-Norvaline	(+)	118	72	10	75	–	–

Reproduced from Teleki et al. [6] with permission. Copyright Elsevier

(b) Adjust capillary voltages to ±4000 V (for both ionization modes).

(c) Adjust nebulizer pressure to 30 psi.

Metabolite-specific MRM transitions and associated MS/MS parameters are optimized for 60 common intracellular metabolites by direct injection of pure standard solutions [6]. If the used triple quadrupole platform corresponds to the Agilent 6400 mass spectrometer series, specified parameters are directly applicable (*see* Table 1). To establish optimal MS/MS performance for the metabolites of interest, the following steps are required:

3. Choose the direct injection mode (column bypassing) with a solvent flow rate of 0.2 mL/min as HPLC configuration.
4. Apply a solvent composition which largely corresponds to expected elution condition of the targeted metabolite (e.g., 50% (vv^{-1}) acetonitrile).
5. Perform a series of 5 μL injections of a prepared (*see* Subheading 3.3) pure standard solution (e.g., 50 μM) in the positive (ESI+) or negative ionization mode (ESI−) within different operation modes:

(a) In the *single stage full scan* mode with adapted mass ranges (e.g., 50–2000 m/z), select the resulting precursor ions with respect to maximum signal intensity and analyte stability (usually singly protonated $[M + H]^+$ or deprotonated $[M - H]^-$ species).

(b) Analyze CID fragmentations (collision-induced dissociation) of suitable precursor ions in *product ion scan* mode under varying applied collision energies (e.g., 0–200 V).

(c) Fine-tune the applied collision energies (CE) and fragmentor voltages (*F*) for every mass transition in the MRM mode.

To ensure absolute quantifications of metabolites in a complex sample matrix, such as intracellular metabolite extracts, a standard-based evaluation of chromatographic performances, linear dynamic ranges, repeatabilities, and detection limits is crucial (*see* Table 2). To establish good chromatographic performance for the metabolites of interest, the following steps are required:

6. Choose the column injection mode with a solvent flow rate of 0.2 mL/min as HPLC configuration.
7. Set the following gradient program (*see* **Note 13**) with a constant column temperature of 40 °C:
   (a) Isocratic hold 0% B for 1 min.
   (b) Linear gradient from 0% B to 75% B for 30 min.
   (c) Linear gradient from 75% B to 100% B for 4 min.
   (d) Wash with 100% B for 5 min.
   (e) Linear gradient from 100% B to 0% B for 10 min.
   (f) Re-equilibrate 0% B for 15 min.
8. Perform a series of 5 μL injections of a prepared (*see* Subheading 3.3) pure standard solution (e.g., 50 μM) or standard mixes with preoptimized MRM transitions and associated MS/MS parameters.
   (a) Evaluate metabolite-specific retention time reproducibility, peak width and shape (*see* **Note 14**).
   (b) Evaluate metabolite-specific linear dynamic range with adapted calibration ranges and levels (e.g., 0.5–200 μM based on 12 non-equidistant levels) (*see* **Note 15**).
   (c) Optionally evaluate metabolite-specific detection limits with adapted calibration ranges and levels (e.g., 1 nM to 5 μM based on 12 non-equidistant levels) (*see* **Note 16**).

Table 2
Overview of obtained linearity ranges ($\leq 5\ \mu M$) with related regression coefficients, method detection limits and retention times of investigated low-molecular weight metabolites measured under alkaline (pH 9.2) or acid mobile phase conditions (pH 5.6) (parts 1 and 2)

		Linearity range						Method detection limit				Retention time	
		Alkaline			Acidic			Alkaline	Acidic	Alkaline	Acidic		
Metabolite	Polarity	Min (nM)	Max (nM)	R^2	Min (nM)	Max (nM)	R^2	(nM)		pmol (on column)		Alkaline (min)	Acidic (min)
(a) *Part 1*													
aKB	(−)	50	5000	1.0000	50	5000	0.9998	20.7	4.2	0.10	0.02	8.02 ± 0.04	8.63 ± 0.07
Fum	(−)	50	5000	0.9999	500	5000	0.9951	33.8	606.0	0.17	3.03	22.07 ± 0.02	22.57 ± 0.01
Suc	(−)	500	5000	0.9999	500	5000	0.9976	117.1	138.0	0.59	0.69	20.59 ± 0.02	20.12 ± 0.01
Mal	(−)	20	5000	0.9998	20	5000	0.9998	2.2	8.5	0.01	0.04	21.83 ± 0.01	22.14 ± 0.01
aKG	(−)	50	5000	0.9974	100	5000	0.9916	9.4	59.6	0.05	0.30	21.42 ± 0.02	21.77 ± 0.01
Cit[a]	(−)	50	5000	0.9938	1000	5000	0.9838	439.6	513.3	2.20	2.57	24.15 ± 0.01	23.97[c] ± 0.02
IsoCit[a]	(−)	20	5000	0.9953	500	5000	0.9945	21.5	454.1	0.11	2.27	24.83 ± 0.01	24.75[c] ± 0.02
cis-Aco	(−)	20	1000	0.9899	50	5000	0.9854	3.9	41.6	0.02	0.21	23.08 ± 0.01	22.76 ± 0.01
Xu5P	(−)	50	5000	0.9884	200	5000	0.9953	30.1	181.9	0.15	0.91	21.15 ± 0.01	21.14 ± 0.01
6PG	(−)	1000	5000	0.9903	1000	5000	0.9950	744.3	1131.1	3.72	5.66	24.03 ± 0.01	24.53 ± 0.01
E4P[b]	(−)	–	–	–	2000	5000	0.9972	–	1713.6	–	8.57	–	21.73 ± 0.11
R5P[a]	(−)	20	5000	0.9973	50	5000	0.9956	154.5	67.5	0.77	0.34	21.11 ± 0.02	–
Ru5P[a]	(−)	20	5000	0.9973	50	5000	0.9956	154.5	67.5	0.77	0.34	21.27 ± 0.01	–
F6P	(−)	50	5000	0.9954	100	5000	0.9990	35.1	53.2	0.18	0.27	21.63 ± 0.02	21.81 ± 0.01
G6P	(−)	50	5000	0.9972	100	5000	0.9992	33.5	79.3	0.17	0.40	22.49 ± 0.02	22.72 ± 0.01
FbP	(−)	500	5000	0.9998	2000	5000	0.9980	360.8	2949.3	1.80	14.75	24.99 ± 0.01	25.93 ± 0.07

(continued)

Table 2 (continued)

Metabolite	Polarity	Linearity range						Method detection limit				Retention time	
		Alkaline			Acidic			Alkaline	Acidic	Alkaline	Acidic		
		Min (nM)	Max (nM)	R^2	Min (nM)	Max (nM)	R^2	(nM)		pmol (on column)		Alkaline (min)	Acidic (min)
DHAP	(−)	200	5000	0.9990	500	5000	0.9985	53.9	423.1	0.27	2.12	20.77 ± 0.01	20.78 ± 0.01
PEP	(−)	50	5000	0.9904	500	5000	0.9994	86.5	532.7	0.43	2.66	23.61 ± 0.02	24.07 ± 0.02
2PG[a]	(−)	100	5000	0.9874	500	5000	0.9984	87.8	148.4	0.44	0.74	22.86 ± 0.02	23.32 ± 0.02
3PG[a]	(−)	100	5000	0.9874	500	5000	0.9984	87.8	148.4	0.44	0.74	23.11 ± 0.02	23.61 ± 0.02
NAD	(+)	10	5000	0.9982	20	5000	0.9963	7.6	13.1	0.04	0.07	19.64 ± 0.01	20.03 ± 0.01
NADP	(+)	20	2000	0.9901	500	5000	0.9847	31.4	1495.0	0.16	7.47	23.15 ± 0.01	23.51 ± 0.01
AMP	(−)	50	5000	0.9995	100	5000	0.9987	13.4	144.1	0.07	0.72	19.41 ± 0.01	19.33 ± 0.01
ADP	(−)	100	5000	0.9928	1000	5000	0.9832	40.1	1428.0	0.20	7.14	21.31 ± 0.01	22.08 ± 0.03
ATP	(−)	100	5000	0.9925	1000	5000	0.9836	59.6	1130.4	0.30	5.65	22.80 ± 0.02	23.75 ± 0.05
GMP	(−)	100	5000	0.9990	100	5000	0.9960	49.9	198.5	0.25	0.99	22.07 ± 0.01	22.11 ± 0.01
GDP	(−)	100	5000	0.9844	1000	5000	0.9975	70.9	2274.7	0.35	11.37	23.95 ± 0.02	24.91 ± 0.03
GTP	(−)	500	5000	0.9915	2000	5000	0.9971	102.4	1525.2	0.51	7.63	25.45 ± 0.01	26.73[c] ± 0.12
CMP	(−)	100	5000	0.9990	100	5000	0.9974	49.9	160.8	0.25	0.80	21.38 ± 0.01	21.34 ± 0.01
(b) *Part 2*													
CDP	(−)	50	5000	0.9905	1000	5000	0.9938	56.3	1030.1	0.28	5.15	23.21 ± 0.01	24.02 ± 0.01
CTP	(−)	200	5000	0.9915	1000	5000	0.9842	51.9	2004.1	0.26	10.02	24.66 ± 0.01	25.90 ± 0.17
UMP	(−)	100	5000	0.9991	100	5000	0.9985	4.7	123.4	0.02	0.62	20.54 ± 0.01	20.40 ± 0.01

UDP	(−)	50	5000	0.9858	500	5000	0.9840	39.2	1094.0	0.20	5.47	22.57 ± 0.01	23.33 ± 0.02
UTP	(−)	100	5000	0.9968	2000	5000	0.9850	72.8	1711.8	0.36	8.56	24.11 ± 0.01	25.24 ± 0.01
cAMP	(−)	100	5000	0.9970	100	5000	0.9988	59.3	375.8	0.30	1.88	15.35 ± 0.01	15.65 ± 0.01
AIBA	(+)	20	5000	0.9996	50	5000	0.9938	41.7	3.6	0.21	0.02	15.86 ± 0.01	16.26 ± 0.01
GABA	(+)	10	5000	0.9994	20	5000	0.9920	12.7	18.2	0.06	0.09	17.64 ± 0.01	18.02 ± 0.01
HSer	(+)	20	5000	0.9990	50	5000	0.9963	10.2	22.0	0.05	0.11	17.51 ± 0.01	17.92 ± 0.01
*O*AcSer	(+)	20	5000	0.9996	200	5000	0.9986	33.3	286.0	0.17	1.43	15.76 ± 0.01	15.00 ± 0.01
*O*SHS	(+)	10	5000	0.9991	50	5000	0.9956	14.0	38.3	0.07	0.19	18.69 ± 0.01	19.08 ± 0.01
Cysth	(+)	10	5000	0.9953	50	5000	0.9938	4.9	22.6	0.02	0.11	21.45 ± 0.01	22.10 ± 0.01
Gly	(+)	200	5000	0.9992	500	5000	0.9963	28.5	99.2	0.14	0.50	17.96 ± 0.01	18.37 ± 0.01
Ala	(+)	100	5000	0.9990	200	5000	0.9956	41.0	167.5	0.20	0.84	16.91 ± 0.01	17.32 ± 0.01
Ser	(+)	100	5000	0.9988	200	5000	0.9964	9.9	140.9	0.05	0.70	18.45 ± 0.01	18.90 ± 0.01
Pro	(+)	10	5000	0.9992	50	5000	0.9968	9.1	66.8	0.05	0.33	14.37 ± 0.01	14.76 ± 0.01
Val	(+)	20	5000	0.9991	100	5000	0.9964	26.4	49.7	0.13	0.25	14.54 ± 0.01	14.94 ± 0.01
Thr	(+)	50	5000	0.9993	100	5000	0.9967	43.1	97.6	0.22	0.49	17.14 ± 0.01	17.58 ± 0.02
Leu	(+)	20	5000	0.9993	20	5000	0.9965	7.9	35.9	0.04	0.18	12.74 ± 0.01	13.15 ± 0.01
Ile	(+)	20	5000	0.9990	20	5000	0.9966	17.2	22.2	0.09	0.11	13.22 ± 0.01	13.64 ± 0.01
Asn	(+)	50	5000	0.9991	100	5000	0.9964	15.0	100.6	0.08	0.50	18.35 ± 0.01	18.79 ± 0.01
Asp	(+)	50	5000	0.9989	200	5000	0.9972	31.7	192.0	0.16	0.96	20.05 ± 0.01	20.52 ± 0.01
Gln	(+)	20	5000	0.9987	50	5000	0.9959	3.4	54.6	0.02	0.27	17.95 ± 0.01	18.37 ± 0.01
Lys	(+)	20	5000	0.9987	100	5000	0.9933	20.7	36.3	0.10	0.18	24.40 ± 0.01	24.78 ± 0.01
Glu	(+)	100	5000	0.9991	100	5000	0.9965	37.9	49.6	0.19	0.25	19.75 ± 0.01	20.18 ± 0.01
Met	(+)	20	5000	0.9990	20	5000	0.9966	16.8	21.9	0.08	0.11	14.01 ± 0.01	14.43 ± 0.01

(continued)

Table 2 (continued)

Metabolite	Polarity	Linearity range						Method detection limit				Retention time	
		Alkaline			Acidic			Alkaline	Acidic	Alkaline	Acidic		
		Min (nM)	Max (nM)	R^2	Min (nM)	Max (nM)	R^2	(nM)		pmol (on column)		Alkaline (min)	Acidic (min)
Phe	(+)	20	5000	0.9989	20	5000	0.9965	4.1	4.9	0.02	0.02	12.69 ± 0.01	13.14 ± 0.01
Arg	(+)	20	3000	0.9936	100	3000	0.9869	6.1	49.9	0.03	0.25	24.65 ± 0.01	25.01 ± 0.01
Trp	(+)	20	5000	0.9991	50	5000	0.9983	65.5	67.9	0.33	0.34	13.84 ± 0.01	14.27 ± 0.01
His[b]	(+)	–	–	–	200	5000	0.9860	–	106.3	–	0.53	–	18.35 ± 0.02

Reproduced from Teleki et al. [6] with permission. Copyright Elsevier
[a]Peaks were not completely separated
[b]Insufficient stability under alkaline condition (pH 9.2)

3.5 Intracellular LC-MS/MS Analysis

1. Check MS performance as mentioned above (*see* Subheading 3.4).
2. Adjust source and chromatographic parameters as mentioned above (*see* Subheading 3.4).
3. Create an adapted MS/MS program for simultaneous quantification of targeted metabolites (*see* Fig. 1):
   (a) Divide the HPLC run in suitable measuring time segments with respect to evaluated retention times and peak geometries (*see* **Note 17**).
   (b) Include the ionization mode (ESI±), selected MRM transitions (for analytes and internal standards), scan widths (e.g., 1 *m/z*) and desired signal amplifications (electron multiplier) into corresponding segments (*see* **Note 17**).
   (c) Adjust the dwell time (detection period) to 100 ms for each MRM transition.
   (d) Check and optionally adjust the overall program by column injection of 5 μL prepared standard mixes (e.g., 50 μM).
4. Transfer prepared (*see* Subheading 3.3) sample vials into the autosampler cooled at 5 °C (*see* **Note 18**).
5. Create a sample measuring sequence in the column injection mode (5 μL) (*see* **Note 19**):
   (a) Start with a blank run.
   (b) Followed by an initial test run of adapted standard mixes.
   (c) Measure optionally standard-based external calibration curves starting with the lowest level (*see* Subheading 3.6).
   (d) Followed by experimental samples.
   (e) Frequently measure standard mixes as quality control (e.g., 50 μM, every ten samples).
   (f) End with a blank run.

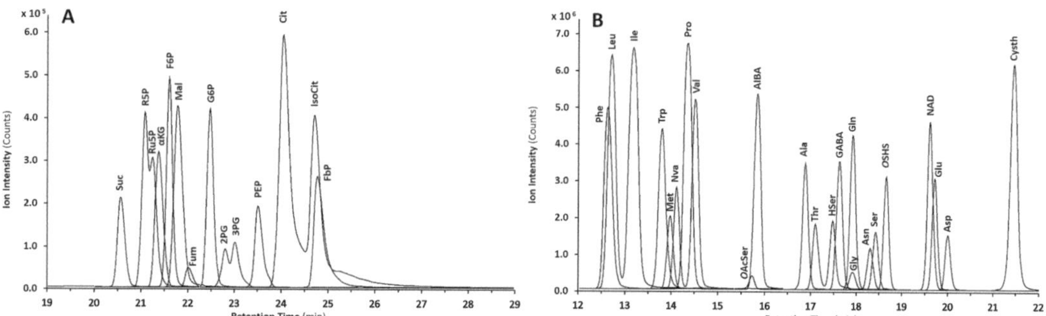

Fig. 1 Standard-based representative extracted ion chromatograms (MRM) obtained with alkaline ZIC®-pHILIC method (pH 9.2) in negative (**a**) and positive ionization mode (**b**). Depicted abbreviations of metabolite compounds are listed in Table 1

3.6 Data Analysis and Quantification Strategies

1. Open LC-MS/MS data with compatible analysis software (e.g., MassHunter Qualitative or Quantitative Analysis).
   (a) Extract MRM transitions from acquired raw data (*Total Ion Chromatogram*, TIC).
   (b) Use retention times to identify corresponding metabolite peaks and potentially interfering peaks (isobaric analytes) (*see* **Note 9**).
   (c) Check and optionally correct peak integrations (e.g., base lines, integration limits).

To obtain reliable absolute quantifications in complex intracellular sample matrices, a quantification bias caused by co-eluting untargeted compounds must be considered (e.g., ion suppression) [23]. Matrix effects can be accounted for by isotope dilution [24] as common calibration strategy (*see* **Note 20**).

2. Isotope dilution mass spectrometry (IDMS) enables a time-saving absolute quantification of targeted metabolites by conventional external calibration (*see* **Note 8**):
   (a) Prepare multicomponent ^{12}C standard mixes for external calibration.
   (b) Adjust metabolite-specific calibration ranges (at least seven levels) based on previously estimated concentrations (*spiking*, *see* **Note 20**).
   (c) Add constant levels of $U^{13}C$-labeled algal extracts as internal standard solution to calibration standards and samples.
   (d) Prepare metabolite-specific external calibration curves by linear regression of peak areas normalized by internal standard areas ($U^{13}C$ analog) against applied concentration levels.
3. Determine absolute intracellular metabolite concentrations c_i with respect to the applied extraction method (*see* Subheading 3.1).
   (a) Calculate initial extract concentrations c_e considering dilutions in sample preparation.
   (b) Normalize resulting concentrations with applied biomass concentrations c_x:

$$c_i\ [\mu\text{mol}/g_{CDW}] = c_e[\mu\text{mol/L}]/c_x[g_{CDW}/\text{L}]$$

   (c) Optionally, resulting intracellular concentrations could be converted to volumetric pool concentrations considering actual or assumed intracellular volumes [25].

$$c_i\ [\mu\text{mol/L}] = c_i\ [\mu\text{mol}/g_{CDW}]/2.3 \times 10^{-3}[\text{mL}/g_{CDW}]$$

To demonstrate the quality of the overall method, exemplary results of selected metabolites obtained by three different calibration strategies (external calibration, isotope dilution, and standard addition) are summarized in Fig. 2 and Table 3 (*see* **Note 20**).

4 Notes

1. HILIC and normal stationary phases tend to have a continuous accumulation of metallic trace elements including iron (III) ions [26]. Interactions with metal ions have a negative impact on retention characteristics (e.g., peak tailing) of several anionic metabolites (e.g., fructose-1,6-bisphosphate, citrate) with related sensitivity reductions. The application of PEEK (polyether-ether-ketone) capillaries and a metal-free bioinert autosampler, largely eliminates exposed metal surfaces within the HPLC system. Routine injections of high-concentration EDTA solutions (e.g., 1 g/L, 100 μL) additionally reduce residual on-column metal ions and significantly improve resulting chromatographic peak performances [26].
2. Listed parameters refer to shake-flask cultivations of *E. coli* during exponential growth. Fermentations in bioreactors enable the installation of (semi)automated volumetric devices for more rapid sampling [27].
3. Due to comparatively low extraction volumes (50–200 μL) an unwanted carryover of supernatants may result in a significant bias of calculated concentrations. Obtained intracellular pool concentrations can be slightly corrected considering actual intercellular volumes of washed cell pellets (*see* also **Note 6**).
4. Simultaneous extracellular sampling enables the additional quantification of relevant uptake and excretion rates (e.g., metabolic *stimulus response experiments*, SRE) [27].
5. Metabolome studies of microbial overproducers face the inherent challenge that intracellular and extracellular metabolite amounts may differ by orders of magnitude (e.g., product metabolite). Sequential sampling strategies such as *fast centrifugation treatment* (FCT) enable the implementation of isotonic washing steps and avoid unfavorable leakage-related differential approaches (e.g., *cold methanol quenching*, CMQ) [28].
6. Variations during sample preparations could be checked and corrected by addition of defined amounts of suitable global internal standard (e.g., L-norvaline). In order to increase the predictive significance of the metabolic-specific measurement parameters (e.g., linear dynamic range) and calibration strategies, global internal standards were only considered for

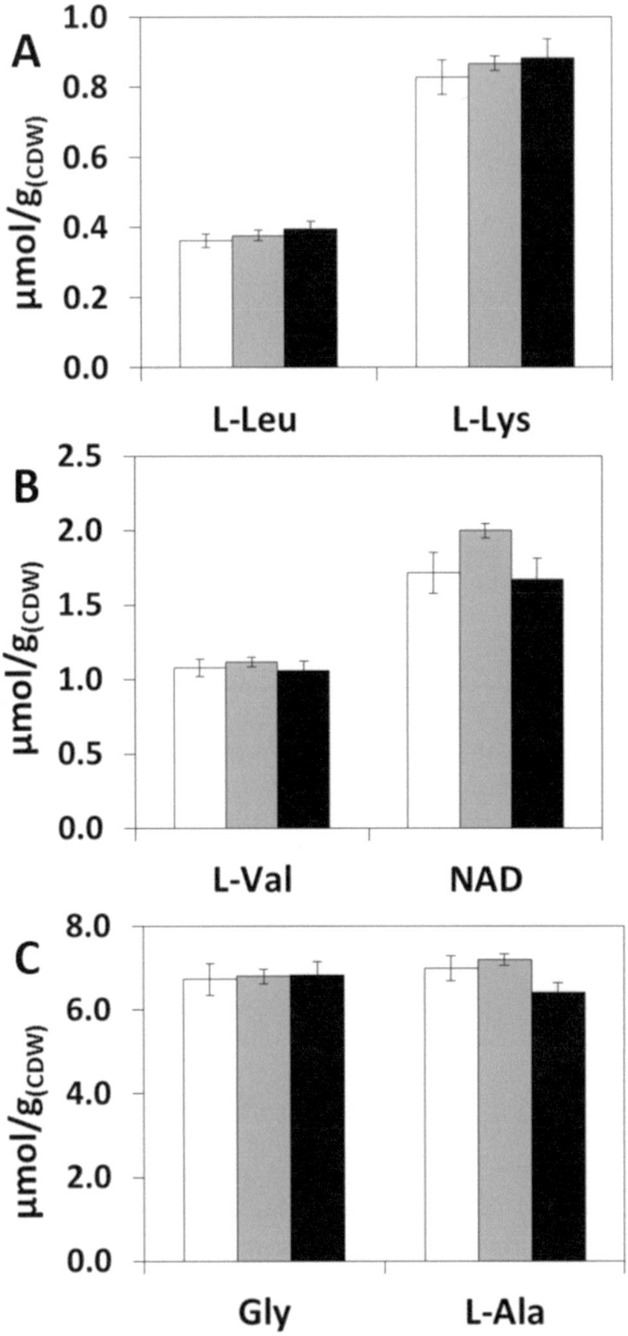

Fig. 2 Absolute intracellular concentrations of selected metabolites in exponentially growing *E. coli* K-12 MG1655 quantified by HILIC-MS/MS under alkaline conditions (pH 9.2). Measurements are based on standard-based external calibration (white), isotope dilution (grey) and standard addition (black). Error bars describe the standard deviation of four replicates measurements. Results are in good accordance with previously published data [31–34]. Reproduced from Teleki et al. [6] with permission. Copyright Elsevier

Table 3
Calibration functions and corresponding correlation coefficients of investigated metabolites (*see also* Figs. 1 and 2). Reproduced from Teleki et al. [6] with permission. Copyright Elsevier

			External calibration			Isotope dilution			Standard addition			One-way ANOVA	
Metabolite	RT [min]		Slope	Intercept	R^2	Slope	Intercept	R^2	Slope	Intercept	R^2	F	P value
L-Leucine	13.17	±0.04	3.20*E*+05	3.80*E*+04	0.9910	4.01*E*−01	4.50*E*−02	0.9913	3.28*E*+05	2.55*E*+05	0.9986	3.560	0.0726
L-Lysine	24.94	±0.05	2.41*E*+05	2.90*E*+04	0.9975	5.08*E*−02	4.10*E*−03	0.9969	2.35*E*+05	4.09*E*+05	0.9995	1.622	0.2503
L-Valine	14.98	±0.04	1.42*E*+05	4.23*E*+03	0.9969	2.59*E*−01	3.40*E*−03	0.9975	1.45*E*+05	2.99*E*+05	0.9989	1.379	0.3002
NAD	20.06	±0.05	4.25*E*+04	−2.74*E*+03	0.9978	1.31*E*+00	−8.93*E*−02	0.9982	4.40*E*+04	1.36*E*+05	0.9994	9.193	0.0067
Glycine	18.42	±0.05	1.91*E*+04	7.34*E*+03	0.9994	7.77*E*−02	−1.96*E*−02	0.9991	1.95*E*+04	2.60*E*+05	0.9999	0.121	0.8876
L-Alanine	17.36	±0.05	6.93*E*+04	−5.56*E*+04	0.9977	5.03*E*−02	−3.82*E*−02	0.9984	7.06*E*+04	8.90*E*+05	0.9997	1.778	0.2235

$F_{(\mathrm{crit})} = 4.256$

monitoring of instrumental fluctuations but not for normalization of obtained peak areas (*see* Tables 2 and 3).

7. Hot water extractions enable well-defined and effective endo-metabolome extractions with comparatively low experimental effort. However, for thermally labile metabolites (e.g., nucleotides) we recommend more gentle approaches like cold chloroform-methanol extractions (CME) [28].
8. The application of $U^{13}C$-metabolites as internal standards (IDMS) ensures the accurate compensation for analyte losses during sample generation, processing and ion suppression in various MS ionization sources. However, such compounds are either a significant cost factor or unavailable at all. Commercially available $U^{13}C$ algal extracts provide a convenient and cost-effective alternative for a broad range of common intracellular metabolites [15].
9. Despite comparatively high chromatographic selectivity and retention time stability, accurate peak integration within complex sample matrices may be hampered by interfering isobaric peaks. In cases of doubt, selective addition of metabolite standards to the sample preparation (i.e., *spiking*) usually provides clarification.
10. Adaption of metabolite extracts (60% (vv^{-1}) acetonitrile) to hydrophobic HILIC starting conditions (90% (vv^{-1}) acetonitrile) may lead to unwanted precipitation of salts and charged or highly polar metabolites. In these cases, we recommend a sufficient pre-dilution of corresponding extracts.
11. Irrespective of applied ionization modes, standard-based HILIC-ESI-MS/MS studies reveals significantly improved chromatographic performances, detection limits, linearity ranges and ESI-MS responsivities for 59 intracellular key metabolites under alkaline mobile phase conditions (*see* Table 2) [6]. Unexpected ionization behaviors (e.g., amino acids in positive ionization mode under alkaline conditions) are controversially discussed and are defined as *wrong-way-round ionization* (WWR) [29].
12. To ensure measurements with optimal sensitivity and signal stability the implementation of cleaning protocols according to manufacturer specifications is mandatory. We recommend routine cleaning of removable ESI components with abrasive paper (4000 grit) and flushing of the nebulizer with 50% (vv^{-1}) aqueous isopropanol solution on a weekly basis. Residual contaminations are easily removable by a subsequent ultrasonification procedure.
13. To ensure sufficient hydration of HILIC stationary phase particles, maintain at least 3% (vv^{-1}) of water in the used mobile phases [22]. The composition of eluent A (90% acetonitrile,

10% water) has been adjusted accordingly. A minimum proportion of water additionally prevents a blockage of the pump valves based on a potential polymerization of pure acetonitrile.

14. Retention time reproducibility within the described chromatographic conditions is fairly high. Ninety-eight percentage of analyzed metabolites showed absolute standard deviations of less than 2 s under alkaline mobile phase conditions [6]. We recommend an initial equilibration of at least 50 column volumes after column installation and a re-equilibration of 8–10 column volumes between sample injections.
15. ESI-MS response linearities are specific for each metabolite, sample matrix and eluent condition. Reliable calibration strategies are based on pre-evaluated operating ranges. By adding $U^{13}C$-analogues as internal standards (IDMS), corresponding external calibration ranges can be significantly extended.
16. Methodical detection limits were calculated as the injected amount of each metabolite that results in a statistical significant peak above background noise level. Calculations were based on standard deviations of triplicates at the lower linearity range boundary multiplied by the expansion coefficient ($t = 6.96$) for defining the 99% confidence interval [30].
17. Comprehensive quantifications of a broad spectrum of metabolites require the division of chromatographic runs in suitable measuring time segments. LC-MS resolutions are limited by the metabolite-specific peak widths (usually 0.5–2 min) and corresponding cycle times (sum of dwell and switching times). To ensure sufficient peak resolution we recommend at least 15–20 run cycles (measuring points) during elution of the targeted analyte. Provided that the dwell time is adjusted to 100 ms, a maximum cycle time of 2 s is acceptable for accurate quantitative integrations.
18. Direct injection experiments via prepared standard solutions of studied metabolites reveals sufficient measurement stability under alkaline conditions (signal decrease <20%) for 24 h. However, L-histidine and erythrose-4-phosphate showed higher degradation rates and should be analyzed by isotope dilution to compensate for analyte losses.
19. In case of a significant signal break down (e.g., quality control samples) the current sample sequence should be stopped. A quick cleaning of the nebulizer and ESI source usually solves the problem. If $U^{13}C$ analogues are used as internal standards the sample sequence could be continued without any problems.
20. Matrix effects in multicomponent intracellular extracts have led to the almost obligatory application of isotope dilution (IDMS) for absolute metabolite quantifications. Conventional

external calibration may lead to a quantification bias caused by interfering untargeted compounds (e.g., ion suppression). Alternatively, the standard addition can be used, which is based on a time-consuming multiple addition (e.g., $n = 4$) of adequate amounts of metabolite standard directly to the sample aliquots (internal calibration). However, exemplary results of selected metabolites revealed irrespective of applied calibration strategy high similarity (<10%) and confirm the quality of chromatographic separation with a sufficient reduction of ESI-MS-based matrix effects (*see* Fig. 2 and Table 3).

Acknowledgments

The work was funded by the Bundesministerium für Bildung und Forschung (grant number: 0315867, BMBF, Berlin, Germany) in cooperation with Evonik Industries AG.

References

1. Oldiges M, Lütz S, Pflug S, Schroer K, Stein N, Wiendahl C (2007) Metabolomics: current state and evolving methodologies and tools. Appl Microbiol Biotechnol 76:495–511
2. Fei F, Bowdish DM, McCarry BE (2014) Comprehensive and simultaneous coverage of lipid and polar metabolites for endogenous cellular metabolomics using HILIC-TOF-MS. Anal Bioanal Chem 406:3723–3733
3. Oldiges M, Takors R (2005) Applying metabolic profiling techniques for stimulus-response experiments: chances and pitfalls. Adv Biochem Eng Biotechnol 92:173–196
4. Magnus JB, Oldiges M, Takors R (2009) The identification of enzyme targets for the optimization of a valine producing Corynebacterium glutamicum strain using a kinetic model. Biotechnol Prog 25:754–762
5. Dunn WB, Bailey NJ (2005) Measuring the metabolome: current analytical technologies. Analyst 130:606–625
6. Teleki A, Sánchez-Kopper A, Takors R (2015) Alkaline conditions in hydrophilic interaction liquid chromatography for intracellular metabolite quantification using tandem mass spectrometry. Anal Biochem 475:4–13
7. Dunn WB, Ellis DI (2005) Metabolomics: current analytical platforms and methodologies. Trends Anal Chem 24:285–294
8. Mashego MR, Rumbold K (2007) Microbial metabolomics: past, present and future methodologies. Biotechnol Lett 29:1–16
9. Griffin JL, Williams HJ, Sang E, Clarke K, Rae C, Nicholson JK (2001) Metabolic profiling of genetic disorders: a multitissue H-1 nuclear magnetic resonance spectroscopic and pattern recognition study into dystrophic tissue. Anal Biochem 293:16–21
10. Dell'mour M, Jaitz L, Oburger E, Puschenreiter M, Koellensperger G, Hann S (2010) Hydrophilic interaction LC combined with electrospray MS for highly sensitiveanalysis of underivatized amino acids in rhizosphere research. J Sep Sci 33:911–922
11. Kvitvang HF, Andreassen T, Adam T, Villas-Bôas SG, Bruheim P (2011) Highly sensitive GC/MS/MS method for quantitation of amino and nonamino organic acids. Anal Chem 83:2705–2711
12. Ramautar R, Somsen GW, de Jong GJ (2013) CE-MS for metabolomics: developments and applications in the period 2010–2012. Electrophoresis 34:86–98
13. Lane AN, Fan TW, Higashi RM (2008) Isotopomer-based metabolomic analysis by NMR and mass spectrometry. Methods Cell Biol 84:541–588
14. Edwards JL, Chisolm CN, Shackman JG, Kennedy RT (2006) Negative mode sheathless capillary electrophoresis electrospray ionization-mass spectrometry for metabolite analysis of prokaryotes. J Chromatogr A 1106:80–88
15. Vielhauer O, Zakhartsev M, Horn T, Takors R, Reuss M (2011) Simplified absolute metabolite quantification by gas chromatography-isotope

dilution mass spectrometry on the basis of commercially available source material. J Chromatogr B 879:3859–3870
16. Nguyen HP, Schug KA (2008) The advantages of ESI-MS detection in conjunction with HILIC mode separations: fundamentals and applications. J Sep Sci 31:1465–1480
17. Buszewski B, Noga S (2012) Hydrophilic interaction liquid chromatography (HILIC)—a powerful separation technique. Anal Bioanal Chem 402:231–247
18. Preinerstorfer B, Schiesel S, Lämmerhofer M, Lindner W (2010) Metabolic profiling of intracellular metabolites in fermentation broths from beta-lactam antibiotics production by liquid chromatography-tandem mass spectrometry methods. J Chromatogr A 1217:312–328
19. Nguyen HP, Yang SH, Wigginton JG, Simpkins JW, Schug KA (2010) Retention behavior of estrogen metabolites on hydrophilic interaction chromatography stationary phases. J Sep Sci 33:793–802
20. Plassmeier J, Barsch A (2007) Investigation of central carbon metabolism and the 2-methylcitrate cycle in Corynebacterium glutamicum by metabolic profiling using gas chromatography-mass spectrometry. J Biotechnol 130:354–363
21. Bolten CJ, Kiefer P, Letisse F, Portais JC, Wittmann C (2007) Sampling for metabolome analysis of microorganisms. Anal Chem 79:3843–3890
22. Merck-Sequant (2008) A practical guide to HILIC – including ZIC®-HILIC applications, 5th edn. Merck-Sequant, Umea
23. Kiefer P, Portais JC, Vorholt JA (2008) Quantitative metabolome analysis using liquid chromatography-high-resolution mass spectrometry. Anal Biochem 382:94–100
24. Mashego MR, Wu L, van Dam JC, Ras C, Vinke JL, van Winden WA, van Gulik WM, Heijnen JJ (2004) MIRACLE: mass isotopomer ratio analysis of U-13C-labeled extracts. A new method for accurate quantification of changes in concentrations of intracellular metabolites. Biotechnol Bioeng 85:620–628
25. Wang L, Zhou YJ, Ji D, Zhao ZK (2013) An accurate method for estimation of the intracellular aqueous volume of Escherichia coli cells. J Microbiol Methods 93:73–76
26. Pesek JJ, Matyska MT, Fischer SM (2011) Improvement of peak shape in aqueous normal phase analysis of anionic metabolites. J Sep Sci 34:3509–3516
27. Teleki A, Rahnert M, Bungart O, Gann B, Ochrombel I, Takors R (2017) Robust identification of metabolic control for microbial l-methionine production following an easy-to-use puristic approach. Metab Eng 41:159–172
28. de Koning W, van Dam K (1992) A method for the determination of changes of glycolytic metabolites in yeast on a subsecond time scale using extraction at neutral pH. Anal Biochem 204:118–123
29. Mansoori BA, Volmer DA, Boyd RK (1997) "Wrong-way-round" electrospray ionization of amino acids. Rapid Commun Mass Spectrom 11:1120–1130
30. Wells G, Prest H, Russ CW IV (2011) Technical note: signal, noise, and detection limits in mass spectrometry. Agilent Technologies, Santa Clara, CA 5990-7651EN
31. Hiller J, Franco-Lara E, Weuster-Botz D (2007) Metabolic profiling of Escherichia coli cultivations: evaluation of extraction and metabolite analysis procedures. Biotechnol Lett 29:1169–1178
32. Yuan J, Fowler WU, Kimball E, Lu W, Rabinowitz JD (2006) Kinetic flux profiling of nitrogen assimilation in Escherichia coli. Nat Chem Biol 2:529–530
33. Bennett BD, Kimball EH, Gao M, Osterhout R, van Dien SJ, Rabinowitz JD (2009) Absolute metabolite concentrations and implied enzyme active site occupancy in Escherichia coli. Nat Chem Biol 5:593–599
34. Buchholz A, Takors R, Wandrey C (2001) Quantification of intracellular metabolites in Escherichia coli K12 using liquid chromatographic-electrospray ionization tandem mass spectrometric techniques. Anal Biochem 295:129–137

Chapter 11

Liquid Chromatography and Mass Spectrometry Analysis of Isoprenoid Intermediates in *Escherichia coli*

Edward E.K. Baidoo, George Wang, Chijioke J. Joshua, Veronica Teixeira Benites, and Jay D. Keasling

Abstract

Isoprenoids are a highly diverse group of natural products with broad application as high value chemicals and advanced biofuels. They are synthesized using two primary building blocks, namely, isopentenyl diphosphate (IPP) and dimethylallyl diphosphate (DMAPP) that are generated via the mevalonate (MVA) or deoxy-D-xylulose-5-phosphate (DXP) pathways. Isoprenoid biosynthetic pathways are prevalent in eukaryotes, archaea, and bacteria. Measurement of isoprenoid intermediates via standard liquid chromatography–mass spectrometry (LC-MS) protocols is generally challenging because of the hydrophilicity and complex physicochemical properties of the molecules. In addition, there is currently no reliable analytical method that can simultaneously measure metabolic intermediates from MVA and DXP pathways, including the prenyl diphosphates. Therefore, we describe a robust hydrophilic interaction liquid chromatography time-of-flight mass spectrometry (HILIC-TOF-MS) method for analyzing isoprenoid intermediates from metabolically engineered *Escherichia coli* strains.

Key words Isoprenoid intermediates, Metabolomics, LC-MS, HILIC, Mevalonate, Deoxy-D-xylulose-5-phosphate

1 Introduction

Isoprenoids are a diverse groups of natural products that serve as sterols, carotenoids, quinones and hormones, which are involved in defense and communication in most organisms [1]. There are over 35,000 distinct types of isoprenoids that are synthesized from IPP and DMAPP via the MVA or DXP pathways. The MVA pathway is generally prevalent in eukaryotes and archaea, while the DXP pathway is more abundant in eubacteria [1]. The MVA pathway consists of six enzymes that catalyze the transformation of acetyl-CoA (Ac-CoA) to IPP (Fig. 1). The IPP is subsequently isomerized to DMAPP by isopentenyl diphosphate isomerase (IDI). In contrast, the DXP pathway consists of seven enzymes that transform

Edward E.K. Baidoo (ed.), *Microbial Metabolomics: Methods and Protocols*, Methods in Molecular Biology, vol. 1859, https://doi.org/10.1007/978-1-4939-8757-3_11,

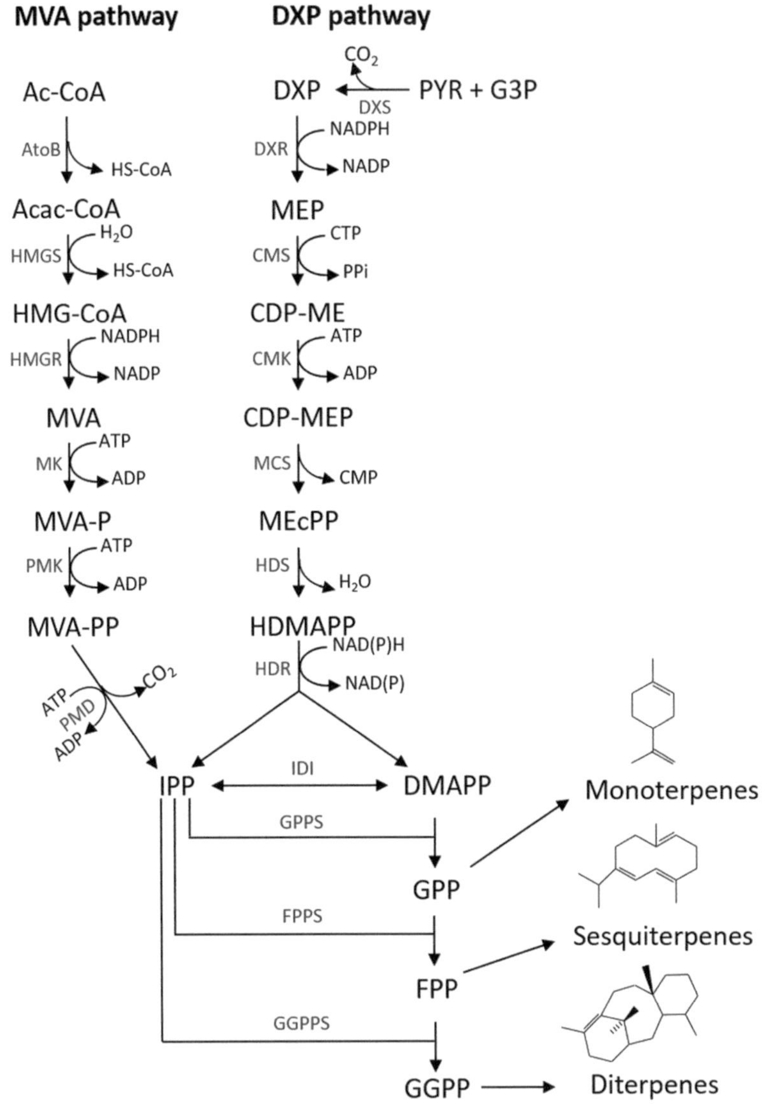

Fig. 1 MVA and DXP pathways. The MVA pathway utilizes six enzymes that catalyze the transformation of Ac-CoA to IPP, which is isomerized to DMAPP. The DXP utilizes seven enzymes that catalyze the transformation of PYR and G3P to IPP and DMAPP. Metabolite abbreviations—MVA-P (mevalonate-5-phosphate), MVA-PP (mevalonate diphosphate), MEP (2-C-methylerythritol 4-phosphate, CDP-ME (4-diphosphocytidyl-2-C-methylerythritol), CDP-MEP (4-diphosphocytidyl-2-C-methyl-D-erythritol 2-phosphate), MEcPP (2-C-methyl-D-erythritol 2,4-cyclodiphosphate), HDMAPP ((E)-4-Hydroxy-3-methyl-but-2-enyl diphosphate), GPP (geranyl diphosphate), FPP (farnesyl diphosphate) and geranylgeranyl diphosphate (GGPP). All other metabolite abbreviations are described in the abstract and introduction. Enzyme abbreviations—AtoB (acetoacetyl-CoA thiolase), HMGS (HMG-CoA synthase), HMGR (HMG-CoA reductase), MK (mevalonate kinase), PMK (phosphomevalonate kinase), DXS (DXP synthase), DXR (DXP reductoisomerase), CMS (CDP-ME synthase), CMK (4-diphosphocytidyl-2-C-methylerythritol synthase), MCS (MEcPP synthase), HDS (HDMAPP synthase), HDR (HDMAPP reductase), IDI (IPP isomerase), GPPS (GPP synthase), FPPS (FPP synthase), and GGPPS (GGPP synthase)

glyceraldehdye-3-phosphate (G3P) and pyruvic acid (PYR) to IPP and DMAPP at a ratio of ~6:1 respectively (Fig. 1).

Quantitative evaluation of metabolic intermediates from heterologous expression of the MVA pathway in *Escherichia coli* is often challenged by the existence of an endogenous DXP pathway in the organism. In addition, metabolic intermediates from isoprenoid biosynthetic pathways are difficult to analyze because the molecules are highly hydrophilic, anionic (mostly phosphorylated) and labile. These intermediates are also comprised of compounds such as prenyl diphosphates, CoAs, and organic acids with diverse molecular characteristics that make chromatographic resolution extremely challenging.

One can divide the MVA pathway into upper and lower segments, splitting the pathway at MVA. The upper portion of the MVA pathway involves the following reactions: Ac-CoA → acetoacetyl-coenzyme A (Acac-CoA) → 3-hydroxy-3-methylglutaryl-coenzyme A (HMG-CoA) that are readily analyzed using reversed phase LC-MS methods for measuring acyl-CoAs [2–4]. Several studies have also reported well characterized LC-MS methods for measuring glyceraldehyde-3-phosphate (G3P) and pyruvate (PYR), as well as the precursors of the DXP pathway [5–7]. Some analytical methods convert the phosphorylated intermediates to their respective alcohols by dephosphorylating the molecules with potato acid phosphatase [8]. The alcohols were subsequently analyzed via LC with reversed phase C18 [9]. Currently, there are relatively few validated analytical methods for simultaneously profiling metabolites from MVA and DXP pathways, especially the prenyl diphosphates. Most recently reported methods are effective for measuring the lower MVA pathway intermediates [10, 11] and the prenyl phosphates [10–13], or the DXP pathway intermediates [12, 14–16]. The two most effective LC-MS strategies for measuring isoprenoid intermediates employ reversed phase chromatography [10–12, 16] and HILIC based techniques [14]. In this chapter, we present a robust HILIC-TOF-MS method for measuring prenyl diphosphates and metabolic intermediates from MVA and DXP pathways (Fig. 2). HILIC retention of the analytes is mediated by hydrogen bonding with a thin film of water bound to the ZIC®-pHILIC stationary phase and weak electrostatic interaction with its zwitterionic functional groups. This chromatographic separation is coupled TOF mass spectrometry, which provides high resolution accurate mass measurements of deprotonated ions in the negative ion mode. Herein, we present the application of the HILIC-TOF-MS method to the measurement of isoprenoid intermediates from *Escherichia coli* strain MBIS2 [17] that was engineered to produce the antimalarial drug precursor amorpha-4,11-diene via the MVA pathway (Fig. 3).

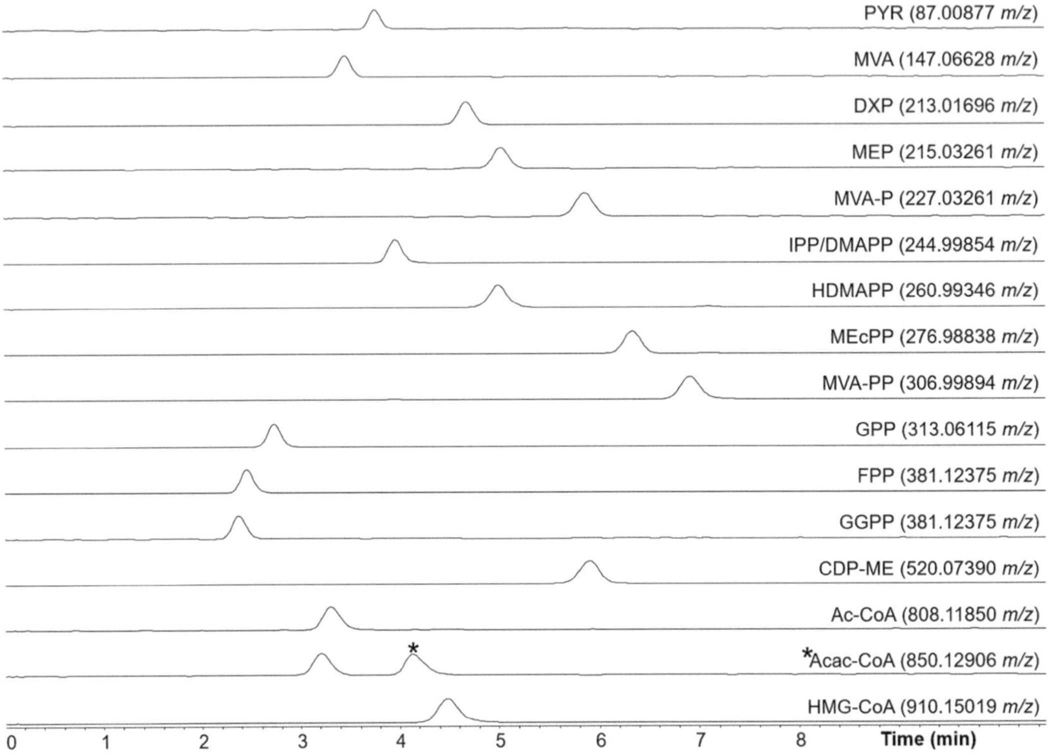

Fig. 2 HILIC-TOF-MS extracted ion chromatograms of isoprenoid intermediates. A 6.25 μM chemical standards mixture was used

2 Materials

Prepare all solutions using HPLC-grade solvents or greater. Prepare and store all reagents at room temperature (unless indicated otherwise). Diligently follow all appropriate waste disposal regulations when disposing of waste materials. Researchers should also follow all safety guidelines prescribed by their institution and funding sources.

2.1 Sample Preparation and Standard Solution Chemicals

1. Solvents used: methanol, water, and acetonitrile.
2. Quenching and extraction solvents: methanol and water (*see* **Note 1**).
3. Sample medium and reconstitution solution: acetonitrile–methanol–water (6:1:3, v/v/v).
4. Internal standards (ISD) solution: 5–10 μM methionine sulfone, deuterated d_4-succinic acid, and deuterated d_4-citric acid in acetonitrile–methanol–water (6:1:3, v/v/v) (*see* **Note 2**). This solution is briefly mixed by vortexing.

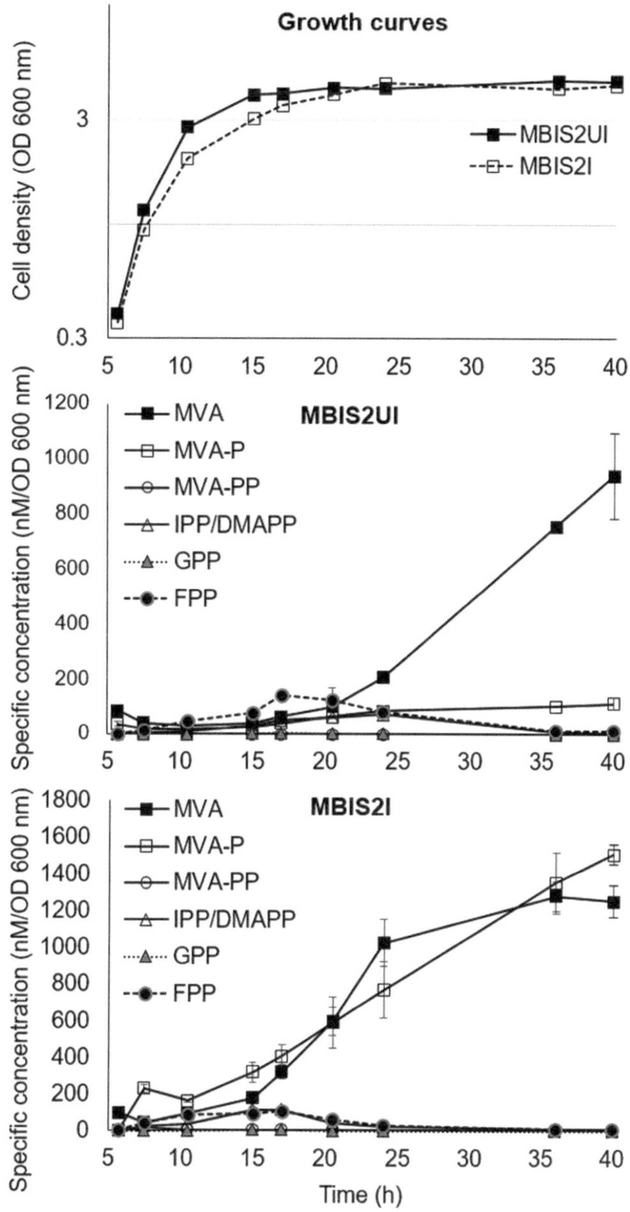

Fig. 3 Growth and intracellular metabolite levels of *E. coli* DH1 harboring the MBIS2 MVA pathway construct, grown on 0.4% glucose in LB medium. *E. coli* harboring MevT (which produces enzymes in the upper portion of the MVA pathway, i.e., AtoB, HMGS, and HMGR) and MBIS2 (which produces enzymes in the lower portion of the MVA pathway, i.e., MK, PMK, and PMD, as well as IDI, IspA (also known as FPPS) and amorphadiene synthase) plasmids, which were driven by P_{trc} promotors, was induced at $OD_{600nm} = \sim 0.37$ by the addition of 1 mM IPTG [17]. Growth was monitored for up to 40 h and liquid culture was sampled at nine time points. Metabolite concentration was normalized to cell density. UI and I are abbreviations for uninduced and IPTG induced, respectively

5. Quality control (QC) sample: Lyophilized *E. coli* wild-type extract reconstituted in a solution containing chemical standards (at 10 μM) of target intermediates and ISD in acetonitrile–methanol–water (6:1:3, v/v/v). This solution is briefly mixed by vortexing (*see* Subheading 3.1 for sample preparation procedure).
6. Calibration curve solution: Chemical standards of target intermediates and ISD in acetonitrile–methanol–water (6:1:3, v/v/v). This solution is briefly mixed by vortexing.

2.2 Sample Preparation Procedure Equipment

1. Vortexer.
2. Balance.
3. 1.7–2 mL polypropylene microcentrifuge tubes with screw caps containing 'O' rings.
4. 1.7–2 mL amber polypropylene microcentrifuge tubes (*see* **Note 3**) with snap lock lids.
5. Lyophilizer.
6. Speedvac.
7. −20 °C freezer.
8. −80 °C freezer.
9. Weighing boats.
10. 3000 Da molecular weight cutoff (MW/CO) centrifuge filter tubes (Amicon Ultra from MilliporeSigma) (*see* **Note 4**).
11. Temperature controlled benchtop centrifuge.

2.3 Analytical Methodology (e.g., LC–MS) Chemicals

1. Chemicals used: water, acetonitrile, methanol, MVA and DXP pathway chemical standards (Echelon Biosciences Incorporated, Salt Lake City, UT USA).
2. HILIC mobile phase solution: The mobile phase is composed of 10 mM ammonium carbonate (Fluka Analytical, Sigma-Aldrich, St. Louis, MO, USA) and 118.4 mM ammonium hydroxide (14.8 M ammonium hydroxide solution, 30–33% NH3 basis, EMD Chemicals, Billerica, MA, USA) in acetonitrile–water (60.2:39.8, v/v). Ammonium carbonate is prepared from a 1 M stock solution in water, which is stored at 4 °C (*see* **Notes 5** and **6**).
3. HILIC column storage solution: 5 mM ammonium acetate in acetonitrile–water (8:2, v/v).
4. Needle wash solvents: water, methanol, and acetonitrile (*see* **Note 7**).
5. MS reference mass (RM) solution: This is a mixture of (1) 5 μM trifluoroacetic acid ammonium salt (CF3CO2N1H4, CAS NO. 3336-58-1) (i.e., TFA) and (2) 5 μM hexakis(1H, 1H,

3H-tetrafluoropropoxy)phosphazine ($C_{18}H_{18}F_{24}N_3O_6P_3$, 921.23 Da, CAS NO.: 58943-98-9) (i.e., HTFP) in methanol–water (1:1, v/v). The original stock solutions were provided by Agilent Technologies.

6. MS calibrant (CB) solution: This is the Agilent Technologies ESI-L Low Concentration Tuning Mix (100 mL) containing (1) betaine ($C_5H_{11}N_1O_2$, 117.15 Da, CAS NO.: 107–43-7), (2) TFA, (3) hexamethoxyphosphazine ($C_{10}H_{19}NO_6P$, CAS NO.: 957–13-1), (4) hexakis(2,2-difluoroethoxy)phosphazine ($C_{12}H_{18}F_{12}N_3O_6P_3$, 621.19 Da, CAS NO.: 186817-57-2), (5) HTFP, (6) hexakis(1H, 1H, 5H-octafluoropentoxy) phosphazine ($C_{30}H_{18}F_{48}N_3O_6P_3$, 1521.334 Da, CAS NO.: 16059-16-8), (7) hexakis(1H, 1H, 7H-dodecafluoroheptoxy) phosphazine ($C_{42}H_{18}F_{72}N_3O_6P_3$, 2121.42 Da, CAS NO.: 3830-74-8), (8) hexakis(1H, 1H, 9H-perfluorononyloxy) phosphazine ($C_{54}H_{18}F_{96}N_3O_6P_3$, 2721.51 Da, CAS NO.: 186043-67-4), (9) hexakis(1H, 1H,4H-hexafluorobutyloxy) phosphazine ($C_{24}H_{18}F_{36}N_3O_6P_3$, 1221.28 Da, CAS NO.: 186406-47-3), (10) hexakis(1H, 1H, 6H-decafluorohexyloxy)phosphazine ($C_{36}H_{18}F_{60}N_3O_6P_3$, 1821.37 Da, CAS NO.: 186406-48-4), (11) hexakis(1H, 1H, 8H-tetradecafluorooctyloxy)phosphazine ($C_{48}H_{18}F_{84}N_3O_6P_3$, 2421.46 Da, CAS NO.: 186406-49-5), (12) tris(trifluoromethyl)-1, 3, 5-triazine ($C_6F_9N_3$, 285.07 Da, CAS NO.: 368-66-1), and (13) tris(heptafluoropropyl)-1, 3, 5-triazine ($C_{12}F_{21}N_3$, 585.11 Da, CAS NO.: 915-76-4). These compounds are all present at <0.01% and at a purity of 97% or greater in HPLC grade acetonitrile–deionized water (95:5, v/v).

2.4 Analytical Equipment

1. Amber 1.5 mL glass LC vials with 250 μL glass inserts and PTFE septa poly propylene screw caps.
2. HPLC system information: an Agilent Technologies 1200 Series Rapid Resolution HPLC system (Agilent Technologies, Santa Clara, CA, USA). The HPLC system is equipped with red 0.12 mm internal diameter (i.d.) stainless steel capillary tubing. The sample injection volume, autosampler tray temperature and column compartment are set to 3 μL, 6 °C (*see* **Note 8**) and 35 °C (*see* **Note 9**), respectively.
3. LC column information: a SeQuant® ZIC®-pHILIC guard column (20-mm length, 2.1-mm internal diameter, and 5-μm particle size, from EMD Millipore, Billerica, MA, USA) in series with a long SeQuant® ZIC®-pHILIC column (150-mm length, 4.6-mm internal diameter, and 5-μm particle size). Thus, the mobile phase flow path is as follows: guard column → analytical column → TOF-MS.
4. The autosampler is configured to use two 54 well trays (*see* **Note 10**).

5. Needle wash information: vials 1, 2, and 3 contain water, methanol, and acetonitrile, respectively.
6. MS system information: The HPLC system is coupled to an Agilent Technologies 6210 time-of-flight mass spectrometer (LC-TOF-MS) by a 1:3 post-column split (*see* **Notes 11** and **12**).
7. Electrospray ionization (ESI) sprayer.
8. LC-MS acquisition software information: Agilent MassHunter Workstation (version 6) software was used acquired the data.
9. LC-MS data analysis software information: Agilent MassHunter Qualitative Analysis (version 6) and Profinder (version 6) software (*see* **Note 13**).

3 Methods

3.1 Sample Preparation Procedure

1. Transfer 1.5 mL of *E. coli* cell culture (from a culture tube, shake flask, etc., at up to $OD_{600nm} = 5$) to a 1.7 mL microcentrifuge tube (*see* **Note 14**).
2. Separate biomass from broth by centrifuging the cell culture at 13,000 × *g* for 1 min at room temperature. For extracellular metabolite analysis, transfer 195 μL of the supernatant to a microcentrifuge tube containing 65 μL of ice cold methanol and mix by vortexing. Then store the resulting 260 μL extracellular solution at −20 °C until ready to proceed with **step 7** (*see* **Note 15**). Proceed directly to **step 3** if extracellular metabolites are not required.
3. Quickly decant the rest of the supernatant and proceed to **step 4** for intracellular metabolite extraction from the cell pellet.
4. Add 250 μL of ice cold methanol to the resulting pellet and mix thoroughly by vortexing. Store the samples at −20 °C until you are ready to proceed with the next step.
5. Add 250 μL of water to the MeOH lysate and mix thoroughly by vortexing.
6. Centrifuge the methanol–water lysate at 13,000 × *g* for 5 min at 4 °C.
7. Transfer 450 μL of the intracellular supernatant to a 3000 Da MW/CO centrifuge filter tube (*see* **Notes 16** and **17**). For extracellular metabolite analysis, transfer 260 μL of the extracellular solution from **step 2** to a 3000 Da MW/CO centrifuge filter tube.
8. Filter the intracellular samples by centrifuging at 13,000 × *g*, at −2 °C, for up to 90 min. Filter extracellular samples by centrifuging at 13,000 × *g*, at −2 °C, for up to 52 min. The filter

may contain trace amounts of glycerin. If glycerin affects downstream applications, it can be removed from the filter by flushing it with 0.5 mL of LC-MS grade water via centrifugation at room temperature. The filter should then be turned upside down in the tube and centrifuged to remove residual water. This should be conducted before sample transfer.

9. Transfer 400 μL of intracellular filtrate to a 2 mL microcentrifuge tube. For extracellular metabolite analysis, transfer 150 μL of the extracellular filtrate to a 0.6 mL microcentrifuge tube containing 300 μL acetonitrile and mix by vortexing. The extracellular metabolite sample is now prepared. If the extracellular samples are too dilute, then transfer them to a 2 mL microcentrifuge tube and proceed to **steps 10–12**).
10. Add 500 μL of water to the resulting filtrate.
11. Freeze with liquid nitrogen and lyophilize for 1–2 days.
12. Reconstitute the dried extracts in the reconstitution solution from a volume of 100–400 μL depending on the abundance of target metabolites and the sensitivity of the mass spectrometer used.

3.2 LC–MS Data Acquisition

1. Purge the LC system with the mobile phase.
2. Load the LC-MS method via the appropriate software (Tables 1, 2, and 3).

Table 1
Sample injection and column compartment parameters

Autosampler
1. Set autosampler tray temperature to 6 °C
2. Set the autosampler to the 54-well tray configuration (*see* **Note 18**)
3. Set blank, standard, sample, QC, and needle wash vial positions
Sample injection
1. Set sample injection volume to 3 μL
2. Set draw speed to 200 μL/min
3. Set eject speed to 200 μL/min
4. Set draw position to 0 for standards and up to 5 mm for samples (*see* **Note 19**)
Injector program
1. Needle into vial 1 (water), default offset (*see* **Note 20**)
2. Needle into vial 2 (methanol), default offset
3. Needle into vial 3 (water), default offset
4. Needle into vial 4 (acetonitrile), default offset (*see* **Note 21**)
5. Draw default amount from sample, with default speed and default offset
6. Inject sample
Column compartment
1. Set the column compartment to 35 °C (*see* **Note 22**)
2. Make sure that both sides of the compartment utilize the above temperature
3. Choose the correct column compartment configuration (*see* **Note 23**)

Table 2
Binary pump parameters

Time (min)	Mobile phase (%)	Flow rate (mL/min)	Maximum back pressure (bar)
0	100	0.45	200
6	100	0.45	200
6.5	100	0.605	200
12.5	100	0.605	200

Table 3
MS parameters

MS parameters	Values
Ionization mode	Negative
Mass range (*m/z*)	65–1100
Spectra/s	0.86
MS/spectrum	1162.8
Transients/spectrum	15,922
Dry gas temperature (°C)	300
Dry gas flow rate (L/min)	11
Nebulizer pressure (psi)	30
Fragmentor voltage (V)	100
Skimmer voltage (V)	50
OCT 1 RF Vpp (V)	170
MS detection window (ppm)	50
Minimum peak height (counts)	200
TFA fragment, CF_3, RM (*m/z*)	68.995758
TFA anion RM (*m/z*)	112.985587
HTFP + TFA anion adduct RM (*m/z*)	1033.98811

3. Make sure all the method parameters are correct and switch on the LC modules.
4. Condition the LC column by flushing with the HILIC mobile phase solution at flow rate of 100 μL/min. Once the LC system backpressure stabilizes, gradually increase the flow rate incrementally until the starting mobile phase flow rate of 450 μL/min is reached.

Table 4
Sequence worklist

Order of sequence
1. Run three to five consecutive blank samples (*see* **Note 25**)
2. Run a seven-point calibration curve from 1.5625 to 50 μM (*see* **Note 26**)
3. Run blank sample
4. Run QC sample (*see* **Note 27**)
5. Run blank sample
6. Run first set of samples (*see* **Note 28**)
7. Run blank sample
8. Run QC sample
9. Run blank sample
10. Run second set of samples
11. Run blank sample
12. Run QC sample
13. Run second calibration curve (*see* **Note 29**)
14. Run blank sample
15. Run column storage method (i.e., flow storage solution through the HILIC column at. 0.45 mL/min for 15 min)

5. Switch on the MS system and connect the calibrant delivery system to the ESI sprayer.
6. Perform an MS check tune (*see* **Note 24**).
7. Load the LC-MS sequence work list information (Table 4).
8. Run the LC-MS sequence work list.

3.3 LC-MS Data Analysis Procedure

1. Extract metabolite features with accurate $[M - H]^- \leq \pm\ 0.02$ *m/z* and retention time information via the data analysis software.
2. Obtain integration list with compound names, retention times, peak areas, ppm error, etc.
3. Identify compounds via accurate mass, retention time, and isotopic distribution information. If tandem mass spectrometry is available, MS/MS fragmentation spectra can also be used for identification.
4. Normalize measurements with internal standards (*see* **Note 30**) and calculate concentrations via external calibration curves.
5. Normalize data via OD_{600nm} measurements or cell counts.

6. Perform relevant statistical analysis (e.g., Student's *t*-test to determine whether the replicate data is significantly different). More elaborate statistical analyses may be required depending on the outcome of the interpretation step.

7. For data interpretation, compare metabolite profiles to growth and physiology information (e.g., Fig. 3). Pathway bottlenecks can be identified by observing the accumulation of intermediate metabolites (which may have occurred due to reduced activity of the proceeding enzyme). Toxic intermediates can be determined by correlating metabolite data with growth and physiological information. The method is also capable of monitoring dephosphorylated pathway intermediates (e.g., IP/DMAP and FP), glucose, adenylate energy charge, reducing equivalents, and central carbon metabolites of the glycolysis pathway and the tricarboxylic acid (TCA) cycle in the negative ion mode of MS (i.e., $[M-H]^-$). Incorporation of quantitative central carbon metabolite and cofactor data allows the researcher to assess the impact the engineered pathway has on metabolism and physiology. Furthermore, the high-resolution capability of TOF-MS allows untargeted metabolomic measurements to be conducted in parallel with targeted workflows and offers the possibility of discovery experiments.

4 Notes

1. Methanol is used to quench metabolic activity, lyse cells, and extract metabolites. Water is used for cell lysis and metabolite extraction.

2. Add a small volume (e.g., 1–2 μL) of the concentrated ISD stock solution to the sample (20–50 μL) to give a final concentration between 5 and 10 μM. The volume of the ISD should be kept small to minimize the dilution of the sample and to ensure that the ISD experiences the full effect of the sample matrix. The final concentration of the ISD can be reduced when a more sensitive mass spectrometer is used.

3. Amber polypropylene microcentrifuge tubes prevent the degradation of light-sensitive compounds by reducing the amount of light that enters the tube. The same is true for amber glass vials.

4. Choose filters with low binding properties. This can be tested by conducting metabolite recovery studies on prospective filters. That is, Recovery (%) = (Peak area of filtered standard/ Peak area of unfiltered standard) × 100.

5. The mobile phase is prepared in a volumetric flask and mixed by a sequential combination of ten flask inversions followed by sonication for ~5 min.

6. The volatile nature of the mobile phase requires that it is prepared freshly each day. Alternatively, a mobile phase composed of 10 mM ammonium acetate and 118.4 mM ammonium hydroxide in acetonitrile/water (54.9:45.1, v/v) can be used (instead of ammonium carbonate). For this mobile phase, the flow rate profile should be changed to the following: hold at 0.45 mL/min for 6 min, linearly increase the flow rate to 0.55 mL/min in 0.5 min and hold at 0.55 mL/min for 5.5 min. The total run time is 12 min. This mobile phase was found to be compatible with the Agilent Technologies 1290 Infinity II LC system (i.e., UHPLC). When using the ammonium acetate mobile phase, the sample medium can be composed of methanol–water (1:1, v/v). This can simplify the sample preparation procedure in Subheading 3.1 by avoiding the use of acetonitrile.
7. The wash solvents remove contaminants from the outside of the LC autosampler injector needle to prevent the carryover of highly abundant metabolites.
8. The autosampler tray temperature should be maintained ideally as cold as possible to prevent thermally labile metabolites from degrading significantly during the run sequence. 4 °C is normally the lowest operational temperature used on the Agilent Technologies 1200 Series Rapid Resolution HPLC system.
9. Mobile phases with high organic solvent contents can cause significant chromatographic peak broadening for HILIC measurements at elevated temperatures. Peak broadening can be reduced by lowering the temperature of the column compartment. However, a reduction in temperature can lead to increased mobile phase viscosity and, hence, back pressure. Therefore, it is imperative that the operator is aware of the maximum allowable back pressure limits for the chromatographic column and HPLC system and does not exceed them. The maximum back pressures for the ZIC®-pHILIC column and Agilent Technologies 1200 Series Rapid Resolution HPLC system are 200 and 600 bar, respectively.
10. 96 well plates can be used if the number of samples exceeds 108.
11. For a flow rate of 450 μL/min, the splitter ensures that 150 μL/min of the LC column effluent is being sent to the MS. This is to ensure optimal working performance of the LC-MS system for experiments with long acquisition times resulting from the analysis of many samples. Alternatively, the full flow can be sent to the MS system if sensitivity is an issue. This normally involves directly connecting the PEEK tubing that delivers the LC column effluent to a divert valve system situated at the front of the MS system. In this case, the

following approach is typically used: divert to waste (at the start of the run to remove unretained compounds) → MS system (where data is acquired) → divert to waste (after the final compound has eluted and during the column equilibration phase).

12. Contact between both instrument set-ups is established by using a LAN card to trigger the MS into operation after initiating a run cycle from the Agilent MassHunter Workstation software (Agilent Technologies, CA, USA).
13. Agilent MassHunter Qualitative Analysis is used to confirm the identities of chemical standards and determine their retention times. Agilent MassHunter Profinder is used to curate metabolite data.
14. *E. coli* is grown in the culture medium of choice (e.g., M9 minimal media, lysogeny broth (LB), terrific broth (TB), etc.).
15. Here are two alternative approaches for measuring extracellular metabolites: (a) If sensitivity is not an issue and the supernatant has been determined to be relatively clean with little-to-no requirement for filtration (e.g., from minimal medium), then transfer 150 μL of the supernatant to a microcentrifuge tube containing 300 μL of ice cold acetonitrile and 50 μL of methanol. This will produce a final solution composition of acetonitrile–methanol–water (6:1:3, v/v/v), which should be mixed by vortexing. Centrifuge this solution at 13,000 × *g* for 5 min to sediment/precipitate undissolved macromolecules or salt if possible. Then decant the supernatant and analyze for extracellular metabolites. (b) If filtration is required and no relevant extracellular enzymes are present, then centrifuge 200 μL of *E. coli* culture at 13,000 × *g*. Transfer 150 μL of the supernatant to a 3000 Da MW/CO centrifuge filter tube at 4 °C and centrifuge at 13,000 × *g* for up to 52 min. Transfer 90 μL of the filtrate to a tube containing 180 μL of acetonitrile and 30 μL of methanol. This will produce a final solution composition of acetonitrile–methanol–water (6:1:3, v/v/v), which should be mixed by vortexing. If precipitation is observed (resulting from culture medium salts, etc.), the extract can be recentrifuged or refiltered to remove precipitants.
16. The purpose of the filtration step is to remove protein and other macromolecules. This reduces the complexity of the sample and makes the analysis process easier.
17. The Amicon Ultra centrifuge filter has a chemical compatibility of $\leq$60% methanol and $\leq$20% acetonitrile.
18. If 96-well plates are to be used, make sure they are chemically compatible with the organic solvents in the sample medium.

19. If there are precipitants at the bottom of the LC vials, it may be important to raise the needle to the draw position (to a predetermined depth) to avoid injecting them.
20. The default needle offset is typically 0–3 mm from the bottom of the vial. Each manufacturer will have their own default value.
21. If speed is required and little-to-no carryover is observed, then wash the needle with a single solution of water–methanol–acetonitrile (5:2.5:2.5, v/v/v) instead. This solution can be mixed by vortexing and briefly (<20 s) degassed by sonication.
22. The temperature range of this column compartment is from 10 °C below ambient to 80 °C. In addition to this, the ZIC®-pHILIC column has a maximum upper temperature limit of 60 °C.
23. When a column switching valve is employed, make sure to choose the configuration that delivers the mobile phase and sample to the column (i.e., column position 1 or 2).
24. The check tune quickly checks the MS parameters and calibrates the MS mass axis.
25. This equilibrates the column with the mobile phase and provides stable retention times from the first run onward.
26. The calibration curve is a twofold serial dilution of 50 μM → 25 μM → 12.5 μM → 6.25 μM → 3.125 μM → 1.5625 μM → 0.7825 μM.
27. QC is a low positive test sample used to validate consistent laboratory analysis within predetermined limits of precision. QC samples are typically defined by each laboratory under a strict set of criteria.
28. Three to five biological replicates are normally adequate for metabolic engineering experiments.
29. This calibration curve will be averaged with the first. The averaged calibration curve will more accurately reflect the detector response from the beginning to the end of the sequence worklist.
30. Internal standards can be used to reduce error in cases where there is considerable variation in peaks areas. They are also used to identify metabolites via retention time indices.

Acknowledgments

The authors would like to acknowledge that this work was part of the DOE Joint BioEnergy Institute (http://www.jbei.org) supported by the US Department of Energy, Office of Science, Office of Biological and Environmental Research, through contract DE-AC02-05CH11231 between Lawrence Berkeley National Laboratory and the US Department of Energy.

References

1. Kirby J, Keasling JD (2009) Biosynthesis of plant isoprenoids: perspectives for microbial engineering. Annu Rev Plant Biol 60:335–355
2. Pitera DJ, Paddon CJ, Newman JD et al (2007) Balancing a heterologous mevalonate pathway for improved isoprenoid production in Escherichia coli. Metab Eng 9:193–207
3. Gilibili RR, Kandaswamy M, Sharma K et al (2011) Development and validation of a highly sensitive LC-MS/MS method for simultaneous quantitation of acetyl-CoA and malonyl-CoA in animal tissues. Biomed Chromatogr 25:1352–1359
4. Hayashi O, Satoh K (2014) Determination of acetyl-CoA and malonyl-CoA in germinating rice seeds using the LC-MS/MS technique. Biosci Biotechnol Biochem 70:2676–2681
5. Bartek T, Blombach B, Lang S et al (2011) Comparative 13C metabolic flux analysis of pyruvate dehydrogenase complex-deficient, L-valine-producing Corynebacterium glutamicum. Appl Environ Microbiol 77:6644–6652
6. Paczia N, Nilgen A, Lehmann T et al (2012) Extensive exometabolome analysis reveals extended overflow metabolism in various microorganisms. Microb Cell Factories 11:122
7. Kato H, Izumi Y, Hasunuma T et al (2012) Widely targeted metabolic profiling analysis of yeast central metabolites. J Biosci Bioeng 113:665–673
8. Fujii H, Koyama T, Ogura K (1982) Efficient enzymatic hydrolysis of polyprenyl pyrophosphates. Biochim Biophys Acta 712:716–718
9. Martin VJJ, Pitera DJ, Withers ST et al (2003) Engineering a mevalonate pathway in Escherichia coli for production of terpenoids. Nat Biotechnol 21:796–802
10. Henneman L, van CAG, Denis SW et al (2008) Detection of nonsterol isoprenoids by HPLC-MS/MS. Anal Biochem 383:18–24
11. Henneman L, van CAG, Kulik W et al (2011) Inhibition of the isoprenoid biosynthesis pathway; detection of intermediates by UPLC-MS/MS. Biochim Biophys Acta 1811:227–233
12. Zou R, Zhou K, Stephanopoulos G et al (2013) Combinatorial engineering of 1-deoxy-D-xylulose 5-phosphate pathway using cross-lapping in vitro assembly (CLIVA) method. PLoS One 8:e79557
13. Zhang D, Poulter CD (1993) Analaysis and purification of phosphorylated isoprenoids by reversed-phase HPLC. Anal Biochem 213:356–361
14. Li Z, Sharkey TD (2013) Metabolic profiling of the methylerythritol phosphate pathway reveals the source of post-illumination isoprene burst from leaves. Plant Cell Environ 36:429–437
15. Zhou K, Zou R, Stephanopoulos G et al (2012) Metabolite profiling identified methylerythritol cyclodiphosphate efflux as a limiting step in microbial isoprenoid production. PLoS One 7:e47513
16. Bongers M, Chrysanthopoulos PK, Behrendorff JBYH et al (2015) Systems analysis of methylerythritol-phosphate pathway flux in E. coli: insights into the role of oxidative stress and the validity of lycopene as an isoprenoid reporter metabolite. Microb Cell Factories 14:193
17. Nowroozi FF, Baidoo EEK, Ermakov S et al (2014) Metabolic pathway optimization using ribosome binding site variants and combinatorial gene assembly. Appl Microbiol Biotechnol 98:1567–1581

Chapter 12

Determining the Mode of Action of Antimalarial Drugs Using Time-Resolved LC-MS-Based Metabolite Profiling

Simon A. Cobbold and Malcolm J. McConville

Abstract

Methods for assessing the mode of action of new antimalarial compounds identified in high throughput phenotypic screens are needed to triage and facilitate lead compound development and to anticipate potential resistance mechanisms that might emerge. Here we describe a mass spectrometry-based approach for detecting metabolic changes in asexual erythrocytic stages of *Plasmodium falciparum* induced by antimalarial compounds. Time-resolved or concentration-resolved measurements are used to discriminate between putative targets of the compound and nonspecific and/or downstream secondary metabolic effects. These protocols can also be coupled with ^{13}C-stable-isotope tracing experiments under nonequilibrative (or nonstationary) conditions to measure metabolic dynamics following drug exposure. Time-resolved ^{13}C-labeling studies greatly increase confidence in target assignment and provide a more comprehensive understanding of the metabolic perturbations induced by small molecule inhibitors. The protocol provides details on the experimental design, *Plasmodium falciparum* culture, sample preparation, analytical approaches, and data analysis used in either targeted (pathway focused) or untargeted (all detected metabolites) analysis of drug-induced metabolic perturbations.

Key words Malaria, Plasmodium, Antimalarial, Metabolomics, Metabolism, LC-MS

1 Introduction

Malaria remains a significant global health problem and the emergence of resistance to existing frontline drugs continues to undermine efforts to control this devastating disease. The identification and development of new antimalarial drugs that target key developmental stages of *P. falciparum*, the major cause of human malaria, therefore remains a high priority. In a significant advance, collaborative efforts between pharmaceutical companies and academic or not-for-profit groups have led to the identification of thousands of new compounds and chemical scaffolds which show potent and selective antimalarial activity in high throughput untargeted phenotypic screens [1–4]. However, the further development and optimization of many of these compounds have been

Edward E.K. Baidoo (ed.), *Microbial Metabolomics: Methods and Protocols*, Methods in Molecular Biology, vol. 1859, https://doi.org/10.1007/978-1-4939-8757-3_12,

hampered by the lack of information on their mode of action [5]. Recent advances in whole genome sequencing of parasite strains that are resistant to individual compounds have been invaluable in identifying the targets of new antimalarials [6–9]. However, this approach is time-consuming and may not always distinguish between the real target of the compound and gene mutations that confer resistance by drug efflux, detoxification, or metabolic bypass. Alternatively, focussed antimalarial drug libraries have been screened for activity against individual protein targets in vitro or live cell assays [10–12]. While these targeted approaches can be used to rapidly screen many compounds they are, by definition, unsuitable for defining any off-target or characterizing drugs with pleiotropic affects.

In a third approach, described here, global metabolite profiling is used to identify direct and downstream impacts of antimalarials on parasite metabolism. A rationale for this approach is that the majority of antimalarial compounds, for which information on mode of action are available, target metabolic processes directly or indirectly. These include metabolic enzymes (e.g., dihydrofolate reductase) and metabolite transporters (e.g., hexose transporter) [13–15]. Techniques that monitor metabolism are therefore likely to detect suitable and novel drug targets in the malaria parasite and complement resistance screening and in vitro assays.

Metabolite profiling via mass spectrometry is increasingly being used to measure the metabolic perturbations induced by antimalarial compounds (both new and existing) and infer which enzyme or pathway is inhibited. For example, targeted metabolite profiling was used to define the mode of action of antimalarial compounds that target polyamine biosynthesis, the nonmevalonate pathway of isoprenoid biosynthesis, and the mitochondrial respiratory chain [16–18]. Targeted metabolite profiling approaches are particularly useful when the mode of action of an antimalarial is suspected, allowing for focussed and highly sensitive analysis of drug-induced changes. In contrast, untargeted metabolite analysis can be employed without any existing knowledge of the putative target and is capable of indicating both the mechanism of action and also whether multiple targets might be involved [19]. Targeted and untargeted metabolite profiling approaches were used to identify a number of compounds in the "Malaria Box" antimalarial library that inhibited pyrimidine biosynthesis without any prior information of targets [20, 21]. Inferred targets identified in this way were subsequently validated using genetic approaches [21].

While metabolite profiling approaches can identify putative antimalarial targets from large screens (i.e., hundreds of compounds), one limitation is the false-negative rate inherent to single dimension screening. Put simply, it is often difficult to determine which changes are specific to the drug's mode of action and which are nonspecific or caused by cell death? This issue is partly overcome here by using either time- or drug concentration-resolved experimental approaches which provide information on pleiotropic compounds and secondary effects, the speed of drug uptake, and the speed of action for any particular antimalarial [19]. The inclusion of time-resolved metabolite profiling was successfully used to identify two Malaria Box compounds that were unexpectedly found to inhibit both the New Permeability Pathways in *P. falciparum* infected erythrocytes and pyrimidine biosynthesis in intracellular parasites [22]. Time-resolved metabolite profiling requires more cell material, resources, and instrument time and is therefore not suited for screening hundreds of compounds. However, time-resolved ^{13}C-stable isotope profiling is suited to efficiently validate candidates that have emerged from large-scale screening approaches. This is exemplified by screening of the Malaria Box for inhibitors of the parasite lactate transporter and subsequent validation using a comparable metabolite profiling approach as described here [23].

Here we outline a protocol for antimalarial metabolite profiling that can be readily used to investigate the mode of action of focussed drug sets of ~10 compounds [19], and can be used to complement/validate other approaches, such as resistance screening and targeted in vitro assays. The inclusion of ^{13}C-stable isotope labeling provides additional information on metabolic dynamics and combined with an untargeted profiling approach is capable of capturing the perturbations induced by antimalarial compounds across the metabolic network.

2 Materials

Diligently follow all appropriate waste disposal regulations when disposing of waste materials.

2.1 *P. falciparum* Culture

1. Tissue culture hood.
2. Complete RPMI 1640 media (supplemented with 25 mM HEPES, 100 μM hypoxanthine, 0.5% w/v Albumax II, and 10 mg/L gentamicin sulfate).
3. Human group O, Rh+ erythrocytes.
4. 80% ethanol (v/v) spray.
5. 150 cm^2 vented culture flasks (Falcon™).
6. Incubator 37 °C.
7. Gas mixture (5% CO_2, 1% O_2, and 94% N_2).
8. Autopipette, sterile plastic pipettes (10, 25, and 50 mL).
9. 5% sorbitol (w/v).

2.2 Purification of P. falciparum-Infected Erythrocytes

1. Magnex Cell Separator (Colebrook Bioscience).
2. Four CS magnetic enrichment columns (Miltenyi Biotech).
3. Disposable 50 mL reagent reservoir (Corning).
4. RPMI 1640 complete media.
5. Plastic 2 mL transfer pipettes.
6. 50 mL plastic tube (Falcon™).

2.3 Drug Exposure Experiment

1. Hemocytometer (Fisher Scientific).
2. 25 cm^2 vented culture flask (Falcon™).
3. Gas mixture (5% CO_2, 1% O_2, and 94% N_2).
4. RPMI 1640 complete media.
5. 1 M stock of ^{13}C-U-Glucose (Sigma) in water.
6. 2 mL microcentrifuge tubes (Eppendorf).
7. Stock of antimalarial of interest (e.g., 1 mM atovaquone in DMSO).

2.4 Metabolite Extraction and Sample Preparation

1. Ice-cold 1× phosphate-buffered saline (PBS) (approximately 50 mL).
2. 80% Acetonitrile (99.9% hyper grade (Merck) and Milli-Q® water) + 1 μM ^{13}C^{15}N-aspartate (Cambridge Isotopes).
3. Ice bucket.
4. Cold room or refrigerated microcentrifuge.
5. LC-MS vials (screw cap, clear glass, Agilent), vial inserts (250 μL pulled point glass, Agilent), and screw caps (Agilent).
6. 15 mL centrifuge tube (Falcon™).

2.5 LC-MS Acquisition

1. SeQuant® ZIC®-pHILIC 5 μm polymeric column (150 × 4.6 mm; Merck) with guard column (ZIC®-pHILIC 20 × 2.1 mm; Merck).
2. Hyper-grade acetonitrile (Merck).
3. 20 mM ammonium carbonate (HPLC grade; Fluka) made up in Milli-Q® water.
4. LC-MS system: we use an Agilent 1260 binary pump coupled to an Agilent 6545 Q-TOF but any high mass-accuracy full-scan MS would be suitable.

3 Methods

3.1 P. falciparum In Vitro Cultivation

1. *P. falciparum* 3D7 parasites are continuously cultured in 150 cm^2 tissue culture flasks containing RPMI 1640 medium with O^+ human blood cells (5–10% parasitemia, 2% hematocrit)

under controlled atmospheric conditions (5% CO_2, 1% O_2, and 94% N_2) (*see* **Note 1**).

2. Cultures are synchronized with 5% sorbitol at ring stage using established methods [24].
3. Cultures are typically harvested when parasites reach trophozoite stage (24–36 h). The culture media is replaced with fresh complete RPMI 1640 and the cell suspension transferred to 50 mL centrifuge tubes and incubated at 37 °C in a water bath prior to isolation of infected erythrocytes by magnetic enrichment.

3.2 Purification of *P. falciparum*-Infected Erythrocytes

1. Four CS columns are washed with three volumes (in reference to the reservoir capacity of the column ~2 mL) of Milli-Q® water and three volumes of complete RPMI 1640 prior to use (*see* **Note 2**). The columns are mounted in a Magnex Cell Separator, and fitted with three-way stopcock and a 21 G needle to allow unbound medium to flow into 50 mL reagent reservoirs aligned below the columns (*see* **Note 3**).
2. *P. falciparum*-infected erythrocyte cultures are centrifuged (500 × *g*, 5 min, room temperature) and the supernatant aspirated. Fresh complete RPMI 1640 is added to make a final hematocrit of 6–8% (i.e., a 50 mL culture at 2% hematocrit is resuspended to 12.5–15 mL; *see* **Note 4**).
3. The suspension of infected and uninfected erythrocytes is pipetted onto the CS columns and the three-way stopcock opened until the liquid meniscus reaches the top of the resin. Cell suspensions are continually added until the entire cell suspension is loaded onto each CS column (a maximum of 15 mL 8% hematocrit was added to each column).
4. The CS columns are washed with complete RPMI 1640 (*see* **Note 5**) until the eluent runs clear, and the column washed with one final volume of RPMI 1640. This protocol routinely ensures >98% parasitemia in the bound erythrocytes.
5. The washed CS columns are sequentially removed from the magnetic apparatus and the retained *P. falciparum* trophozoite-infected erythrocytes eluted into fresh 50 mL centrifuge tubes with 20 mL of complete RPMI 1640. The resultant cell suspensions are combined into a single 50 mL centrifuge tube and incubated at 37 °C in a water bath for 0.5–1 h to minimize variability due to any perturbations.
6. The percent parasitemia and number of infected erythrocytes in the eluent is determined by Giemsa-staining smears and hemocytometer counting (typically a 1/10 dilution of the cell suspension is sufficient for accurate hemocytometer counts) during the incubation. The volume is subsequently adjusted to achieve a concentration of 1×10^8 *P. falciparum*-infected erythrocytes per sample replicate.

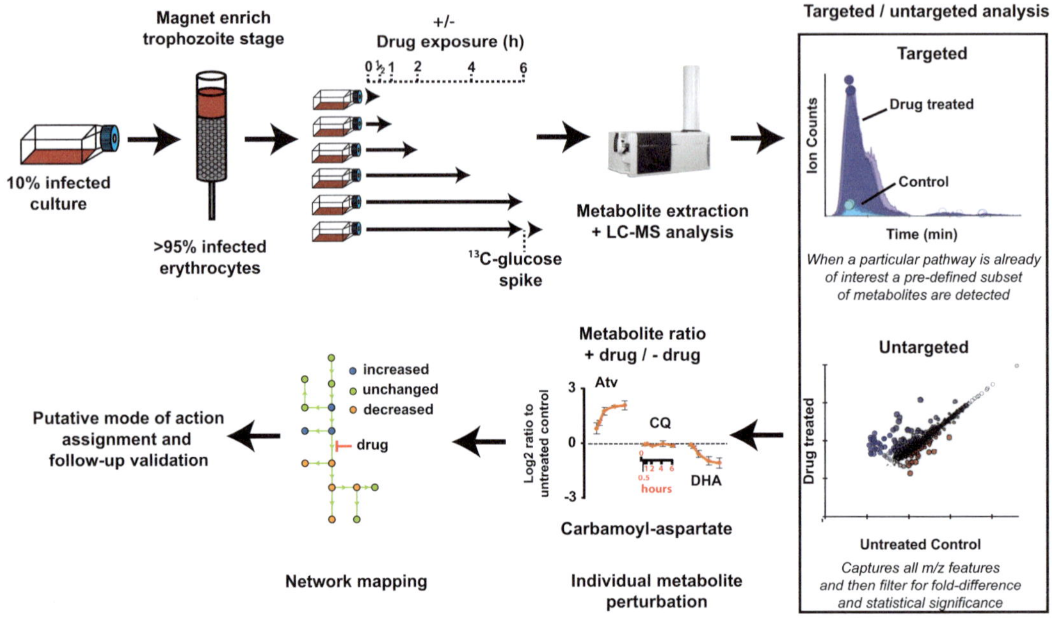

Fig. 1 Metabolite profiling workflow

3.3 Time-Resolved Antimalarial Exposure

1. Complete RPMI 1640 medium (50 mL) containing antimalarial compounds at a concentration of $10 \times IC_{50}$ is freshly prepared. A total of 48 mL is required for four time points (6 mL each, 10^8 infected erythrocytes, in duplicate: 4 time points × 2 technical replicates × 6 mL = 48 mL) (*see* **Note 6**).
2. Purified *P. falciparum*-infected erythrocytes are aliquoted into two fresh 50-mL centrifuge tubes (each corresponding to 8×10^8 cells) and cells pelleted by centrifugation (500 × *g*, 5 min).
3. The supernatant is aspirated and a parallel time-course series initiated with the addition of 48 mL of complete RPMI 1640 containing the antimalarial to one of the tubes and RPMI 1640 without antimalarial to the second tube (untreated control) (Fig. 1). The two cell suspensions (drug treated and untreated) are rapidly aliquoted into 25 cm^2 vented tissue culture flasks (12 mL each flask, corresponding to each time point to be extracted as technical duplicate).
4. The 25 cm^2 culture flask are placed in a culture cabinet and incubated at 37 °C under standard culturing conditions (5% CO_2, 1% O_2, and 94% N_2), and removed at predetermined time intervals (0.5, 1, 3, and 6 h) for extraction and metabolite analysis (*see* Subheading 3.5).
5. Once the time course has been initiated, the remaining cell suspension is centrifuged and resuspended in 12 mL complete RPMI 1640 to generate a time zero point. This suspension is

returned to the water bath for 15 min to allow equilibration before metabolite extraction (*see* **Note 7**).

The example given above is for a time-resolved antimalarial profiling experiment. Refer to **Note 8** for a description of a concentration-resolved antimalarial profiling experiment and the rationale for selecting either approach.

3.4 Nonstationary ^{13}C-U-Glucose Labeling

1. ^{13}C-glucose labeling studies provide additional information on changes in metabolic fluxes induced by different drugs. For these approaches, incubate purified infected erythrocytes with different antimalarials (or as untreated controls) as described in the previous section for between 3 and 6 h prior to commencing ^{13}C-glucose labeling (Fig. 1).
2. Spike parasite cultures with 1 M ^{13}C-U-glucose to give a final 1:1 ratio of ^{13}C-glucose/^{12}C-glucose in the medium (*see* **Note 9**). This protocol minimizes any perturbation of parasite cultures during drug treatment and subsequent labeling period.
3. Following addition of ^{13}C-glucose, return the culture flasks to the culture cabinet for a further 30 min before metabolite extraction.
4. Collect the media and analyze by LC-MS to empirically determine the ratio of ^{13}C-glucose:^{12}C-glucose in the medium (predicted to be 1:1) and the determined values used to adjust maximum labeling of individual metabolites (theoretically 50%) as described in Subheading 3.7.

3.5 Metabolite Extraction

1. The 25 cm^2 flasks containing untreated control and drug-treated *P. falciparum*-infected erythrocytes are removed at specified time points and cell suspension transferred to 15 mL centrifuge tubes and centrifuged (500 × *g*, 5 min).
2. An aliquot of the supernatant (5 μL) is transferred to a fresh microcentrifuge tube containing 200 μL of 80% acetonitrile (containing 1 μM ^{13}C^{15}N-asparate as the internal standard). The remaining supernatant is aspirated and the cell pellet metabolically quenched by addition of 2 mL of ice-cold PBS (*see* **Note 10**). The supernatant sample is further processed as described below for the cell extract samples.
3. The chilled cell suspension is evenly transferred to two microcentrifuge tubes (1 mL each, representing one technical replicate with 1 × 10^8 cells) and centrifuged for 15 s (14,000 × *g*, 4 °C) using either refrigerated centrifuge or a centrifuge in a cold room (*see* **Note 11**).
4. The supernatant is rapidly aspirated and 200 μL of 80% acetonitrile (containing 1 μM ^{13}C^{15}N-asparate as the internal standard) added to each sample and the cell pellet rapidly

resuspended and then vortex mixed or drawn across a microcentrifuge rack to facilitate extraction (*see* **Note 12**).

5. Cell extracts are immediately centrifuged (14,000 × *g*, 5 min) and the supernatant, containing polar metabolites, collected and transferred directly to 250 μL glass inserts in 1.5 mL LC-MS vials (*see* **Note 13**).
6. At each time point, smears are taken from the residual cell suspension left in the 15 mL centrifuge tube in order to monitor parasites morphology throughout the time course.
7. The remaining time points for both untreated and antimalarial treated are quenched and processed as described above. All samples are stored at 4 °C if LC-MS analysis can be performed within 18 h, or at −80 °C if longer storage is required.

3.6 LC Separation and MS Acquisition

1. Separate metabolites on a SeQuant® ZIC®-pHILIC column (5 μM, 150 × 4.6 mm, Millipore) using a 1200 series HPLC system (Agilent). Elute metabolites with solvents A (20 mM ammonium carbonate in water) and B (100% acetonitrile) at a flow rate of 0.3 mL/min using a binary gradient as described in Table 1.
2. MS detection is performed using an Agilent Q-TOF mass spectrometer 6545 operating in negative ESI mode, with the MS tuned to low mass range and optimized for fragile ions. Additional MS settings are described: the scan range is set to 85–1200 *m/z* between 2 and 28 min at 0.9 spectra/second, gas temperature set to 150 °C, drying gas 10 L/min, nebulizer 20 psig, sheath gas temperature 300 °C, sheath gas flow 10 L/min, capillary voltage 2500 V and the nozzle voltage 200 V. Additional LC parameters include autosampler temperature 4 °C, injection volume 5–15 μL, and data collected in centroid mode with MassHunter Workstation software (Agilent).
3. An internal reference ion solution is continually run throughout the chromatographic separation and the 119.03632 (purine) and 981.99509 (hexakis adduct) *m/z* ions used to maintain mass accuracy.
4. The order of sample injection is randomized by sample treatment and length of incubation, with pooled biological quality control (PBQC) samples generated by mixing 5 μL of every sample to be analyzed into a single pooled sample. PBQC were run throughout the sequence (approximately every ten samples) to monitor instrument performance while 80% acetonitrile blanks were run every ten injections to ensure minimal carryover between sample runs.
5. A collection of 150 pure metabolite standards grouped into five samples is then sequentially injected onto the LC-MS

Table 1
LC gradient

Time (min)	Flow rate (mL/min)	Solvent A (20 mM ammonium carbonate)	Solvent B (acetonitrile)
0	0.3	20	80
0.5	0.3	20	80
15.5	0.3	50	50
17.5	0.3	70	30
18.5	0.3	95	5
21	0.3	95	5
23	0.3	20	80
29.5	0.3	20	80

instrument and used to define retention time for each metabolite present and aid metabolite identification.

6. In many cases, it is necessary to confirm metabolite identification using MS/MS. This is done when a metabolite of interest has no corresponding pure standard or when ambiguous results are obtained (e.g., when an *m/z* feature corresponds to multiple possible metabolites and cannot be chromatographically resolved). Auto MS/MS can be performed with a preferred *m/z* target list (including the retention time observed during the initial LC-MS analysis). MS/MS spectra for each preferred *m/z* feature are typically generated with collision energies of 10, 20, and 40 V, with a maximum of two *m/z* features at any one time and a minimum ion intensity of 10,000.

3.7 Data Processing

1. LC-MS .d files are converted to .mzXML files using MS convert and uploaded into the MAVEN software package [25].
2. Samples are aligned using a maximum of ten iterations using a polynomial degree of 5, the minimum ion intensity threshold set to 30,000, each group must contain at least six good peaks and limit total number of groups in the alignment to 1000.
3. Feature (*m/z*) extraction is performed using the following settings: mass domain resolution = 10 ppm, time domain resolution = 10 scans, minimum number of good peaks = 6, minimum signal/baseline ratio = 5, minimum peak width = 5 scans, and the minimum peak intensity = 8000.
4. A targeted metabolite analysis is performed by searching a predefined metabolite library containing approximately 150 metabolites with specific retention times which correspond to the same set of metabolites run in Subheading 3.6. Here

metabolite identification is performed using exact mass (10 ppm mass resolution) and retention times (0.5 min window), compared to pure standards run during the same acquisition. The area top for each positively assigned metabolite is integrated and used to calculate the ratio of drug treated to untreated controls. Isotopologue data can also be extracted for the ^{13}C-U-glucose labeled samples and the complete dataset exported as a .csv file.

5. The log 2 ratios (+drug/−drug) across the time series are calculated and statistical testing performed. For time-resolved experiments it is preferable to use a one-way ANOVA with Tukey post-hoc testing, whereby second order polynomials are fitted to each time series (treated/untreated ratios) and tested for statistical significance at $\Upsilon_{max/min}$. Alternatively, a two-way ANOVA can be used for comparing multiple antimalarials across time, with multiple hypothesis testing controlled for using the Benjamini correction via Metaboanalyst 3.0 [26]. The second approach has the benefit of capturing any interaction between the two variables (drug treatment and time) and requires less data processing.
6. In some situations it is desirable to perform an untargeted analysis. Here, pairwise comparisons (untreated vs treated per time point) are performed in the MAVEN software package with a minimum fold-difference of 1.5, missing values set to 300, and significance determined using an adjusted P value < 0.01 using the Benjamini post-hoc test for multiple hypothesis testing. Statistically significant features are then visually inspected and coeluting adducts, in-source fragments, and isotopologues are removed. The mass of each parental ion is then searched using the METLIN database using M−H and M+Cl adducts and metabolite assignment confirmed with pure standards where possible or MS/MS performed as described in Subheading 3.6. MS/MS spectra are analyzed with the Agilent MassHunter Qualitative software package and MS/MS spectra searched against the METLIN database. Spectra are manually verified and identifications confirmed.
7. Isotopologue data is corrected for the natural abundance of carbon using established procedures [27] with the following formulas:

$$M^{0}_{\text{real}} = \frac{M^{0}_{\text{measured}}}{0.989^{N}}$$

$$M^{+1}_{\text{real}} = \frac{M^{+1}_{\text{measured}}}{0.989^{N-1}} - 0.011\,N \times M^{0}_{\text{real}}$$

$$M^{+2}_{\text{real}} = \frac{M^{+2}_{\text{measured}}}{0.989^{N-2}} - 0.011^2 \times \frac{N}{2(N-1)} \times M^0_{\text{real}} - 0.011 \times (N-1) \times M^{+1}_{\text{real}}$$

where M^0 is the amount of the monoisotopic compound, M^{+1} is the amount of the compound with one ^{13}C atom, M^{+2} is the amount of the compound with two ^{13}C atoms. N is the number of carbon atoms in the molecule, "real" refers to values corrected natural isotope abundance and "measured" refers to raw values determined by LC-MS.

The maximal percent ^{13}C-labeling is determined by using the empirically determined ratio of ^{13}C-glucose/^{12}C-glucose in each supernatant sample. Percent labeling for each metabolite of interest is corrected for the maximal percent labeling observed. Flux through particular enzymatic reactions can be indirectly inferred by analyzing the percent labeling and total pool size data for each metabolite of interest.

8. Network mapping can be performed via BioCyc (www.biocyc.org) and pathway enrichment can be performed either via mummichog directly or via XCMSonline [28, 29]. The putative target(s) of an antimalarial can be inferred from the perturbation to the metabolic network. Generally, perturbations occur within three enzymatic steps of inhibition [30], with metabolite increases generally occurring upstream of inhibition and metabolite decreases occurring downstream. However, this assumption is commonly confounded by other variables, such as allosteric regulation, pathway convergence, and metabolite compartmentalization and should be used with caution. The inclusion of nonstationary stable isotope labeling therefore provides more information on how metabolism is altered under drug treatment and provides a greater insight into the underlying mechanism causing the observed metabolic perturbation.

4 Notes

1. Typically 0.5–1.0 × 10^9 purified trophozoite-infected erythrocytes can be obtained from a 50 mL culture (8–10% parasitemia, 2% hematocrit) when estimating the amount of culture required for a particular experiment. In this protocol we use a time-resolved pairwise comparison (drug treated vs untreated, five time points, performed in duplicate) which equates to twenty individual samples. 1 × 10^8 purified infected erythrocytes is required per sample, equating to 100–150 mL of culture (8–10% parasitemia, 2% hematocrit) for each drug combination.
2. Complete RPMI 1640 contains sodium bicarbonate which causes the pH of the medium to rise during magnetic purification of infected erythrocytes. While the increase in pH during a

standard enrichment (~15 min) is minimal, it is recommended that sodium bicarbonate free RPMI be used if parasites are left in the medium for any period of time.

3. In order to handle the volumes of cell culture medium used in these experiments, it is recommended to use the Colebrook Bioscience magnetic enrichment apparatus. This apparatus has four times the capacity of the VarioMACs enrichment device and allows for faster enrichment and parallel enrichment of different parasite lines. Note that the design of the Colebrook Bioscience apparatus is open access and can be easily downloaded at https://3dwarehouse.sketchup.com/collection.html?id=268370ca88a21c547e1b0af8b0e461e8, 3D printed and assembled on site. It is feasible to use the VarioMACs enrichment device, but this restricts the size of experiments achievable.

4. We have observed that recovery of infected erythrocytes is reduced when hematocrits exceed 8%, due to the increased viscosity of the suspension and/or saturation of the enrichment column.

5. It is recommended to "condition" the silica bed of the CS column by allowing the initial washes with RPMI 1640 media to fully enter the bed of the column (but not to run dry), and to then repeat by adding small volume of medium (0.25 mL), until the media on the silica bed is translucent. The reservoir is then filled with RPMI 1640 to begin washing the columns. This approach speeds up the washing process and improves the final purity of the enriched *P. falciparum*-infected erythrocytes.

6. Purified infected erythrocytes need to be diluted in large volumes (approximating 0.5% hematocrit) to ensure the parasites do not become nutrient starved/limited and that the pH is maintained around 7.2–7.4. If trophozoite stage parasites are incubated at a hematocrit higher than 0.5% they will become metabolically stressed and the experiment will be compromised. A general rule of thumb is 1×10^8 trophozoite stage cells need to be incubated in at least 1 mL of RPMI 1640 per hour of incubation.

7. In this example a single parasite line is tested, it is therefore appropriate to collect a single time zero sample (in technical duplicates) to represent the initial metabolite profile of both untreated and treated conditions.

8. Time-resolve vs concentration-resolved.

   Here we present the method for performing a time-resolved analysis of antimalarial induced metabolite perturbations. An alternative approach would be to perform a concentration-resolved analysis, whereby a range of antimalarial drug concentrations are tested over a fixed length of incubation (e.g., from $0.5 \times IC_{50}$ to $20 \times IC_{50}$ for 3 h).

Both approaches have advantages in helping to resolve direct antimalarial induced metabolic perturbations over non-specific and/or secondary effects. Information about the speed of action of a drug, if available, can help establish which approach would be preferable. For example, antimalarials like chloroquine take some time to reach the site of action and need to accumulate before inducing metabolic changes. For drugs with this mode of action it is preferable to use a concentration-resolved experimental approach.

9. ^{13}C-glucose is commonly used due to its rapid incorporation into many intermediates in central carbon metabolism. However, other $^{13}C/^{15}N$ labeled precursors (i.e. glutamine or bicarbonate) can be used if a drug is suspected to target specific pathways or pathways not connected to glucose metabolism. For example, labeling studies with ^{13}C-bicarbonate can be used to measure pyrimidine biosynthesis and the activity of inhibitors that affect this pathway (e.g., such as atovaquone inhibiting dihydroorotate dehydrogenase indirectly).

   We mention the use of nonstationary stable isotope labeling. This means selecting an incubation period at which most biosynthetic intermediates have not yet reached maximal percent labeling and equilibrated. From previous work, 30 min provides sufficient labeling for detection via LC-MS, while still being within the dynamic/nonstationary phase of ^{13}C-incorporation for most labeled metabolites. The use of labeled (^{13}C) and unlabeled (^{12}C) precursors is advantageous as it leads to the generation of complex isotopic incorporation of metabolites, which provides useful information of the operation of specific metabolic pathways in central carbon metabolism.

10. The timing of this step needs to be rapid and consistent across all samples. While the metabolic quenching and extraction steps are not technically demanding, emphasis is on the reproducibility of the time taken to process each sample and thereby minimize any variation in metabolism.

11. The number of samples that can be processed simultaneously is 2–4. This is to maintain reproducibility of the time taken to quench and extract each sample across the experiment. As such it is recommended that if more than two conditions are being tested in a single experiment (i.e., more than one antimalarial tested) then different time points need to be staggered by at least 5 min so that four samples at a maximum are processed at any given time.

12. It is necessary to ensure that all of the supernatant is removed before metabolite extraction is performed to ensure a consistent final volume is maintained.
13. It has been observed that when cell pellets are left for longer than 10 min in the extraction solvent, hemoglobin and other particulate matter are also solubilized, which can lead to the presence of precipitates in subsequent extracts (particularly after freeze–thaw cycles), which may necessitate additional centrifugation/filtration cleanup steps prior to LC-MS analysis.

Acknowledgments

This work was supported by grants from the Australian National Health and Medical Research Council (NHMRC). M.J.M. is a NHMRC Principal Research Fellow, and S.A.C. is a University of Melbourne Early Career Research Fellow.

References

1. Gamo FJ et al (2010) Thousands of chemical starting points for antimalarial lead identification. Nature 465(7296):305–310
2. Guiguemde WA et al (2010) Chemical genetics of Plasmodium falciparum. Nature 465 (7296):311–315
3. Meister S et al (2011) Imaging of Plasmodium liver stages to drive next-generation antimalarial drug discovery. Science 334 (6061):1372–1377
4. Spangenberg T et al (2013) The open access malaria box: a drug discovery catalyst for neglected diseases. PLoS One 8(6):e62906
5. Van Voorhis WC et al (2016) Open source drug discovery with the malaria box compound collection for neglected diseases and beyond. PLoS Pathog 12(7):e1005763
6. Herman JD et al (2014) A genomic and evolutionary approach reveals non-genetic drug resistance in malaria. Genome Biol 15(11):511
7. McNamara CW et al (2013) Targeting Plasmodium PI(4)K to eliminate malaria. Nature 504 (7479):248–253
8. Rottmann M et al (2010) Spiroindolones, a potent compound class for the treatment of malaria. Science 329(5996):1175–1180
9. Spillman NJ et al (2013) Na(+) regulation in the malaria parasite Plasmodium falciparum involves the cation ATPase PfATP4 and is a target of the spiroindolone antimalarials. Cell Host Microbe 13(2):227–237
10. Lehane AM et al (2014) Diverse chemotypes disrupt ion homeostasis in the malaria parasite. Mol Microbiol 94(2):327–339
11. Pretzel J et al (2016) Characterization and redox regulation of Plasmodium falciparum methionine adenosyltransferase. J Biochem 160(6):355–367
12. Wu W et al (2015) A chemical rescue screen identifies a Plasmodium falciparum apicoplast inhibitor targeting MEP isoprenoid precursor biosynthesis. Antimicrob Agents Chemother 59(1):356–364
13. Ferone R, Burchall JJ, Hitchings GH (1969) Plasmodium berghei dihydrofolate reductase. Isolation, properties, and inhibition by antifolates. Mol Pharmacol 5(1):49–59
14. Foote SJ, Galatis D, Cowman AF (1990) Amino acids in the dihydrofolate reductase-thymidylate synthase gene of Plasmodium falciparum involved in cycloguanil resistance differ from those involved in pyrimethamine resistance. Proc Natl Acad Sci U S A 87 (8):3014–3017
15. Joet T et al (2003) Validation of the hexose transporter of Plasmodium falciparum as a novel drug target. Proc Natl Acad Sci U S A 100(13):7476–7479
16. Biagini GA et al (2012) Generation of quinolone antimalarials targeting the Plasmodium falciparum mitochondrial respiratory chain for

the treatment and prophylaxis of malaria. Proc Natl Acad Sci U S A 109(21):8298–8303

17. van Brummelen AC et al (2009) Co-inhibition of Plasmodium falciparum S-adenosylmethionine decarboxylase/ornithine decarboxylase reveals perturbation-specific compensatory mechanisms by transcriptome, proteome, and metabolome analyses. J Biol Chem 284(7):4635–4646
18. Zhang B et al (2011) A second target of the antimalarial and antibacterial agent fosmidomycin revealed by cellular metabolic profiling. Biochemistry 50(17):3570–3577
19. Cobbold SA et al (2016) Metabolic dysregulation induced in Plasmodium falciparum by dihydroartemisinin and other front-line antimalarial drugs. J Infect Dis 213(2):276–286
20. Allman EL et al (2016) Metabolomic profiling of the malaria box reveals antimalarial target pathways. Antimicrob Agents Chemother 60 (11):6635–6649
21. Creek DJ et al (2016) Metabolomics-based screening of the malaria box reveals both novel and established mechanisms of action. Antimicrob Agents Chemother 60 (11):6650–6663
22. Dickerman BK et al (2016) Identification of inhibitors that dually target the new permeability pathway and dihydroorotate dehydrogenase in the blood stage of *Plasmodium falciparum*. Sci Rep 6:37502
23. Hapuarachchi SV et al (2017) The malaria parasite's lactate transporter PfFNT is the target of antiplasmodial compounds identified in whole cell phenotypic screens. PLoS Pathog 13 (2):e1006180
24. Lambros C, Vanderberg JP (1979) Synchronization of Plasmodium falciparum erythrocytic stages in culture. J Parasitol 65(3):418–420
25. Clasquin MF, Melamud E, Rabinowitz JD (2012) LC-MS data processing with MAVEN: a metabolomic analysis and visualization engine. Curr Protoc Bioinformatics 14:11
26. Xia J, Wishart DS (2016) Using MetaboAnalyst 3.0 for comprehensive metabolomics data analysis. Curr Protoc Bioinformatics 55:14-10-1–14-10-91
27. Yuan J, Bennett BD, Rabinowitz JD (2008) Kinetic flux profiling for quantitation of cellular metabolic fluxes. Nat Protoc 3(8):1328–1340
28. Gowda H et al (2014) Interactive XCMS online: simplifying advanced metabolomic data processing and subsequent statistical analyses. Anal Chem 86(14):6931–6939
29. Li S et al (2013) Predicting network activity from high throughput metabolomics. PLoS Comput Biol 9(7):e1003123
30. Fuhrer T et al (2017) Genomewide landscape of gene-metabolome associations in Escherichia coli. Mol Syst Biol 13(1):907

Chapter 13

Use of Liquid Chromatography–Mass Spectrometry-Based Metabolomics to Identify Biomarkers of Tuberculosis

Juntuo Zhou and Yuxin Yin

Abstract

Liquid chromatography–mass spectrometry (LC-MS) based metabolomics has proven to be a powerful analytical tool for biomarker screening. Here we describe two workflows which employ untargeted metabolomics to study serum biomarkers in tuberculosis patients. Expression profiles for samples of hydrophilic metabolites and hydrophobic metabolites (lipids) may be obtained by this method.

Key words LC-MS, Metabolomics, Serum, Tuberculosis, Hydrophilic metabolite, Lipid

1 Introduction

Metabolomics aims to comprehensively study all metabolites in a given sample [1]. This technique has not achieved coverage of all metabolites in all samples, as low abundance metabolites are harder to detect. However, metabolomics has proven to be a powerful tool for the study of metabolism due to the extensive coverage of analyzed metabolites (hundreds to thousands of metabolites in a given sample can be analyzed) in a single assay with metabolomics techniques [2]. The information obtained from metabolomic measurements can provide data for both biological study [3, 4] and biomarker screening for clinical use [5, 6].

Nuclear magnetic resonance (NMR) [7], gas chromatography or liquid chromatography coupled with mass spectrometry (GC-MS/LC-MS) [8, 9] are the main platforms currently used for metabolomics analyses. However, all of these platforms have their advantages and shortcomings [10]. Metabolomics has been applied for the study of tuberculosis [11–13], and in a recent study, Cheng et al. used 1H-nuclear magnetic resonance spectroscopy based metabolomic analysis to study tuberculous, malignant, and transudative pleural effusion [14]. Besides, Lijun et al. employed LC-MS based metabolomic analysis to study candidate biomarkers

Edward E.K. Baidoo (ed.), *Microbial Metabolomics: Methods and Protocols*, Methods in Molecular Biology, vol. 1859, https://doi.org/10.1007/978-1-4939-8757-3_13,

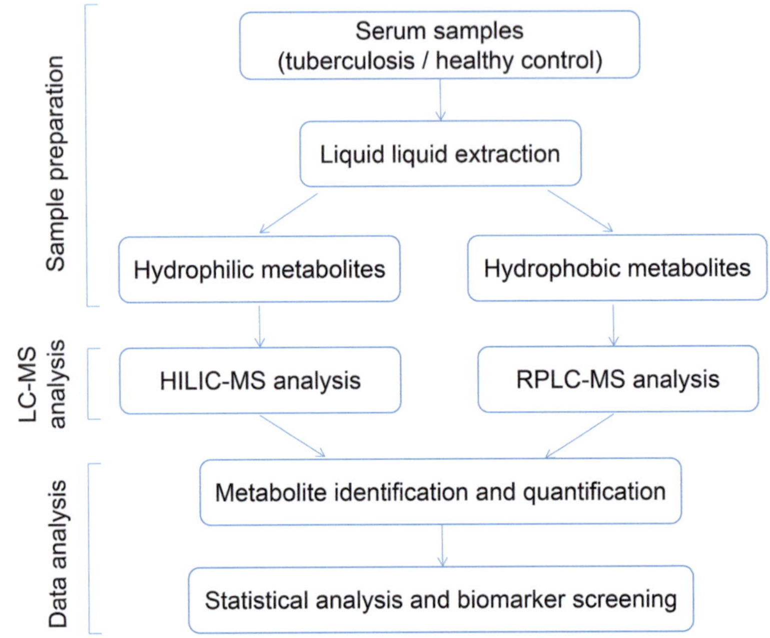

Fig. 1 A schematic flowchart for the LC-MS based untargeted metabolomics

in serum of TB patients [15]. LC-MS based methodology which features high sensitivity and wide metabolite coverage has been proven to be a powerful tool for metabolomics studies. The entire analytical procedure includes sample pretreatment (metabolite extraction), LC-MS data acquisition (hydrophilic interaction liquid chromatography coupled to mass spectrometry (HILIC-MS)) or reversed phase liquid chromatography coupled to mass spectrometry (RPLC-MS)), and data analysis (metabolite identification and quantification). Bioinformatics analysis can also be performed to extract more biological information, when necessary.

Here we present protocols for two separate LC-MS workflows which employ HILIC separation for hydrophilic metabolite analysis and RPLC separation for hydrophobic metabolite analysis. These workflows can be used for metabolomic measurements of serum samples from tuberculosis patients, and the results obtained can be used for clinical biomarker screening or pathologic studies. A schematic flowchart for the LC-MS based untargeted metabolomics is shown in Fig. 1.

2 Materials

2.1 Reagent

All mobile phases and solutions are prepared with HPLC or higher grade solvents. Acetonitrile, methanol, chloroform, and isopropanol may be purchased from Fisher Scientific, and ammonium acetate may be purchased from Sigma-Aldrich. Ultrapure water may be supplied by Milli-Q® water. Diligently follow all appropriate waste disposal regulations when disposing of waste materials.

2.2 Serum Samples

Serum samples of TB patients and healthy controls are collected using a protocol approved by the Ethics Committee at Peking University Health Science Center, and informed consent is obtained from each patient. People in all the groups are well matched in age and gender.

2.3 Solutions for Serum Sample Pretreatment

1. Liquid–liquid extraction solution: chloroform–methanol (2:1, v/v).
2. RPLC residue dissolving solution: chloroform–methanol (1:1, v/v).
3. RPLC sample diluting solution: isopropanol–acetonitrile–H_2O (2:1:1, v/v/v).

2.4 HILIC Separation

1. Liquid chromatograph: U3000 UHPLC (Thermo Fisher Scientific).
2. Mobile phase A: 5 mM ammonium acetate in water with 5% acetonitrile.
3. Mobile phase B: acetonitrile.
4. Column: Xbridge amide column (100 × 4.6 mm i.d., 3.5 μm; Waters) (*see* **Note 1**).

2.5 RPLC Separation

1. Liquid chromatograph: U3000 UHPLC (Thermo Fisher Scientific).
2. Mobile phase A: acetonitrile–water (60/40, v/v), containing 10 mM ammonium acetate and 0.1% formic acid.
3. Mobile phase B: isopropanol–acetonitrile (90/10, v/v), containing 10 mM ammonium acetate and 0.1% formic acid.
4. Column: Acquity CSH C18 column (100 × 4.6 mm i.d., 1.7 μm, Waters).

2.6 Mass Spectrometer

Q-Exactive (hybrid quadrupole-Orbitrap mass spectrometer) coupled with HESI ion source (Thermo Fisher Scientific).

2.7 Other Equipment

Microfuge 22R centrifuge (Beckman Coulter), Vortex-Genie 2 vortex mixer (Scientific Industries) and Micropipette (Eppendorf).

3 Methods

3.1 Preparation of Serum Sample

1. Locate the blood sample in centrifuge tube at room temperature for 1–2 h (*see* **Note 2**).
2. Centrifuge the sample at 3000 rpm (or 1000 × *g*) for 20 min at 4 °C.

3. Transfer the supernatant (serum) to new tubes and store at −80 °C.

3.2 Serum Sample Pretreatment

1. Thaw the serum sample in ice, and vortex for 5 s.
2. Equally combine each sample including both disease and healthy control groups to generate a quality control (QC) sample (*see* **Note 3**).
3. Add fourfold (400 μL) of the liquid–liquid extraction solution to each serum sample (100 μL) and QC sample (100 μL).
4. Vortex the samples for 30 s and then leave to stand for 5 min at room temperature.
5. Centrifuge the mixture at 13,000 rpm (or 14,000 × *g*) for 15 min.
6. Collect the upper aqueous phase (hydrophilic metabolites, used for HILIC-MS) and lower organic phase (hydrophobic metabolites, used for RPLC-MS) in new tubes separately and evaporate at room temperature under vacuum (*see* **Note 4**).
7. For HILIC-MS analysis, dissolve the residue in 100 μL water, and vortex for 30 s. Centrifuge the sample at 12,000 rpm (13,000 × *g*) for 15 min, and transfer the supernatant into vials for LC-MS analysis.
8. For RPLC-MS analysis, dissolve the residue in 100 μL RPLC residue dissolving solution, and vortex for 30 s. Dilute the solution by adding 300 μL of the RPLC sample diluting solution and vortex for 30 s. Centrifuge the sample at 12,000 rpm (13,000 × *g*) for 15 min, and transfer the supernatant into vials for LC-MS analysis (*see* **Note 5**).

3.3 HILIC-MS Analysis

1. Utilize the Ultimate 3000 UHPLC system for HILIC separation. Set the column temperature to 30 °C.
2. Set the linear gradient for separation as follows (also *see* Table 1): 0 min, 90% B; 3 min, 90% B; 15 min, 35% B, 18 min, 35% B; 19 min, 90% B and 24 min, 90% B. The flow rate is 0.5 mL/min.
3. Set the MS to work in the data dependent acquisition (DDA) mode. Each acquisition cycle consists of one survey scan (MS1 scan) at 35,000 resolution from 50 to 750 *m/z*, followed by ten MS/MS scans in HCD mode at 17,500 resolution. Set up MS/MS parameters as follows:
   (a) Set the automatic gain control target (AGC, *see* **Note 6**) to 5e6 (maximum injection time 100 ms) and 5e5 (maximum injection time 64 ms) for the MS1 and MS/MS scan. Set the parameter of fixed first mass (*see* **Note 7**) to 50 *m/z*.
   (b) Set the dynamic exclusion (*see* **Note 8**) to 8 s.

Table 1
The gradient used for HILIC and RPLC separation

Time (min)	Flow rate (mL/min)	Mobile phase A (vol. %)	Mobile phase B (vol. %)
HILIC			
0	0.5	10	90
3	0.5	10	90
15	0.5	65	35
18	0.5	65	35
19	0.5	10	90
24	0.5	10	90
RPLC			
0	0.6	60	40
2	0.6	57	43
2.1	0.6	50	50
12	0.6	40	60
12.1	0.6	25	75
18	0.6	1	99
19	0.6	1	99
20	0.6	60	40
24	0.6	60	40

(c) Set the stepped normalized collision energy (step-NCE) to 10, 20 and 30 (*see* **Note 9**).

(d) Set the spectra recording type to profile mode.

(e) A typical base peak chromatograph for positive and negative ion mode is shown in Fig. 2.

4. Set the HESI ion source parameters as follows: spray voltage, 3.3 kV for positive ion mode and 3.0 kV for negative ion mode; ion source sheath gas, 40; aux gas, 10; capillary temperature, 320 °C; probe heater temperature, 300 °C; S-lens RF level, 55.
5. Analyze the samples in random order and analyze QC samples repeatedly in the acquisition batch. A typical acquisition sequence is shown in Fig. 3 for reference (*see* **Note 10**).
6. Acquire the sample batch in positive ionization mode and negative ionization mode separately (*see* **Note 11**).

3.4 RPLC-MS Analysis

1. Utilize the Ultimate 3000 UHPLC system for RPLC separation. Set the column temperature to 50 °C.

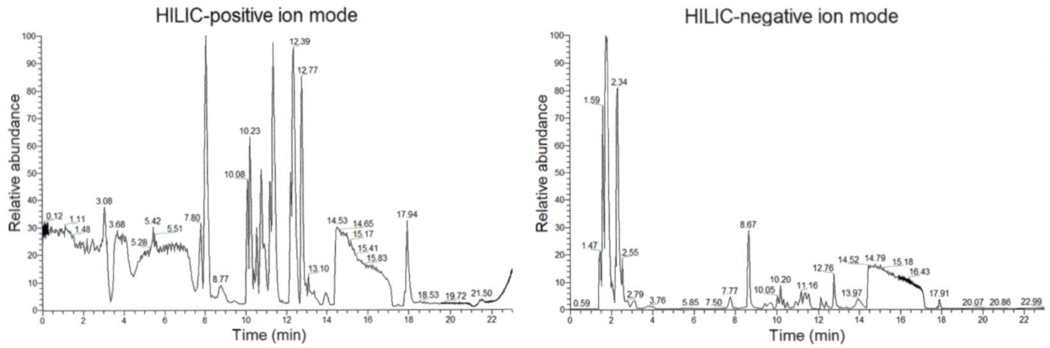

Fig. 2 Typical example of base peak chromatography of untargeted metabolomics analysis of serum sample in HILIC mode. The positive ion mode (left) and negative ion mode (right) are performed separately

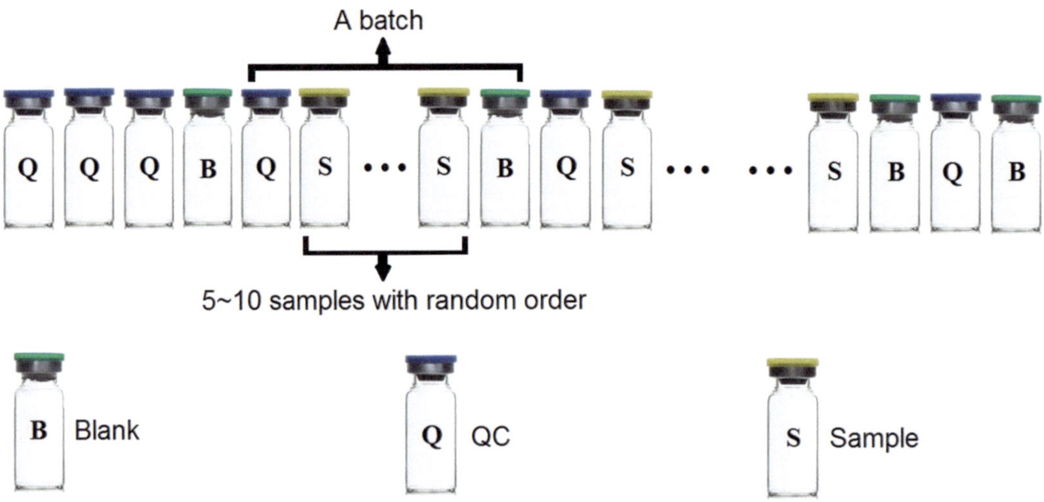

Fig. 3 The general LC-MS acquisition sequence for large numbers of samples. The colors of the bottles represent different type of samples (green, blank; blue, QC; yellow, sample)

2. Set the linear gradient for separation as follows (also *see* Table 1): 0 min, 40% B; 2 min, 43% B; 2.1 min, 50% B; 12 min, 60% B, 12.1 min, 75% B; 18 min, 99% B; 19 min, 99% B; 20 min, 40% B and 24 min, 40% B. The flow rate is 0.6 mL/min.

3. Set the Q Exactive MS to work in data dependent acquisition (DDA) mode. Each acquisition cycle consists of one survey scan (MS1 scan) at 35,000 resolution from 200 to 1200 *m/z*, followed by ten MS/MS scans in HCD mode at 17,500 resolution. Set up MS/MS parameters as follows:

   (a) Set the automatic gain control target (AGC, *see* **Note 6**) to 5e6 (maximum injection time 100 ms) and 5e5 (maximum injection time 64 ms) for the MS1 and MS/MS

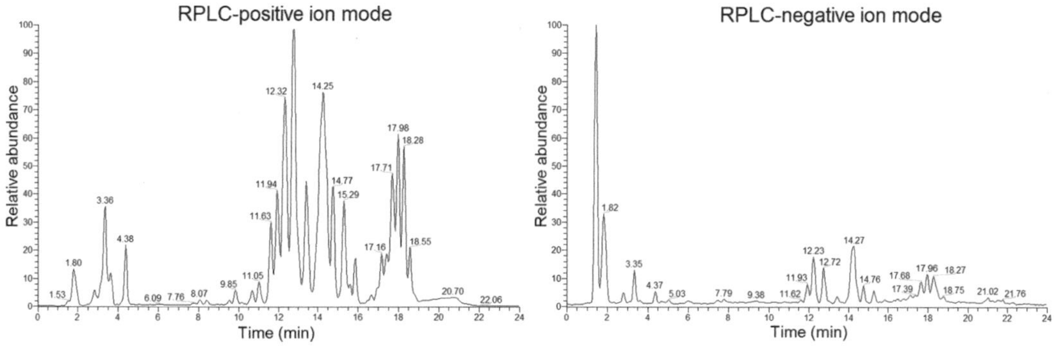

Fig. 4 Typical example of base peak chromatography of untargeted metabolomics analysis of serum sample in RPLC mode. The positive ion mode (left) and negative ion mode (right) are performed separately

scan. Set the parameter of fixed first mass (*see* **Note 7**) to 50 *m/z*.

(b) Set the dynamic exclusion (*see* **Note 8**) to 8 s.

(c) Set the stepped normalized collision energy (step-NCE) to 10, 20, and 30 (*see* **Note 9**).

(d) Set the spectra recording type to profile mode.

(e) A typical base peak chromatograph for positive and negative ion mode is shown in Fig. 4.

4. Set the HESI ion source parameters as follows: spray voltage, 3.3 kV for positive ion mode and 3.0 kV for negative ion mode; ion source sheath gas, 40; aux gas, 10; capillary temperature, 320 °C; probe heater temperature, 300 °C; S-lens RF level, 55.
5. Analyze the samples in random order and analyze QC samples repeatedly in the acquisition batch. A typical acquisition sequence is shown in Fig. 3 for reference (*see* **Note 10**).
6. Acquire the sample batch in positive ionization mode and negative ionization mode separately (*see* **Note 11**).

3.5 Data Analysis

1. Peak extraction. Process the raw MS data using SIEVE software (*see* **Note 12**), and set the parameters as follows: frame time width, 0.5 min; minimum scans across a peak, 9; signal-to-noise ratio, 10; *m/z* step width, 5; peak intensity threshold, 1,000,000; maximum number of frames, 5000; alignment bypass, false; starting retention time, 0 min; and end retention time, 23 min.
2. Statistical Analysis. Calculate the *p*-value and fold change for each metabolite. A metabolite with a *p*-value of <0.05 and fold change of >2 may be considered significantly changed between groups. Determine the *P*-values using the Student's t-test (comparison between two groups) or ANOVA (comparison of three groups or more) (*see* **Note 13**).

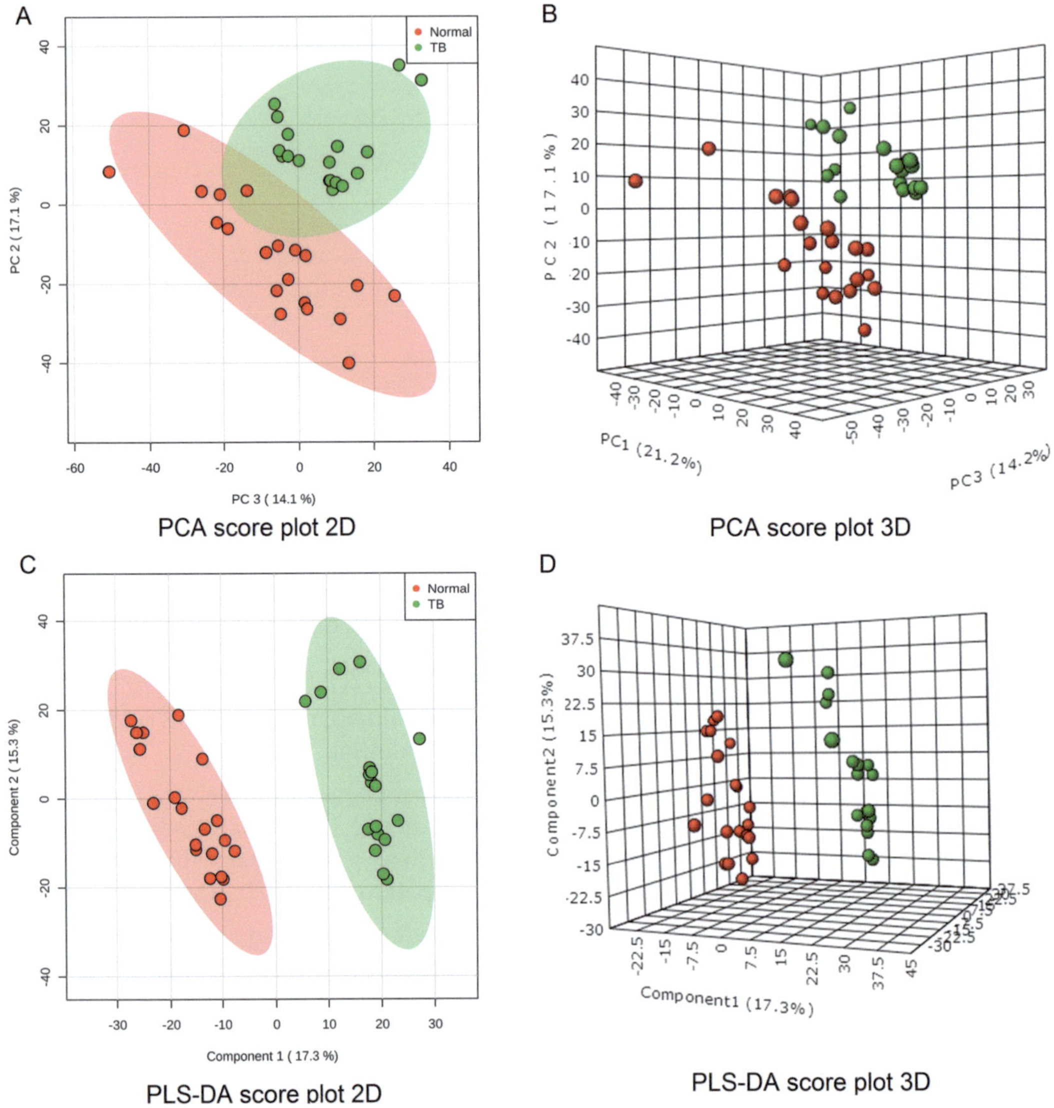

Fig. 5 The analytical result of serum metabolomics of TB patients in the HILIC mode. (**a**) PCA score plot in 2D. (**b**) PCA score plot in 3D. (**c**) PLSDA score plot in 2D. (**d**) PLSDA score plot in 3D

3. Multivariate statistical analysis (MSA). MSA is a commonly used strategy for identifying dysregulated metabolites between groups. Perform the principal component analysis (PCA), partial least-squares discriminant analysis (PLS-DA) and orthogonal projections to latent structures discriminant analysis (OPLS-DA) using the MetaboAnalyst web service [16] (http://www.metaboanalyst.ca/) (Fig. 5). Details of this procedure can be accessed on the website. Evaluate the importance of each metabolite in the PLS–DA model by variable importance in the projection (VIP) score. Metabolites with VIP scores greater than 1.0 may be considered important in this model (*see* **Note 14**).

4. Identification of significantly dysregulated metabolites. Use the MS1 and MS2 spectra of the metabolite to search against public metabolomics databases, including the mzCloud (https://www.mzcloud.org/), the Human Metabolome Database (HMDB; http://www.hmdb.ca/) and LIPID MAPS (http://www.lipidmaps.org/) to find potential candidates (*see* **Note 15**).
5. Bioinformatics analysis. Subject the list of identified dysregulated metabolites to the MetaboAnalyst pathway analysis tool web service (http://www.metaboanalyst.ca/faces/Secure/upload/PathUploadView.xhtml) to investigate the metabolic pathways correlated with the metabolites. Use the HMDB ID of each metabolite as an identifier.

4 Notes

1. The internal diameter (ID) of the LC columns used for HILIC and RPLC separation in our laboratory is 4.6 mm. For metabolomics analysis, a HILIC or C18 column of 2.1 mm ID can also be used. The column with larger ID has higher loading capacity for samples, and should be used at higher flow rate.
2. The centrifuge tube should contain no chemicals which have anticoagulant activity such as EDTA and heparin.
3. Quality control (QC) samples are used for monitoring the analytical performance of the LC-MS platform [17]. The QC sample is repeatedly analyzed after every 5–10 sample injections during the entire LC-MS acquisition batch.
4. The liquid–liquid extraction (LLE) strategy is used to extract metabolites and remove proteins in serum samples. At the end of the last step (centrifugation), the content in the tube is separated into three parts according to their solubility. The upper liquid layer consists of hydrophilic metabolites, the lower liquid layer consists of hydrophobic metabolites, and the undissolved proteins are present in the solid layer in the middle.
5. Chloroform is used to dissolve the hydrophobic lipids, but high concentrations of chloroform are not suitable for LC-MS analysis. The dissolved sample is therefore diluted three-fold before being subjected to LC-MS analysis.
6. The automatic gain control target (AGC) is a parameter that is used to control the number of ions analyzed in each single MS or MS/MS scan.
7. The parameter of fixed first mass is used to define the start of m/z recording range of a MS/MS scan.

8. The parameter of dynamic exclusion is set to prevent each metabolite being repeatedly analyzed in MS/MS mode in a given period (8 s).
9. When step-NCE is enabled, each metabolite will be analyzed in MS/MS mode using collision energy of 10, 20, and 30 separately, and finally a combined MS/MS spectrum is generated for each metabolite.
10. Analyzing the samples in random order can minimize the influence caused by MS signal perturbation. The QC sample can be used to monitor the stability of LC-MS system during the entire analysis.
11. Metabolites generally have signal response in positive ionization mode or negative ionization mode, and some metabolites have response in both modes. When performing metabolomics analysis, we include positive and negative ionization modes for better metabolite coverage. The MS instrument we use is the Q-Orbitrap (Q Exactive), and we usually perform the positive and negative ionization scan modes separately. If the complexity of the sample to be analyzed is low, the positive-negative ion switching mode can also be used.
12. Many types of software can be used for the peak extraction procedure, such as SIEVE, Progenesis QI (commercial software) and MS-DIAL (free of charge). The provided parameters are used for SIEVE. For other software, parameters can be accessed from the user manual.
13. The *p*-value and fold change for each metabolite can be calculated using the Excel software. The web service MetaboAnalyst can also be used for this purpose. Generally, a *p*-value of 0.05 and fold change of 2 are used as cutoffs to find significantly dysregulated metabolites, but cutoffs can be adjusted according to circumstances. For example, a fold change of 1.5 can be used in some cases.
14. Multivariate statistical analysis is a strategy used to retrieve dysregulated metabolites between experimental groups. MSA can be combined with statistical analysis (Subheading 3.5, **step 2**) to narrow down the candidate metabolites which are to be identified in the next step.
15. The identification of a metabolite is based on MS1 spectra (exact *m/z* of the precursor ion) and MS/MS spectra (exact *m/z* of the product ions generated by the precursor ion). The process of searching spectra against the database can be done manually or with software (MS-DIAL).

Acknowledgments

This work was supported by grants from the National Natural Science Foundation of China (7161007, 81430056, 31420103905, 21305005 and 31400695), the National Key Research and Development Program of China (No. 2016YFA0500302), and the Lam Chung Nin Foundation for Systems Biomedicine.

References

1. Fiehn O (2002) Metabolomics—the link between genotypes and phenotypes. Plant Mol Biol 48(1–2):155–171
2. Patti GJ, Yanes O, Siuzdak G (2012) Innovation: metabolomics: the apogee of the omics trilogy. Nat Rev Mol Cell Biol 13(4):263–269
3. Wikoff WR et al (2011) Untargeted metabolomics identifies enterobiome metabolites and putative uremic toxins as substrates of organic anion transporter 1 (Oat1). J Proteome Res 10 (6):2842–2851
4. De Vos RC et al (2007) Untargeted large-scale plant metabolomics using liquid chromatography coupled to mass spectrometry. Nat Protoc 2(4):778–791
5. Gertsman I, Gangoiti JA, Barshop BA (2014) Validation of a dual LC-HRMS platform for clinical metabolic diagnosis in serum, bridging quantitative analysis and untargeted metabolomics. Metabolomics 10(2):312–323
6. Andersen MB et al (2014) Untargeted metabolomics as a screening tool for estimating compliance to a dietary pattern. J Proteome Res 13(3):1405–1418
7. Kim HK, Choi YH, Verpoorte R (2010) NMR-based metabolomic analysis of plants. Nat Protoc 5(3):536–549
8. Allwood JW et al (2009) Inter-laboratory reproducibility of fast gas chromatography-electron impact-time of flight mass spectrometry (GC-EI-TOF/MS) based plant metabolomics. Metabolomics 5(4):479–496
9. Becker S et al (2012) LC-MS-based metabolomics in the clinical laboratory. J Chromatogr B Analyt Technol Biomed Life Sci 883-884:68–75
10. Zhou J, Yin Y (2016) Strategies for large-scale targeted metabolomics quantification by liquid chromatography-mass spectrometry. Analyst 141(23):6362–6373
11. Loots DT (2016) TB or not TB? Improving the understanding and diagnosis of tuberculosis through metabolomics. Biomark Med 10 (10):1025–1028
12. Kumar N, Shreshtha AK, Patra S (2017) The Metabolomic strategy in tuberculosis therapy. Comb Chem High Throughput Screen 20 (3):235–246
13. Preez ID, Luies L, Loots DT (2017) Metabolomics biomarkers for tuberculosis diagnostics: current status and future objectives. Biomark Med 11(2):179–194
14. Wang C et al (2017) Metabolomic analysis based on 1H-nuclear magnetic resonance spectroscopy metabolic profiles in tuberculous, malignant and transudative pleural effusion. Mol Med Rep
15. Zhong L et al (2016) Serum metabolomic study for the detection of candidate biomarkers of tuberculosis. Int J Clin Exp Pathol 9 (3):3256–3266
16. Xia J et al (2015) MetaboAnalyst 3.0--making metabolomics more meaningful. Nucleic Acids Res 43(W1):W251–W257
17. Dunn WB et al (2012) The importance of experimental design and QC samples in large-scale and MS-driven untargeted metabolomic studies of humans. Bioanalysis 4 (18):2249–2264

Chapter 14

Metabolomics Analysis of Leishmania by Capillary Electrophoresis and Mass Spectrometry

David Rojo, Coral Barbas, and Ángeles López-Gonzálvez

Abstract

Capillary electrophoresis coupled to mass spectrometry is an analytical platform ideal for the analysis of ionic or polar metabolites. It constitutes a perfect complement to reversed-phase liquid chromatography, offering a good alternative to polar stationary phases where reproducibility is not guaranteed. Herein, we describe a robust standardized methodology for the fingerprinting analysis of Leishmania, a taxonomic genus which comprises more than 20 protozoa species.

Key words Fingerprinting, Parasite phenotyping, Polar metabolites, Untargeted metabolomics

1 Introduction

The metabolome, defined as the collection of all the metabolites in a biological specimen (molecules up to 1500 Da), is the closest level to the phenotype across the omics cascade. Its analysis is technically challenging due to the broad range of physicochemical properties of the metabolites, especially taking into account that in metabolomics fingerprinting the goal is to maximize the metabolite coverage as well as to ensure the reproducibility of the detected signals. Based on its technical characteristics, capillary electrophoresis (CE) mass spectrometry (MS) is ideal to profile small and ionizable organic molecules.

CE separates analytes based on their charge and size and for that reason can be considered orthogonal to HPLC. In the capillary, the separation is led by two forces: the electroosmotic flow (related to the charge on the surface of the uncoated fused-silica capillary, affecting all compounds) and the electrophoretic mobility (the attraction of charged compounds to the opposite electrode). Electroosmotic flow (EOF) is the movement of the electrolyte solution into the capillary by applying a voltage. EOF occurs due to the silanol groups on the surface of fused-silica capillaries being

Edward E.K. Baidoo (ed.), *Microbial Metabolomics: Methods and Protocols*, Methods in Molecular Biology, vol. 1859,
https://doi.org/10.1007/978-1-4939-8757-3_14, © Springer Science+Business Media, LLC, part of Springer Nature 2019

negatively charged in function of pH. Under these conditions the positive ions of the electrolyte are attracted to the negatively charged surface and a double charge layer is formed. When a potential is applied, the positive ions of the electrolyte are attracted towards the cathode moving all the analytes inside. In addition, every compound will move towards the anode or cathode at different speeds depending on their size and charge according to their electrophoretic mobility (EPH). The overall mobility of a compound is the result of both EOF and EPH and in the normal polarity mode (i.e., positive mode of CE), ions will reach the detector in the following order: positive, neutral, then negative.

MS detection is carried via electrospray ionization (ESI) in which the sample from the CE is complemented with sheath liquid to correct for volume and to increase volatility. The time-of-flight (TOF) mass analyzer used in this work is equipped with a flight tube that separates ions according to their *m/z* ratios, providing high-resolution accurate mass measurements.

Herein, we describe a robust protocol for the analysis of metabolic profiles of *Leishmania*, a genus constituted by a plethora of protozoa species whose life cycles are divided into two forms: the amastigote (inside the macrophages of the vertebrate host) and the promastigote (in the gut tract of the sandflies). So far, we have focused on the analysis of the promastigotes, the most common form in the in vitro cultures of these parasites.

Our methodology has been applied with minor modifications to (1) study the mechanism of action and resistance of two drugs: an antimonial drug in *L. infantum* (1) and miltefosine in *L. donovani* [2]; (2) investigate virulence factors in *L. braziliensis* [3] and (3) to unveil potential new metabolic pathways in *L. amazonensis* [4]. In these studies, 98 metabolic features were detected as statistically significant in different comparisons, among which the most common are amino acids, peptides and conjugates (71%), followed by organic acids (12%), amines (7%), and purines/pyrimidines and conjugates (7%) (Fig. 1).

2 Materials

Prepare all solutions using ultrapure Milli-Q® water (Millipore, Billerica, MA, USA) and MS grade reagents. Store all prepared solutions at 4 °C (unless indicated otherwise). Conduct all volume and weight measurements in volumetric flasks and analytical balances, respectively. All reagents and solvents are purchased from Sigma-Aldrich (Steinheim, Germany) (unless indicated otherwise) and stored at room temperature. Diligently follow all appropriate waste disposal regulations when disposing of waste materials.

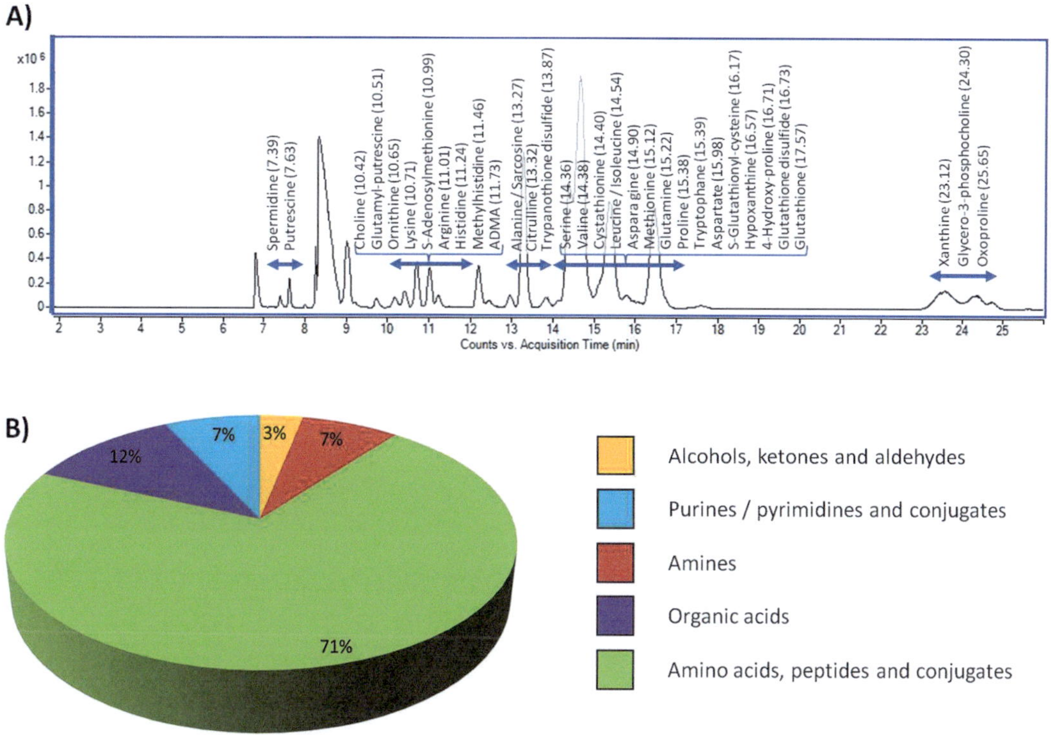

Fig. 1 (**a**) *L. amazonensis*' electropherogram and some metabolites with their migration times in brackets. (**b**) Biochemical classification in percentage of the putatively identified compounds by CE-ESI-TOF-MS analysis of several *Leishmania* spp.

2.1 Metabolite Extraction and Sample Preparation

1. Extraction solvent: MeOH–H_2O (1:1, v/v). For 500 mL of solution, mix 250 mL of MeOH with 250 mL of H_2O.
2. Eppendorf 1.5 mL (Hamburg, Germany).
3. Glass beads, acid-washed: 425–600 μm (sigma-Aldrich, St. Louis, MO, USA).
4. Liquid nitrogen (Air Liquide, Class Alphagaz™ 1).
5. Tissuelyzer LT (Qiagen, Hilden, Germany).
6. Centrifuge 5415 R (Eppendorf, Hamburg, Germany).
7. Analytical vial: Chromacol 03-FIV (300 μL fixed insert vial) (Thermo Scientific, Langerwehe, Germany).
8. Speedvac Concentrator: SpeedVac SPD121P (Thermo Electron Corporation, Milford, MA, USA).
9. Reconstitution solution: Milli-Q® water containing an internal standard (0.2 mM L-methionine sulfone) and 0.1 M formic acid. First, prepare a stock solution of 25 mM L-methionine sulfone by dissolving 0.0453 g of the reagent in 10 mL of H_2O. To 10 mL of the reconstitution solution add 80 μL of 25 mM L-methionine sulfone and 38 μL of formic acid (98% purity, 1.22 g/mL density) up to the final volume of H_2O (*see* **Note 1**).

2.2 CE–MS Analysis

1. Capillary Electrophoresis (7100 Agilent, Waldbronn, Germany) coupled to a TOF Mass Spectrometer (6224 Agilent, Waldbronn, Germany) and controlled by MassHunter Data Acquisition Software (version B.06.01) (Agilent Technologies, Santa Clara, CA, USA).
2. Fused-silica capillary (Agilent, Germany) 100 cm total length, 50 μm internal diameter.
3. Background electrolyte: 0.8 M formic acid in 10% MeOH (v/v). For 1 L, mix 100 mL of MeOH with 30.8 mL of formic acid (98% purity, 1.22 g/mL density) and H_2O up to the final volume. (*see* **Note 1**).
4. New capillary conditioning solution: 1 M NaOH. For 1 L, dissolve 40.0 g of NaOH in H_2O with constant shaking. Be careful to note that this is an exothermic process.
5. Sheath liquid pump: Isocratic pump 1200 series (Agilent, Waldbronn, Germany).
6. Sheath liquid: MeOH–H_2O (1:1) containing 1.0 mM formic acid and purine (detected m/z $[C_5H_4N_4+H]^+$, 121.0509 Da) and hexakis(1H,1H,3H-tetrafluoropropoxy)phosphazine (HP, detected m/z $[C_{18}H_{18}O_6N_3P_3F_{24}+H]^+$, 922.0098 Da) as reference masses for MS analysis (Agilent API-TOF Reference Mass Solution Kit, USA) (*see* **Note 2**). For 100 mL of sheath liquid, first mix 50 mL of MeOH with 50 mL of H_2O. Then, add 10 μL of HP, 5 μL of purine and 5 μL of formic acid (98% purity, 1.22 g/mL density) (*see* **Note 1**).

2.3 Data Treatment

1. MassHunter Qualitative Analysis Software (version B.06.00) (Agilent Technologies, Santa Clara, CA, USA).
2. Mass Profiler Professional (version 13.0) (Agilent Technologies, Santa Clara, CA, USA).
3. CEU Mass Mediator (http://ceumass.eps.uspceu.es/). This is a tool that utilizes accurate mass measurements of metabolites to search against known metabolite databases (KEGG, Metlin, LipidMaps and HMDB). It unifies hits from different sources and includes an advanced search function, which can be used to identify adducts from different ionization modes using the Composite Spectrum if Agilent software is used.

3 Methods

Carry out all procedures at room temperature. Subheadings 3.1 and 3.2 are schematized in Fig. 2.

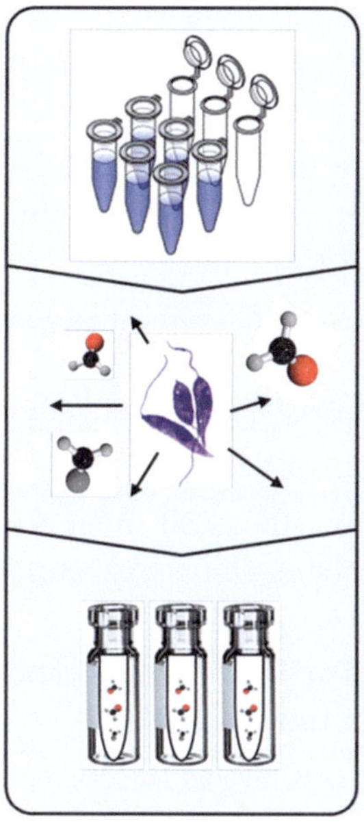

Cellular disruption and metabolite extraction

- 4 x 10^7 promastigotes + 200 µL of MeOH:H_2O and 50 mg of glass beads
- Three cycles of freezing/thawing (liquid N_2 /cold-water bath, 10 s)
- TissueLyser LT (10 minutes at 50 Hz)

Cellular debris precipitation

- Centrifugation (10 minutes at 15,700 x *g*, 4 °C)
- Transfer 100 µL of the resulting supernatant to an analytical vial

Sample conditioning for CE-MS

- Evaporate to dryness in a SpeedVac Concentrator at 35 °C
- Reconstitute in 100 µL of MilliQ water containing an internal standard (0.2 mM L-methionine sulfone) and 0.1 M of formic acid

Quality Control preparation

- Pool and mix equal volumes of each sample

Fig. 2 Main steps for metabolite extraction and sample preparation

3.1 Metabolite Extraction and Sample Preparation

1. Add 200 µL of MeOH–H_2O and 50 mg of glass beads to a 4×10^7 *Leishmania spp.* promastigotes pellet.
2. Physically disrupt the pellets by three cycles of freezing–thawing (liquid N_2–cold-water bath, 10 s) followed by 10 min in a TissueLyser operated at 50 Hz.
3. Pellet the cellular debris and the glass beads by centrifuging for 10 min at 15,700 × *g*, 4 °C.
4. Transfer 100 µL of the resulting supernatant to an analytical vial.
5. Evaporate to dryness in a SpeedVac Concentrator at 35 °C (*see* **Note 3**).
6. Reconstitute in 100 µL of Milli-Q® water containing an internal standard (0.2 mM L-methionine sulfone) and 0.1 M formic acid.

3.2 Quality Control Sample Preparation

Quality control (QC) samples are required at the beginning of the sequence to stabilize the system and throughout the analytical runs at periodic intervals of time to monitor variations in signal across the time. It is recommended to start a new sequence with 5–10 injections of the QC and reinject after every 5–10 samples (depending on the length of the sequence) analyzed in a randomized order.

1. Pool and mix equal volumes of each sample after **step 6** in Subheading 3.1 (*see* **Note 4**).

3.3 CE-MS Analysis

1. All separations are performed in normal polarity with a background electrolyte containing 0.8 M formic acid in 10% MeOH (v/v) at 20 °C.
2. New capillaries are preconditioned with a flush (950 mbar) of 1.0 M NaOH for 30 min, followed by Milli-Q® water for 30 min and the background electrolyte for 30 min.
3. Before each analysis, the background electrolyte content of the inlet home vial is refilled.
4. Before each analysis, the capillary is conditioned by flushing with the background electrolyte for 5 min at 950 mbar.
5. The samples are introduced hydrodynamically at 50 mbar for 50 s. Stacking effect is performed by applying the background electrolyte at 100 mbar afterward for 10 s.
6. The separation conditions are conducted at 30 kV voltage and an internal pressure of 25 mbar for a run time of 30 min.
7. The sheath liquid with two reference masses allows for correction of mass accuracy in the MS. It is continuously infused (at 6 μL/min) while CE separation is ongoing.
8. The MS parameters are: Fragmentor 125 V, skimmer 65 V, octopole 750 V, drying gas temperature 200 °C, nebulizer pressure 10 psi, flow rate 10 L/min and capillary voltage 3500 V. The data are acquired in positive mode with a full scan from *m/z* 85 to 1000 at a rate of 1.41 scan/s.
9. The analytical sequence is set up starting with the analysis of ten QCs followed by the samples; a QC sample is injected in between blocks of samples until the end of the run (*see* **Note 5**).

3.4 Data Treatment Overview

Metabolomics data treatment is a complex matter with multiple paths. For a comprehensive review on the topic, see Godzien et al. (2013) [5]. A brief overview of the main steps is provided in Fig. 3. Note that specific parameters are not reported, they should be set according to the requirements of each study.

1. Deconvolute data with MassHunter Qualitative Analysis Software (or any other with similar performance). Attention should be paid to the possible adducts that can be expected in order to extract the proper metabolic features.
2. Align metabolite features in Mass Profiler Professional (or any other software with similar performance). For successful alignment, attention should be paid to the mass accuracy error window (±10 ppm for a TOF Mass Spectrometer 6224 Agilent) and to the migration time shift (check with the internal standard peak, detected *m/z* $[C_5H_{11}NO_4S+H]^+$, 182.0487 Da).

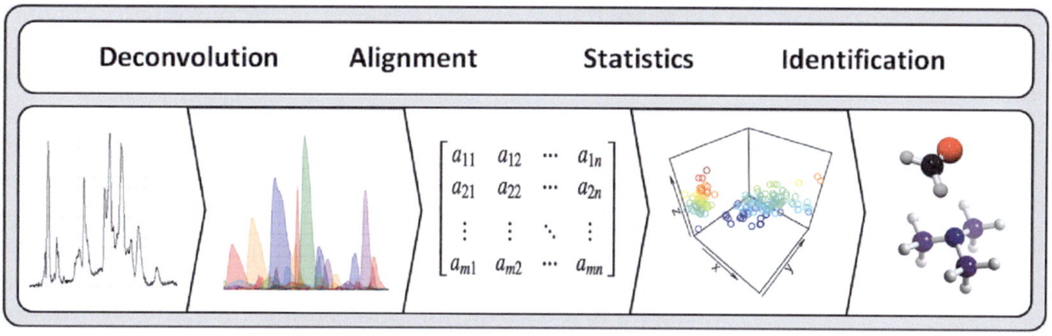

Fig. 3 Main steps in a standard data treatment workflow for metabolomics fingerprinting

3. Statistical analysis for the selection of the potential markers can be done either by multivariate or univariate analysis. There are comprehensive reviews elsewhere [6, 7].
4. The easiest and the most common way to putatively identify a metabolic feature is by linking its accurate mass with a certain structure using public or in house-built databases. For a review about public databases in MS Metabolomics *see* Vinaixa et al. (2015) [8, 9] (*see* **Note 6**). We recommend the use of CEU Mass Mediator.

4 Notes

1. When purity and density are specified, the volumes or weights that are specified in the protocol must be recalculated based on the specification of the solvent or reagent that is available in each laboratory.
2. Agilent API-TOF Reference Mass Solution Kit contains: 5 mM purine in acetonitrile–H_2O (90:10, v/v) and 2.5 mM HP-0921 in acetonitrile.
3. If the samples are not going to be analyzed immediately, the best option is to store them dry in order to minimize their degradation.
4. Independent of the injection volume, the analytical vial should contain enough volume of sample to ensure that the injection end of the capillary is immersed in it. If it is required to prepare the QCs, the extraction volume could be increased up to 300 μL for the same number of parasites.
5. The total number of QC injections, equally split along the sequence, should be enough to ensure a proper assessment of the analytical performance, i.e., 10–15% of the total injections of the analytical run excluding the initial equilibrium.

6. There are four levels of confidence in the metabolites annotation [10]: Level 1 definitively confirmed identifications by the injection of authentic standards; level 2 putatively annotated compounds based on mass and where available a comparison to migration time; level 3 putatively characterized by class; and level 4 unknown compounds. At Subheading 3.4, **step 4** we refer to level 2, named "putative-plus" identification, meaning that identifications are curated with the isotopic pattern matching, migration time evaluation and our own database. For examples of level 2 identifications using CE-ESI-TOF-MS, *see* Godzien et al. (2015) [11] where the electrospray in-source fragmentation pattern is successfully correlated ($r \geq 0.9$) with the MS/MS identification signature.

Acknowledgment

This work has been supported by the funding from the Spanish Ministry of Economy and Competitiveness (CTQ2014-55279-R).

References

1. Rojo D, Canuto GAB, Castilho-Martins EA et al (2015) A multiplatform metabolomic approach to the basis of antimonial action and resistance in Leishmania infantum. PLoS One 10:1–20
2. Canuto GAB, Castilho-Martins EA, Tavares MFM et al (2014) Multi-analytical platform metabolomic approach to study miltefosine mechanism of action and resistance in Leishmania. Anal Bioanal Chem 406:3459–3476
3. Alves-Ferreira EVC, Toledo JS, De Oliveira AHC et al (2015) Differential gene expression and infection profiles of cutaneous and mucosal Leishmania braziliensis isolates from the same patient. PLoS Negl Trop Dis 9:1–1
4. Castilho-Martins EA, Canuto GAB, Muxel SM et al (2015) Capillary electrophoresis reveals polyamine metabolism modulation in Leishmania (Leishmania) amazonensis wild-type and arginase-knockout mutants under arginine starvation. Electrophoresis 36:2314–2323
5. Godzien J, Ciborowski M, Angulo S et al (2013) From numbers to a biological sense: how the strategy chosen for metabolomics data treatment may affect final results. A practical example based on urine fingerprints obtained by LC-MS. Electrophoresis 34:2812–2826
6. Korman A, Oh A, Banks D (2012) Statistical methods in metabolomics. In: Anisimova M (ed) Methods in molecular biology. Springer Science+Business Media, LLC, New York, pp 381–413
7. Steuer R, Morgenthal K, Weckwerth W et al (2007) A gentle guide to the analysis of metabolomic data. In: Weckwerth W (ed) Methods in molecular biology. Humana Press Inc., Totowa, pp 105–126
8. Vinaixa M, Schymanski EL, Neumann S et al (2016) Mass spectral databases for LC/MS- and GC/MS-based metabolomics: state of the field and future prospects. TrAC - Trends Anal Chem 78:23–35
9. Gil de la Fuente A, Grace Armitage E, Otero A, et al (2017), Differentiating signals to make biological sense – a guide through databases for MS-based non-targeted metabolomics
10. Salek RM, Steinbeck C, Viant MR et al (2013) The role of reporting standards for metabolite annotation and identification in metabolomic studies. Gigascience 2
11. Godzien J, Armitage EG, Angulo S et al (2015) In-source fragmentation and correlation analysis as tools for metabolite identification exemplified with CE-TOF untargeted metabolomics. Electrophoresis 36:2188–2195

Part IV

Data Analysis

Chapter 15

A High-Throughput Targeted Metabolomics Workflow for the Detection of 200 Polar Metabolites in Central Carbon Metabolism

Yuping Cai and Zheng-Jiang Zhu

Abstract

Targeted metabolomics aims to analyze a set of pre-selected metabolites from biologically relevant metabolic pathways. The triple quadrupole mass spectrometry (QqQ-MS) based multiple reaction monitoring (MRM) technique is the most widely approach used for targeted metabolomics, and features high selectivity and sensitivity, good reproducibility and wide dynamic range in quantitative analysis. Here, we describe an MRM based targeted metabolomics workflow for the quantitative analysis of 200 polar metabolites in central carbon metabolic pathways, including the data acquisition method and the automated data processing procedures using our in-house R package MRMAnalyzer. The workflow described in this chapter combines a hydrophilic interaction liquid chromatography (HILIC) separation and positive/negative ion polarity switching based MS detection, and is able to acquire data from multiple types of biological samples such as bacteria, cultured mammalian cells, animal tissues and biofluids (e.g., serum and urine). Finally, the MRMAnalyzer software can automatically process the generated large-scale data set with high efficiency. We hope it is a valuable and efficient workflow for researchers to facilitate the respective biological studies using targeted metabolomics.

Key words Targeted metabolomics, Multiple reaction monitoring, Central carbon metabolism, MRMAnalyzer

1 Introduction

Metabolomics quantitatively measures the metabolites and the related metabolic pathways in the biological system, and aims to accurately characterize the physiological and pathological status of the system [1–3]. Metabolomics can give a holistic and deep understanding of the diseases pathogenesis, and facilitates the early diagnosis and treatment of diseases [4, 5]. Untargeted and targeted metabolomics are two commonly used metabolomics approaches

Electronic supplementary material: The online version of this chapter (https://doi.org/10.1007/978-1-4939-8757-3_15) contains supplementary material, which is available to authorized users.

Edward E.K. Baidoo (ed.), *Microbial Metabolomics: Methods and Protocols*, Methods in Molecular Biology, vol. 1859, https://doi.org/10.1007/978-1-4939-8757-3_15, © Springer Science+Business Media, LLC, part of Springer Nature 2019

[1, 6]. Untargeted metabolomics focuses on the comprehensive analysis of all the measureable metabolites in given biological samples including the unknown chemicals. However, this method heavily relies on sophisticated bioinformatics software for data processing and metabolite structural identification [7]. In contrast, targeted metabolomics aims to analyze a set of known and pre-selected metabolites in specific metabolic pathways. The triple quadrupole mass spectrometry (QqQ-MS) based multiple reaction monitoring (MRM) technique is the most widely used approach in targeted metabolomics due to its high selectivity, high sensitivity, good reproducibility and wide dynamic range [8–10]. The data acquisition of known metabolites also significantly reduces the efforts on data processing and metabolite identification [11–14].

The advancements on modern QqQ-MS have significantly improved the capability of targeted metabolomics for metabolic profiling of various types of biological samples. Currently, it is capable of simultaneously analyzing hundreds of metabolites in a single analysis [15, 16]. The MRM technique in targeted metabolomics includes three steps for data acquisition: (1) the selection of a metabolite precursor ion in Q1; (2) the fragmentation of a pre-selected precursor ion in collision cell; and (3) the selection of its product ion in Q3. The combination of a specific pair of precursor ion and product ion, referred to as an MRM transition, represents a targeted metabolite of interest. Here, we describe a well-designed targeted metabolomics workflow using the MRM technique for high-throughput profiling of 200 polar metabolites in central carbon metabolism. The workflow includes sample preparation, LC-MS data acquisition, and automated data processing using an R package MRMAnalyzer [11]. This workflow is capable of acquiring data from various types of biological samples using the standard protocol described in this chapter. The important features of this workflow include the following: (1) A single LC-MS run is needed for the detection of 200 metabolites, and the analysis time is as short as 23 min per injection; (2) the data processing procedure of hundreds and thousands of MRM transitions is fully automated without manual manipulation, and (3) the well-designed workflow is reproducible and suitable for high-throughput analysis for both large-scale biological and clinical samples. In conclusion, the LC-MRM based data acquisition together with automated data processing using MRMAnalyzer has a universal application to profile polar metabolites in biological samples and proffers a unique insight into central carbon metabolism.

2 Materials

All LC-MS/MS solvent and additives used must be of LC-MS grade. The extraction solvent should be stored under −20 °C and kept on ice during the metabolite extraction procedure.

2.1 Metabolite Extraction

1. Methanol (MeOH), LC-MS grade.
2. Acetonitrile (ACN), LC-MS grade.
3. Water (H_2O), LC-MS grade.
4. Eppendorf tube 2 mL.
5. Pipettes (Eppendorf).
6. Glass vial with insert, screw, 300 μL.
7. Vortexer (thermo).
8. Ultrasonic apparatus.
9. Liquid nitrogen.
10. Centrifuge (5424 R, Eppendorf).
11. Vacuum concentrator (Labconco).

2.2 LC-MS Analysis

1. Acetonitrile (ACN), LC-MS grade.
2. Water (H_2O), LC-MS grade.
3. Ammonium acetate and ammonium hydroxide, LC-MS grade.
4. Mobile phase A: 25 mM ammonium acetate and 25 mM ammonium hydroxide in 100% water (*see* **Note 1**).
5. Mobile phase B: acetonitrile.
6. In-line filter (Waters) (*see* **Note 2**).
7. Analytical column, UPLC BEH Amide column, particle size, 1.7 μm; 100 mm (length) × 2.1 mm (i.d.) (Waters).
8. UHPLC system (1290 series, Agilent Technologies).
9. Triple quadrupole mass spectrometer (Agilent 6495 QqQ, Agilent Technologies).

2.3 Data Analysis

1. R programming environment.
2. ProteoWizard software.
3. R package MRMAnalyzer.

3 Methods

3.1 Metabolite Extraction

3.1.1 Escherichia coli

1. Collect 10 mL *Escherichia coli* (*E. coli*) culture into an Eppendorf tube (2 mL) under $OD_{600\ nm} = 1.0$.
2. Proceed with centrifugation at 3461 × *g* for 5 min under 4 °C to remove culture medium and harvest cells. Wash the pellets with 0.6% NaCl three times and centrifuge at 3461 × *g* for 5 min to remove NaCl solution.
3. Then, quickly freeze the pellets in liquid nitrogen and immediately proceed with extraction or store the frozen pellets at −80 °C.
4. Place the *E. coli* pellets on ice and add 1 mL of cold extraction solvent MeOH–ACN–H_2O (2,2:1, v/v/v) to each pellet. The extraction solvent should be stored under −20 °C and kept on ice during the metabolite extraction.
5. Vortex for 30 s and incubate in liquid nitrogen for 1 min.
6. Allow the sample to thaw at room temperature and sonicate for 10 min.
7. This freeze–thaw cycle needs to be repeated three times in total (**steps 3** and **4**).
8. Incubate the samples for 1 h at −20 °C to precipitate proteins.
9. Proceed with 15 min centrifugation at 15,000 × *g* under 4 °C.
10. Transfer the resulting supernatant to a new Eppendorf tube (2 mL), then evaporate to dryness in a vacuum concentrator under 4 °C using vacuum concentrator.
11. Reconstitute the dry extracts with 100 μL of ACN–H_2O (1:1, v/v).
12. Sonicate the reconstitution solution for 10 min, and centrifuge for 15 min at 15,000 × *g* under 4 °C to remove insoluble debris.
13. Transfer the supernatant to an HPLC glass vial and store at −80 °C prior to LC-MS analysis.

3.1.2 Cultured HeLa Cells

1. Remove the culture medium from the cultured HeLa cells (~2×10^6 cells per sample) using pipette.
2. Wash the cells with PBS under 37 °C for three times and remove the PBS. Proceed with trypsinization for 2 min, then add serum medium to stop the reaction.
3. Collect the cells into a new centrifuge tube, and centrifuge at 1153 × *g* for 5 min to remove supernatant.
4. Wash the cell pellets with PBS under 4 °C for three times, and centrifuge at 1153 × *g* for 5 min to remove the PBS.

5. Quickly freeze the cell pellets in liquid nitrogen and immediately proceed with extraction or store the material at −80 °C.

The following extraction procedures for human cells are the same as for *E. coli* samples (Subheading 3.1, *Escherichia coli* 4–13).

3.1.3 Animal Tissues

1. Quickly freeze the animal tissues (e.g., mice liver) in liquid nitrogen immediately after dissection.
2. Cut the tissue on dry ice (~10 mg) into an Eppendorf tube (2 mL).
3. Homogenize the tissue sample with 200 μL of H_2O and five ceramic beads using the homogenizer (Precellys 24, Bertin Technologies, France).
4. Add 800 μL ACN–MeOH (1:1, v/v) to homogenized solution for metabolite extraction.

Compared to metabolite extraction procedures for *E. coli* and cultured cell samples, there is no need to perform freeze–thaw cycles for tissue samples after homogenization. And the following extraction procedures for tissue samples are the same as described for *E. coli* samples (Subheading 3.1, **steps 5–13**).

3.1.4 Biofluid Samples (Plasma, Serum, and Urine) Metabolite Extraction

1. Add 800 μL ACN–MeOH (1:1, v/v) of solvent solution into 200 μL of plasma, serum or urine samples.
2. Vortex the samples for 30 s and sonicate for 10 min.

And also there is no need to perform freeze–thaw cycle for biofluid samples. The following extraction procedures for biofluid samples are the same as described for *E. coli* samples (Subheading 3.1, *Escherichia coli* 6-11).

3.2 Analytical Sample Preparation

In this protocol, several types of samples are required for the targeted metabolomics assay, including 200STD_mix sample, the

Table 1
LC gradient

Time (min)	Flow rate (mL/min)	Eluent A (vol. %)	Eluent B (vol. %)
0.0	0.3	5.0	95.0
1.0	0.3	5.0	95.0
14.0	0.3	35.0	65.0
16.0	0.3	60.0	40.0
18.0	0.3	60.0	40.0
18.1	0.3	5.0	95.0
23.0	0.3	5.0	95.0

pooled QC sample, retention time quality control sample (RTQC), and biological samples. Please refer to the details of each sample type in Subheading 3.4 *Assay for targeted metabolomics.*

1. Prepare the 200STD_mix sample by mixing the stock solutions of 200 metabolite standards. The final concentration for each metabolite is comprehensively optimized (Supplementary Table 1). Prepare the stock solution for each metabolite standard in MeOH–H_2O (1:1, v/v) with a final concentration of 100 ppm (*see* **Note 3**).
2. Prepare the pooled QC sample by mixing aliquots (5–10 μL each) from all biological samples.
3. Prepare the RTQC sample by mixing the stock solutions of 10 metabolite standards. Prepare the stock solution for each metabolite standard in MeOH–H_2O (1:1, v/v). The final concentration of each metabolite is different (*see* **Note 4**).

If one aims to analyze a different set of metabolites of interest, please follow the important instruction (*see* **Note 5**).

3.3 Optimization of MRM Transitions and Measurement of Retention Times

Metabolite MRM transitions and retention times are used for metabolite identification by the R package MRMAnalyzer [11], and are generated using single standard solution of each metabolite (*see* **Note 6**).

1. Analyze each metabolite standard (100 μg/mL) in both ESI positive and negative ion modes via flow injection using the software MassHunter Optimizer (Agilent Technologies). Constant gradient is applied with 50% B to 50% B (0–2 min). The flow rate is 0.1 mL/min and the sample injection volume is 2 μL. In this step, the optimal MRM transition parameters, such as ionization polarity, product ion, collision energy (CE), and fragmentor voltage can be obtained (*see* **Note 7**).
2. Measure the retention time (rt_{single}) of each metabolite by acquiring the corresponding MRM transition individually on a BEH amide column. The sample injection volume is 2 μL. Use the following LC gradient (Table 1):
3. Use MRMAnalyzer for data processing. Metabolite name, ID, MRM transitions and retention time (rt_{single}) are summarized as the metabolite MRM library for the data processing using MRMAnalyzer.

3.4 High-Throughput Targeted Metabolomics Assay

All targeted metabolomics experiments are carried out on an Agilent 6495 QqQ-MS instrument equipped with an Agilent 1290 UHPLC system. The instrument is operated in the positive/negative polarity switching mode.

3.4.1 Liquid Chromatograph

1. Load worklist with method embedded using MassHunter software.
2. Run batch sequence with following LC parameters:
   (a) Wash injection needle one time with needle washing solvent MeOH–H_2O (1:1, v/v).
   (b) Load sample and inject 2 μL of sample.
   (c) Run LC method using the LC gradient as described in Subheading 3.3, **step 2**.

3.4.2 QqQ-MS Parameters

1. Set ESI source conditions to: 350 °C sheath gas temperature; 350 °C dry gas temperature; 12 L/min sheath gas flow rate; 16 L/min dry gas flow rate; capillary voltages of 3000 V and – 2500 V in positive and negative ion modes, respectively; nozzle voltages of 1000 V and 1500 V in positive mode and negative ion modes, respectively; and 40 psi nebulizer pressure.
2. Set the dwell time for each MRM transition to 3 ms. This gives a total cycle time of 1.38 s (*see* **Note 8**).
3. For targeted metabolomic analyses, 200 MRM transitions representing the 200 metabolites are simultaneously monitored, of which there are163 MRM transitions in positive mode and 37 MRM transitions in negative mode. Positive/negative polarity switching is applied.

3.4.3 Assay for Targeted Metabolomics

The design for the targeted metabolomics assay is illustrated in Fig. 1.

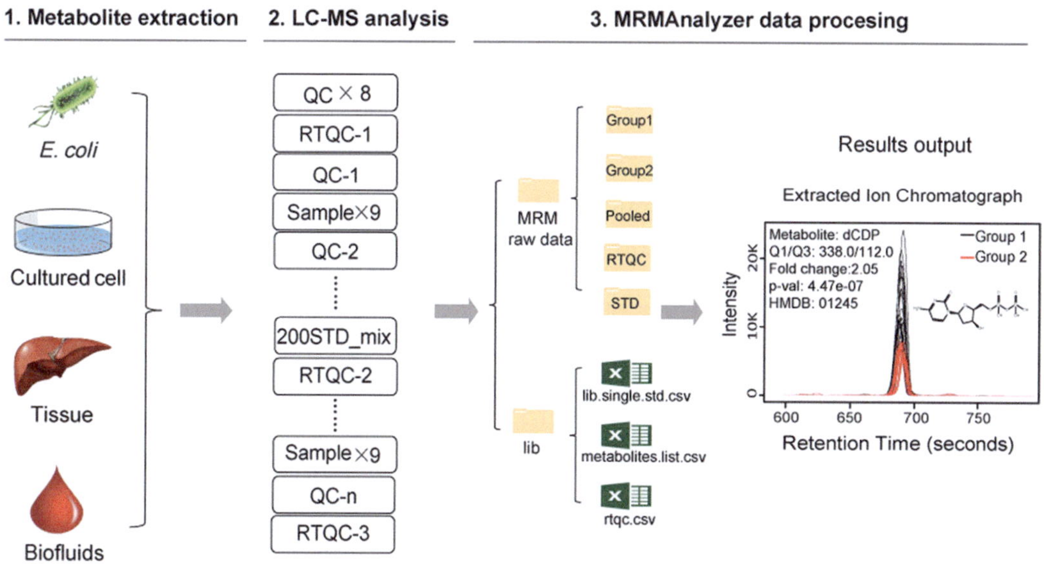

Fig. 1 The designed targeted metabolomics workflow for high-throughput metabolite profiling

1. Analyze the 200STD_mix, which contains chemical standards of the 200 metabolites of interest, together with biological samples from each experiment, and use the measurements from the former as the reference sample for retention time correction and peak alignment.
2. Measure compounds from the pooled QC sample to evaluate the stability and reproducibility of the instrument.
3. The use of RTQC sample is to check the retention time stability of the LC system by comparing the retention times of metabolites in this sample to those in the library. The RTQC sample contains ten metabolites, and the retention times of these distributes almost uniformly throughout the entire chromatographic run (*see* **Note 4**). The maximum tolerance for the retention time shift is 30 s.

   At the beginning of a batch, analyze eight QC samples (and use them for column conditioning). Then measure the QC samples at regular intervals (every nine biological samples) throughout the experiment.
4. Randomize, then analyze all biological samples. 200STD_mix reference sample is analyzed once in the middle of the batch. The RTQC samples are analyzed three times at the beginning, in the middle, and at the end of a batch, respectively (Fig. 1).

3.5 Automated Data Processing of a Large MRM Data Set Using MRMAnalyzer

The previously developed R package MRMAnalyzer is capable of automatically processing the MRM dataset without any manual manipulation [11]. For each targeted metabolomics assay, the MRMAnalyzer is designed to process the RTQC samples, QC samples, 200STD_mix sample and biological samples all altogether. It consists of five steps for data analysis: "pseudo" accurate *m/z* transformation, peak detection and alignment, quality control, metabolite identification, and statistical analyses. Please read our recent publication for the principles of the MRMAnalyzer analysis process [11]. The MRMAnalyzer program and the tutorial are available at our group website (http://www.metabolomics-shanghai.org/software.php).

3.5.1 Installation of MRMAnalyzer

1. Firstly, install the R program on a Windows system (http://www.r-project.org/) since the MRMAnalyzer is developed with R scripts.
2. Install the ProteoWizard software on a Windows system (*see* **Note 9**). (http://proteowizard.sourceforge.net/).
3. Install R packages XCMS, faahKO, baseline, and multtest.
4. Install the MRMAnalyzer package (mrmanalyzer_3.1.1.tar.gz).

Please refer the MRMAnalyzer software tutorial for details about the installation of ProteoWizard and the MRMAnalzyer Package.

3.5.2 Organization of the MRM Raw Data Files

As illustrated in Fig. 1, please create a "MRM raw data" folder. Data files of different sample groups are put into different sub-directories. MRM raw data (.d files) of RTQC sample, QC samples and 200STD_mix sample are put into a subdirectory named "RTQC," "pooled," and "STD," respectively (*see* **Note 10**).

3.5.3 Preparation of Library Files for Data Analysis

Create a "lib" folder so that metabolite information (such as MRM transitions, RT) can be loaded into the MRMAnalyzer (Fig. 1). The "lib" folder requires three .csv files (*see* **Note 11**).

1. Library file-"lib.single.std.csv" (Supplementary Table 2): the file contains in-house ID, MRM transition, and retention time for each defined metabolite.
2. 200STD_mix reference sample file-"metabolites.list.csv" (Supplementary Table 3): the file contains in-house ID, metabolite name, formula, HMDB, METLIN, KEGG, precursor ion, product ion and polarity (*see* **Note 12**).
3. RTQC file-"rtqc.csv" (Supplementary Table 4): the file contains metabolite information for the ten metabolites in RTQC sample including in-house ID, metabolite name, formula, precursor ion, product ion, fragmentor, CE, polarity and RT.

Specifically, the MRMAnalyzer also supports processing other targeted metabolites' MRM data sets. In this case, users are required to modify and update the library files for the metabolites of interest.

3.5.4 Data Processing by MRMAnalyzer

1. Load MRMAnalyzer R package:library(mrmanalyzer).
2. Execute data processing command:

mrmanalyzer (d.in = '/path/to/your/data', pwiz.dir = '/path/to/ProteoWizard', convert = 1, peakwidth = c(20,100), minfrac = 0) (*see* **Note 13**).

Please refer to the MRMAnalyzer software tutorial for the details about the parameter settings.

3.6 Interpretation of Output Results

1. The output files include identified metabolite results (.csv format) and overlapped EICs (.pdf format) of detected metabolites in biological samples.
2. The identified metabolite results (.csv format) consists of metabolite ID information (e.g., metabolite name, formula, HMDB, METLIN, KEGG) and the integrated peak areas of metabolites in biological samples.

3. Relative quantification comparison can be achieved between biological sample conditions based on the peak areas.
4. Subsequent statistical analysis and pathway analysis can also be obtained by performing principal component analysis (PCA) and using MetaboAnalyst [17].
5. The overlapped EICs file (.pdf format) can facilitate further inspection and evaluation of the identified metabolites in samples if necessary.

4 Notes

1. The mobile phase A with 25 mM ammonium acetate and 25 mM ammonium hydroxide in water needs to be prepared freshly to avoid microorganism breeding.
2. It is recommended that an inline filter is used to help extend the lifetime of LC column.
3. The standard mixture sample (200STD_mix) that contains all of the 200 metabolites is measured together with the biological samples, and is used as the reference sample for retention time correction and peak alignment. Each metabolite's concentration in the 200STD_mix sample is comprehensively optimized so that the most intensive peak in a defined retention time region ($rt_{single} \pm 60s$) is the target metabolite. This prerequisite is essential for accurate metabolite identification by the MRMAnalyzer program [11]. Please refer to the previous publication for details. It is worthy to note that only metabolites detected and identified in the 200STD_mix sample by the MRMAnalyzer program could be profiled in biological samples. Therefore, the 200STD_mix sample should be prepared as fresh as possible. According to our experience, the prepared 200STD_mix sample can be stored at −80 °C for 1 year.
4. The RTQC sample contains the chemical standards of ten metabolites of interest with different concentrations: 4-Pyridoxic acid (400 ppb), Purine (400 ppb), Creatinine (400 ppb), Vitamin B2 (400 ppb), Guanosine (400 ppb), N-Acetylglutamine (400 ppb), Sucrose (2 ppm), L-2-Aminoadipic acid (400 ppb), UDP-D-Glucose (2 ppm) and Glutathione disulfide (10 ppm). During LC-MS data acquisition of large sample size, the retention time deviation is different in each chromatographic elution region. Thus, the retention times of the selected metabolites in RTQC sample need to be distributed uniformly throughout the entire chromatographic run to supervise the retention time shift.
5. The customized STD_mix sample should contain all of the targeted metabolites. Each metabolite's concentration in

STD_mix sample should be manually optimized so that the most intensive peak in a defined retention time region (rt_{single} ± 60s) is the target metabolite (*see* **Note 3**). The RTQC sample can be modified according to the customized liquid chromatography system or gradient. The elution times of metabolites in the RTQC sample are recommended to be distribute evenly across the LC gradient.

6. The constructed metabolite MRM library is used for metabolite identification by the MRMAnalyzer. Two important features in metabolite MRM library are MRM transition ("pseudo" accurate *m/z* value) and retention time (rt_{single}). The metabolite identification step is carried out by matching the metabolite features in the metabolite library to the detected peak groups in biological samples. The "pseudo" accurate *m/z* value must be exactly matched (0 ppm error), and the retention time shift tolerance is set as rt_{single} ± 60 s.
7. Firstly, ionization polarity is determined by choosing the condition with higher precursor ion intensity detected in the positive ion mode compared to that produced in the negative mode. Then, the MRM transition is defined by monitoring the most intensive product ion in Q3. Specifically, potential interference between MRM transitions should be taken into consideration when selecting a product ion (Q3 ion). For example, citric acid and isocitric acid have the same molecular weight and cannot be chromatographically separated from each other, so the characteristic product ions of *m/z* 87.0 and *m/z* 73.0 are chosen as the targeted product ions of citric acid and isocitric acid, respectively.
8. The minimum dwell time of 3 ms for each MRM transition is required to assure the detection sensitivity.
9. Note that only the windows version of ProteoWizard software is available at present.
10. The name of the organized folders cannot be changed.
11. The file names in the lib folder should be unique. Otherwise, the package would be stopped as it could not decide which file is the correct library file for data analysis.
12. The file name of the standards mixture should be ended with "metabolites.list.csv" (e.g., "200metabolites.list.csv").
13. The user can optimize some parameters to improve the peak detection according to the LC conditions. The most import one is the "peakwidth," referring to the range of chromatographic peak width and the default value is set to "c (10, 200)". Another parameter is "minfrac," which is set to 0.5 by default. "minfrac" refers to the minimum fraction of

samples necessary in at least one of the sample groups for it to be a valid peak group.

Acknowledgments

The work was financially supported by National Natural Science Foundation of China (Grant No. 21575151) and State High-Tech Development Plan Award (the "863 program", Grant No. 2014AA020526). Z.-J. Z. is supported by Thousand Youth Talents Program.

References

1. Patti GJ, Yanes O, Siuzdak G (2012) Metabolomics: the apogee of the omics trilogy. Nat Rev Mol Cell Biol 13(4):263–269
2. Nicholson JK, Lindon JC (2008) Systems biology: Metabonomics. Nature 455 (7216):1054–1056
3. Fiehn O (2002) Metabolomics - the link between genotypes and phenotypes. Plant Mol Biol 48(1–2):155–171
4. Wishart DS (2016) Emerging applications of metabolomics in drug discovery and precision medicine. Nat Rev Drug Discov 15 (7):473–484
5. Fernie AR, Trethewey RN, Krotzky AJ, Willmitzer L (2004) Metabolite profiling: from diagnostics to systems biology. Nat Rev Mol Cell Biol 5(9):763–769
6. Dettmer K, Aronov PA, Hammock BD (2007) Mass spectrometry-based metabolomics. Mass Spectrom Rev 26(1):51–78
7. Zhu ZJ, Schultz AW, Wang JH, Johnson CH, Yannone SM, Patti GJ, Siuzdak G (2013) Liquid chromatography quadrupole time-of-flight mass spectrometry characterization of metabolites guided by the METLIN database. Nat Protoc 8(3):451–460
8. Dudley E, Yousef M, Wang Y, Griffiths WJ (2010) Targeted metabolomics and mass spectrometry. Adv Protein Chem Struct Biol 80:45–83
9. Kitteringham NR, Jenkins RE, Lane CS, Elliott VL, Park BK (2009) Multiple reaction monitoring for quantitative biomarker analysis in proteomics and metabolomics. J Chromatogr B 877(13):1229–1239
10. Lu W, Bennett BD, Rabinowitz JD (2008) Analytical strategies for LC-MS-based targeted metabolomics. J Chromatogr B 871 (2):236–242
11. Cai Y, Weng K, Guo Y, Peng J, Zhu Z-J (2015) An integrated targeted metabolomic platform for high-throughput metabolite profiling and automated data processing. Metabolomics 11 (6):1575–1586
12. Tsugawa H, Arita M, Kanazawa M, Ogiwara A, Bamba T, Fukusaki E (2013) MRMPROBS: a data assessment and metabolite identification tool for large-scale multiple reaction monitoring based widely targeted metabolomics. Anal Chem 85(10):5191–5199
13. MacLean B, Tomazela DM, Shulman N, Chambers M, Finney GL, Frewen B, Kern R, Tabb DL, Liebler DC, MacCoss MJ (2010) Skyline: an open source document editor for creating and analyzing targeted proteomics experiments. Bioinformatics 26(7):966–968
14. Wong JW, Abuhusain HJ, McDonald KL, Don AS (2012) MMSAT: automated quantification of metabolites in selected reaction monitoring experiments. Anal Chem 84(1):470–474
15. Yuan M, Breitkopf SB, Yang XM, Asara JM (2012) A positive/negative ion-switching, targeted mass spectrometry-based metabolomics platform for bodily fluids, cells, and fresh and fixed tissue. Nat Protoc 7(5):872–881
16. Zhou J, Liu H, Liu Y, Liu J, Zhao X, Yin Y (2016) Development and evaluation of a parallel reaction monitoring strategy for large-scale targeted metabolomics quantification. Anal Chem 88(8):4478–4486
17. Xia J, Psychogios N, Young N, Wishart DS (2009) MetaboAnalyst: a web server for metabolomic data analysis and interpretation. Nucleic Acids Res 37:W652–W660

Chapter 16

Cluster Analysis of Untargeted Metabolomic Experiments

Joshua Heinemann

Abstract

Untargeted metabolite profiling based upon LC-MS methodology can be used to identify unique metabolic phenotypes associated with stress, disease or environmental exposure of cells using mathematical clustering. Here, we show how unsupervised data analysis is a powerful tool for both quality control and answering simple biological questions. We will demonstrate how to format untargeted mass spectrometry data for import into R, a programming language and software environment for statistical computing (R Development Core Team. R: A language and environment for statistical computing, reference index version 2.15. *R Foundation for Statistical Computing*, Vienna, 2012). Using R, we transform untargeted metabolite data using hierarchical clustering and principal component analysis (PCA) to create visual representations of change between biological samples and explore how these can be used predictively, in determining environmental stress, health and metabolic insight.

Key words Clustering, Cluster analysis, Pattern recognition, Untargeted metabolomics, Phenotyping, Data mining

1 Introduction

Untargeted metabolite profile data based upon liquid chromatography-based mass spectrometry (LC-MS) represents dense biological data [1–3]. Using a personal computer, we can format this multidimensional data for computational analysis (*see* **Note 1**). Once the data can be formatted into the appropriate data structure it can be analyzed mathematically in a myriad of different ways to develop underlying biological insight (*see* **Note 2**). Cluster analysis or clustering is a common technique in data mining and used in many fields such as machine learning, data compression and computer graphics [4–7]. It can be used to group a set of objects in a way where more similar objects group closer together than those more distantly related. Clustering is not a single algorithm, but refers to a suite of different algorithms, which differ significantly in their approach to determine relationships. While on the surface clustering may seem to be unrelated to biological analysis, we can use clustering algorithms for data mining untargeted mass

Edward E.K. Baidoo (ed.), *Microbial Metabolomics: Methods and Protocols*, Methods in Molecular Biology, vol. 1859, https://doi.org/10.1007/978-1-4939-8757-3_16,

spectrometry data, both to look at the relationships between cultured microbial samples and their metabolites (*see* **Note 3**). Here we examine the use of Principal Component Analysis and Hierarchical Clustering, two common cluster analysis tools, for digging into multi-dimensional metabolite data.

Principal component analysis or PCA for short is an orthogonal transformation that converts a set of observations of potentially correlated variables into linearly uncorrelated principal components allowing for segregation of the observations based upon these components (*see* **Note 4**) [8–10]. For metabolomics data, observations can refer to samples, and variables to metabolites and their abundance. This means that PCA can cluster samples based upon the dissimilarity of metabolite abundance in the samples, which allows relationships to be observed between samples (*see* **Note 5**).

In Fig. 1 we show a principal component analysis using untargeted metabolite profiles from *Sulfolobus solfataricus. S. solfataricus* is a hyperthermophilic archaea which can be readily grown in culture, one of the few archaea that is able to be cultured [11, 12]. One of its natural habitats is in the boiling acid of the hot spots in Yellowstone Park, and was originally found in the Italian Solfatara volcano. The species grows best in temperatures around 80 °C with pH levels between 2 and 4. These conditions qualify it as an extremophile and it is specifically known as a thermoacidophile. Much effort has been undertaken to understand how this organism responds to viral infection and oxidative stress, as its metabolic systems appear to be unique [11, 12]. In our example, cultures of *S. solfataricus* were grown in two different medias and subjected to two different levels of H_2O_2 (30 μM and 100 μM) stress. These were undertaken with the goal of visualizing a metabolic response that was both reproducible and predictable. Metabolite profiles were then compared using PCA. This example shows how metabolite data can be plotted three-dimensionally to look for common regulatory patterns (*see* **Note 6**). In this example, the data structure is separated into biological replicates (Table 1). Figure 1a–c compares three different classes (C_i) of instantaneous phenotypes of *S. solfataricus* [13, 14]. Where each C_i is associated with initial time point 0 min (t_0) and 30 min (t_1) after stress and using two different medias (m_1) and (m_2), where m_1 is yeast extract media, and m_2 is the same media with supplementary glucose added, and using two different stress levels 30 μM H_2O_2 (s_{30}) or 100 μM H_2O_2 (s_{100}). Overall, there are six different classes of *S. solfataricus* compared in Fig. 1d. The classes (C_i) can be spatially separated using PCA, which becomes more apparent when R's PCA analysis prcomp is used to visualize the data (*see* **Note 7**). It should be noted that PCA is an unsupervised analysis, so it resists human bias by not retaining any class (C) information unlike supervised machine learning (*see* Chapter 17 for more on this topic) [18]. This makes PCA ideal for quality control, as we can see in Fig. 1 prcomp

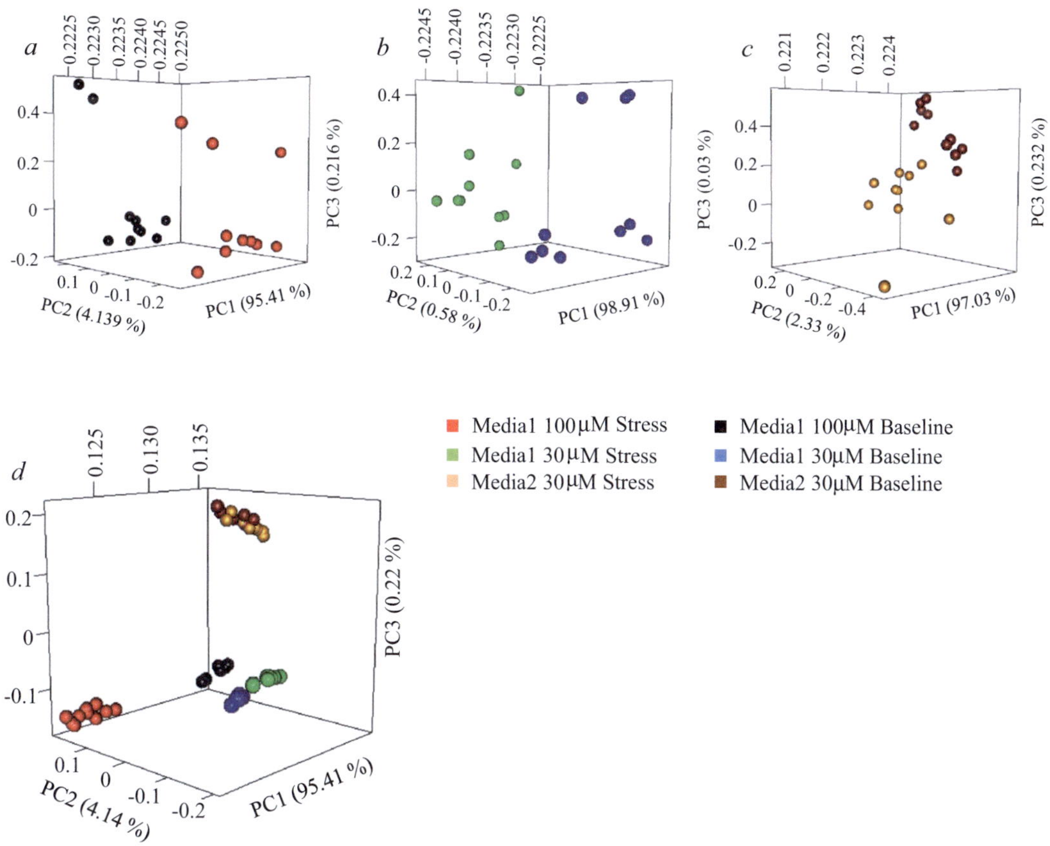

Fig. 1 PCA plots of first three principal components from ten biological replicates of *S. solfataricus* before and after H_2O_2 stress, three variants including (**a**) media 1 before and after 30 min 100 mM H_2O_2, (**b**) Media 1 before and after 30 min 30 μM H_2O_2, (**c**) media 2 before and after 30 min 30 μM H_2O_2, and (**d**) all experiments combined

has no problem grouping without any help. By visual inspection you can see that in experiments of s_{30} stress replicates remain grouped closely at both t_0 and t_1. This is not surprising based upon previous studies that show 30 μM H_2O_2 is a gentle stress for our extremophilic *S. solfataricus* [11]. However, when we look at the higher 100 μM H_2O_2 the results could not be more apparent. The spatial distance between s_{100}, t_0 and t_1 show a much greater distance apart by principal component 1 (PC1). This is because *S. solfataricus* metabolites abundance change more extremely when exposed to the higher concentration of H_2O_2.

A second analysis using the R package is hierarchical clustering (HC) hclust(), which is another visually attractive way to analyze metabolite data [19]. HC creates a distance matrix of relationship between the replicates [4–7]. R creates this distance matrix using dist(), applies the HC using hclust(), then plots the dendrogram. In our example, HC is much better about specifying exact

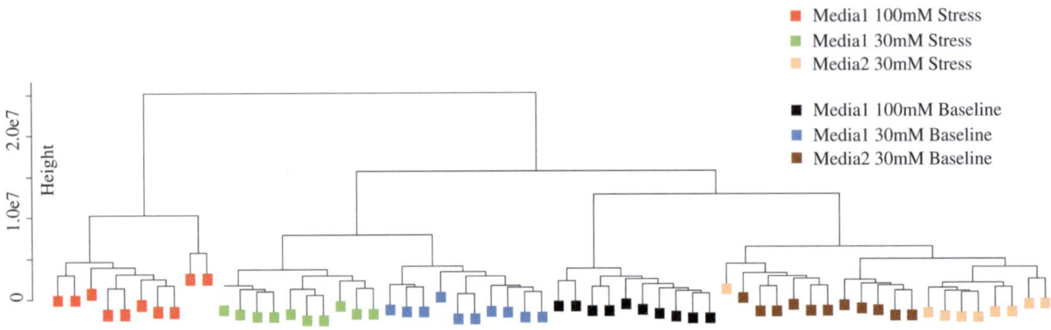

Fig. 2 Hierarchical clustering (HC) distance matrix of relationship of *S. solfataricus* before and after H_2O_2 stress

relationships as the distance matrix allows us a greater resolution than PCA, case and point $m_2t_0s_{30}$ in Fig. 2 which classifies one of the replicates in the wrong group, a relationship which cannot be deduced using PCA. Otherwise, while the analysis is different the same conclusions can be drawn from the data that were observed in the PCA.

Following is a description of how to format your data for import into R and how to analyze this data using both PCA and HC. Example data has been included in the supplementary material.

2 Materials

Installation and data analysis using R was performed on a Windows 7 Operating system, but should be applicable in general to both OSX and various Linux builds compatible with R. R and all the clustering libraries should be compatible on most modern computers. Please refer to R documentation for guidance. Recommended configuration would be Processor (CPU) Intel Core i5-6xxx or equivalent, Operating System Microsoft Windows 7 or higher, Memory: 8 GB RAM, Storage: 500 GB internal hard drive, Monitor/Display: 14″ LCD monitor, resolution of 1600 × 900 or better.

3 Methods

3.1 Data Preparation

Both of these data analysis algorithms can be easily used side by side to answer basic questions about your metabolomics data. However, before this can be done your data must be converted into the appropriate format for import.

We use a comma-separated version (.csv) of the data where each row represents the abundance of a single metabolite, and each column represents a different replicate (Table 1). This should be available for any liquid chromatography mass spectrometry run where metabolites are separately eluted from a chromatography column,

Table 1
Data format for import of metabolite data in R. This contains the first four columns and ten rows from the dataset Clustering_CSV.csv example file. The first column contains the name of the metabolites from XCMS. M100T143 means mass of 100 *m/z* at retention time 143 s etc. Each following column contains the relative intensity of that metabolite in each of the replicates, (Media1_100 μM_Baseline_1...10, etc.)

Metabolite	Media1_100 μM M_Baseline_1	Media1_100 μM _Baseline_2	Media1_100 μM _Baseline_3
M100T143	7931.862018	8006.213351	7790.159829
M101T1396	303380.4001	287845.7038	301148.8986
M101T165	1434666.182	1590029.759	1502503.19
M101T19	80803.03973	68269.51801	81182.25396
M101T291	86876.47482	100728.9849	92428.268
M102T1384	65716.93178	109618.4603	65023.54204
M102T46	90517.28081	91617.47102	91866.27286
M103T1405	39396.86696	35676.64597	39683.62704
M103T172	34170.08346	33365.15348	36185.35018

whether hydrophilic interaction (HILIC) or reversed-phase (C18). XCMS works to extract the data from the mass spectrometric data file by grouping all masses with similar retention into a bin for overall intensity quantification (*see* **Note 2**). This requires the file to be converted to an mzml or mzXML open source file. This version is an eXtensible Markup Languange (XML), which is a common file format for proteomics mass spectrometric data. Typically instrument manufacturers have proprietary software which can handle the conversion or alternatively researchers can use proteowizard or xcms online [15–17]. In our example the data is generated by xcms online, you can set up a free account and with some minor modification your data is in the appropriate format. For instructions on use of xcms please *see* ref. 11.

1. From results.zip file downloaded from xcms online open "XCMS.diffreport.xlsx" in excel.
2. Delete first 22 columns, which are the XCMS differential report values and keep only the raw data, which should be labeled with the name of the replicates.
3. Using excel save this file as filename.csv file for import in R.
4. Download and install the latest version of R for your operating system (https://www.r-project.org/).
5. Open R console.
6. Open Misc toolbar and choose "change working directory...".

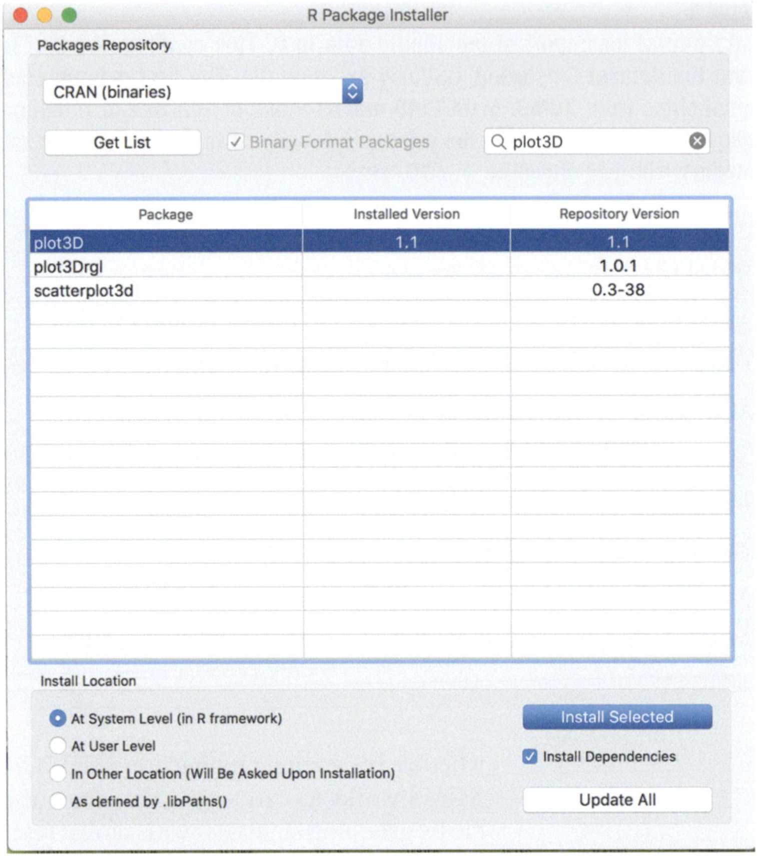

Fig. 3 R Package Installer

7. Navigate to the folder which contains your .csv with your data.
8. To import your .csv data into R use the following command: mydata <- read.table(file="filename.csv", header=TRUE, sep=",").

3.2 Install Dependencies

Principal component analysis in R, as mentioned relies on the prcomp() function which comes in the standard installation (*see* Fig. 3). This function performs a PCA on the given data matrix and returns the results as an object of class prcomp [18]. The calculation is based upon a singular value decomposition of the data matrix. In most cases this is the preferred method for numerical accuracy. The print method for these objects prints the results of the components on the console and plot method produces a 2D screen plot. For 3D plots plot3D will need to be installed and to save a snapshot of the plot rgl must be installed [19].

1. Open R console.
2. From the Packages & Data Menu select R package installer.
3. Enter plot3D in the search bar.

4. Press “Get List” button.
5. Highlight plot3D.
6. Check install dependencies.
7. Press “Install Selected” button.
8. Enter rgl into search bar.
9. Select rgl from package list.
10. Press “Install Selected” button.
11. Close and return to console.
12. Load plot3d with >library(plot3D).
13. Load rgl with > library(rgl).

3.3 PCA Analysis on Datasets

With the dependencies installed we are now ready to perform the PCA. Import your formatted data into the R console as mydata to begin. Using example data set M1_100.csv run the following commands in the console. This should reproduce the data in Fig. 1a. With rgl you can rotate the plot to get the best angle for the snapshot.

1. mydata.pca <- prcomp(mydata, retx=TRUE, center=TRUE, scale.=TRUE)
2. summary(mydata.pca)
3. plot3d((mydata.pca$rotation), xlab="PC1(%)", ylab="PC2 (%)", zlab="", type="s", col=c("black", "black", "black", "black", "black", "black", "black", "black", "black", "black", "red", "red", "red", "red", "red", "red", "red", "red", "red", "red"), size=2, lwd=15, box=F) (*see* **Note 8**)
4. rgl.snapshot("M1_100.png", fmt="png")

Using example data set M1_30.csv run the following commands in the console. This should reproduce the data in Fig. 1b.

1. mydata <- read.table(file="M1_30.csv", header=TRUE, sep=",")
2. mydata.pca <- prcomp(mydata, retx=TRUE, center=TRUE, scale.=TRUE)
3. summary(mydata.pca)
4. plot3d((mydata.pca$rotation), xlab="PC1(%)", ylab="PC2 (%)", zlab="", type="s", col=c("blue", "blue", "blue", "blue", "blue", "blue", "blue", "blue", "blue", "blue", "green", "green", "green", "green", "green", "green", "green", "green", "green", "green"), size=2, lwd=15, box=F)
5. rgl.snapshot("M1_30.png", fmt="png")

Using example data set M2_30.csv run the following commands in the console. This should reproduce the data in Fig. 1c.

1. mydata <- read.table(file="M2_30.csv", header=TRUE, sep=",")
2. mydata.pca <- prcomp(mydata, retx=TRUE, center=TRUE, scale.=TRUE)
3. summary(mydata.pca)
4. plot3d((mydata.pca$rotation), xlab="PC1(%)", ylab="PC2 (%)", zlab="", type="s", col=c("brown", "brown", "brown", "brown", "brown", "brown", "brown", "brown", "brown", "brown", "orange", "orange", "orange", "orange", "orange", "orange", "orange", "orange", "orange", "orange"), size=2, lwd=15, box=F)
5. rgl.snapshot("M2_30.png", fmt="png")

Using example data set H_2O_2_all.csv run the following commands in the console. This should combine all of the data and reproduce the plot in Fig. 1d.

1. mydata <- read.table(file="H_2O_2_all.csv", header=TRUE, sep=",")
2. mydata.pca <- prcomp(mydata, retx=TRUE, center=TRUE, scale.=TRUE)
3. summary(mydata.pca)
4. plot3d((mydata.pca$rotation), pch=20, cex=10, size=10, xlab="PC1(%)", ylab="PC2(%)", zlab="PC3(%)", col=c ("black", "black", "black", "black", "black", "black", "black", "black", "black", "black", "blue", "blue", "blue", "blue", "blue", "blue", "blue", "blue", "blue", "blue", "brown", "brown", "brown", "brown", "brown", "brown", "brown", "brown", "brown", "brown", "red", "red", "red", "red", "red", "red", "red", "red", "red", "red", "green", "green", "green", "green", "green", "green", "green", "green", "green", "green", "orange", "orange", "orange", "orange", "orange", "orange", "orange", "orange", "orange", "orange"))
5. rgl.snapshot("H_2O_2_all.png", fmt="png")

3.4 HC Analysis

The good news is HC analysis doesn't require any additional libraries and can be run very similar to the PCA analysis, only with different commands [20]. The output is different, but the execution is very similar. In this example only one dataset will be demonstrated to show the relationship of all the samples in H_2O_2_all.csv. These commands should reproduce the results of Fig. 2; however additional colors were added to the figure to make the results easier to understand. In this analysis the data must be transposed before executing using the R transposition function t(). A dissimilarity structure then using dist(), hclust is performed on the results of d, then a dendrogram is built to display the output.

6. mydata <- read.table(file="H_2O_2_all.csv", header=TRUE, sep=",")
7. mydata.t <- t(mydata)
8. d <- dist(as.matrix(time.t))
9. hc <- hclust(d)
10. png('hc_H_2O_2.png')
11. plot(hc)
12. dev.off()

4 Notes

1. Dense data refers to each replicate having hundreds to thousands of different variables or molecular feature metabolites, which are represented quantitatively. This is not common to all data types as often each replicate only contains a few variables. This distinction is important as metabolic experiments often have a limited number of experimental replicates or samples and a large number of molecular features. In many cases the number of molecular features in each LC-MS run outnumbers the samples by several orders of magnitude (i.e., 10 control vs 10 stress, with each LC-MS run having 1000 metabolites in each run).
2. It is important to note that data structures are specialized formats for organizing and storing data. General data structure types include arrays. Any data structure is designed to organize the data to suit a specific purpose so that it can be accessed and manipulated homogenously.
3. Orthogonal transformation is a linear transformation which preserves a symmetric inner product. In general, an orthogonal transformation preserves the lengths of vectors and angles between vectors.
4. PCA can be used to map metabolites by their similarity, however we are using them to group replicates in our example. The former results in more of a cloud as there are thousands of metabolites in the data files.
5. Principal components are arranged in a way where the first component represents the majority of the variance between the samples, the second component the next most variance and so on. Plotting a three-dimensional analysis is often not

required to find grouping and often PCA visualization is only undertaken on two components, which can be sufficient if variance is low.

6. princomp(x...) is another algorithm available in R for PCA, however it has trouble plotting data sets with high numbers of samples.
7. Unsupervised analysis is much better for getting an objective view of your data as you do not feed any class information to the algorithm.
8. Different values can be given for plot3D variables to allow you to customize the plot how you like visually. Please refer to R documentation for more details [19].

Acknowledgments

The authors would also like to acknowledge that this work was part of the DOE Joint BioEnergy Institute (http://www.jbei.org) supported by the US Department of Energy, Office of Science, Office of Biological and Environmental Research, through contract DE-AC02-05CH11231 between Lawrence Berkeley National Laboratory and the US Department of Energy.

References

1. Development Core Team R (2012) R: A language and environment for statistical computing, reference index version 2.15.1. R Foundation for Statistical Computing, Vienna
2. Patti GJ, Tautenhahn R, Siuzdak G et al (2012) Meta-analysis of untargeted metabolomic data from multiple profiling experiments. Nat Protoc 7(3):508–516
3. Patti GJ, Yanes O, Shriver LP, Courade J, Tautenhahn R, Manchester M, Siuzdak G et al (2012) Metabolomics implicates altered sphingolipids in chronic pain of neuropathic origin. Nat Chem Biol 8(3):232–234
4. Everitt B (1974) Cluster analysis. Heinemann Educational Books, London
5. Hartigan JA (1975) Clustering algorithms. Wiley, New York
6. Anderberg MR (1973) Cluster analysis for applications. Academic Press, New York
7. Murtagh F (1985) Multidimensional Clustering Algorithms. In: COMPSTAT Lectures 4. Physica-Verlag, Wuerzburg
8. Becker RA, Chambers JM, Wilks AR (1988) The new S language. Wadsworth & Brooks/Cole Advanced Books & Software, Monterey
9. Mardia KV, Kent JT, Bibby JM (1979) Multivariate analysis. Academic Press, London
10. Venables WN, Ripley BD (2002) Modern applied statistics with S. Springer-Verlag, Berlin
11. Heinemann J, Hamerly T, Maaty WS, Movahed N, Steffens JD, Reeves BD, Hilmer JK, Therien J, Grieco PA, Peters JW, Bothner B et al (2014) Expanding the paradigm of thiol redox in the thermophilic root of life. Biochim Biophys Acta 1840:80–85
12. Maaty WS, Wiedenheft B, Tarlykov P, Schaff N, Heinemann J, Robison-Cox J, Valenzuela J, Bothner B et al (2009) Something old, something new, something borrowed; how the thermoacidophilic archaeon Sulfolobus solfataricus responds to oxidative stress. PLoS One 4(9): e6964
13. Gordon AD (1999) Classification. Chapman and Hall / CRC, London
14. McQuitty LL (1966) Similarity analysis by reciprocal pairs for discrete and continuous data. Educ Psychol Meas 26:825–831
15. Kessner D, Chambers M, Burke R, Agus D, Mallick P et al (2008) ProteoWizard: open source software for rapid proteomics tools

development. Bioinformatics 24 (21):2534–2536

16. Tautenhahn R, Böttcher C, Neumann S et al (2008) Highly sensitive feature detection for high resolution LC/MS. BMC bioinformatics 9:504
17. Yanes O, Tautenhahn R, Patti GJ, Siuzdak G et al (2011) Expanding coverage of the metabolome for global metabolite profiling. Anal Chem 83(6):2152–2161
18. https://stat.ethz.ch/R-manual/R-devel/library/stats/html/prcomp.html
19. https://www.rdocumentation.org/packages/rgl/versions/0.97.0/topics/plot3d
20. http://stat.ethz.ch/R-manual/R-devel/library/stats/html/hclust.html

Chapter 17

Machine Learning in Untargeted Metabolomics Experiments

Joshua Heinemann

Abstract

Machine learning is a form of artificial intelligence (AI) that provides computers with the ability to learn generally without being explicitly programmed. Machine learning refers to the ability of computer programs to adapt when exposed to new data. Here we examine the use of machine learning for use with untargeted metabolomics data, when it is appropriate to use, and questions it can answer. We provide an example workflow for training and testing a simple binary classifier, a multiclass classifier and a support vector machine using the Waikato Environment for Knowledge Analysis (Weka), a toolkit for machine learning. This workflow should provide a framework for greater integration of machine learning with metabolomics study.

Key words Machine learning, Untargeted metabolomics, Supervised learning

1 Introduction

Machine learning (ML) came into being during the quest for artificial intelligence, within early AI research, examples used symbolic methods, as well as artificial neural networks [1]. Today ML is widely recognized as the programming of computers to optimize decision-making based upon example data and past experience [2–4]. Learning is necessary when we cannot write a computer program to solve a problem, but instead need a way to use pattern recognition to analyze complex data or experience. One case where learning is necessary is in computer based facial recognition that is, recognizing someone's identity based on differences in the positions and shapes of nose, ears, eyes, hair, and outline. People have a unique conglomeration of attributes, which makes them identifiable, however directly programming an algorithm to evaluate all those features simultaneously is difficult. A learning algorithm is ideal for solving this problem as studying example images and

Electronic supplementary material: The online version of this chapter (https://doi.org/10.1007/978-1-4939-8757-3_17) contains supplementary material, which is available to authorized users.

Edward E.K. Baidoo (ed.), *Microbial Metabolomics: Methods and Protocols*, Methods in Molecular Biology, vol. 1859, https://doi.org/10.1007/978-1-4939-8757-3_17,

comparing them to repositories of negative controls can provide better results [5, 6]. A learning algorithm can be trained to distinguish images containing ML and data mining often utilizes similar methods, however the former focuses on prediction and the latter on finding new information [1].

Machine learning can be loosely separated into three different categories: supervised learning, unsupervised learning, and reinforcement [7]. Supervised learning involves the computer and is presented with example inputs and their desired outputs. In supervised learning a "teacher" trains the algorithm to associate a general rule to map variables with an outcome. In unsupervised learning, no data label is given to the data and the machine is forced to find hidden patterns within the data without a "teacher" [8, 9]. Reinforcement learning is when a learning algorithm is forced to perform a goal within that environment, and the program is given rewards and punishments based on performance [10, 11].

Machine learning can also be grouped upon the desired output of the algorithm. Outputs of machine learning algorithms can be grouped into classification, regression, clustering, and dimensionality reduction. A classifier identifies which set of categories (sub-populations) a new observation belongs to, based upon training received from a training set. Where the training set contains observations (or instances) of known classification [12]. Regression analysis is the process of estimating the relationships between variables. This technique focuses on the relationship between dependent and one or more independent variables. Regression is typically used for prediction and forecasting of the relationships between the variables by looking at multiple observations [1, 12]. Cluster analysis or clustering is the grouping of objects so that objects in the same group (cluster) are more similar to one another, than to members of different clusters (*see* **Note 1**). Exploratory data mining relies heavily on clustering because of the ability of the algorithm to find patterns in dense data [13–15].

Machine learning has a wide reach and there are many variations of use for algorithms. Ultimately the machine learner is evaluated on its accuracy or ability to correctly place known data into groups (*see* **Note 2**). A core objective of a learner is to generalize from experience. Generalization in the context of ML, the learner should be able to accurately make decisions about unseen data from previous experience [1–5]. For best performance (accuracy), the complexity of the analysis should match the complexity of the data (*see* **Note 3**). If it does not there is risk of overfitting the data, which fails to give the algorithm the ability to generalize, the core objective. When analyzing untargeted metabolomics data, there is substantial risk of overfitting because the number of variables (metabolites) is greater than the number of instances (replicates) (*see* **Note 4**), increasing the likelihood that the algorithm can only predict accurately within those data sets [4]. Prediction of new datasets will confound the machine-learning algorithm because it over fit its model to the training data.

To resist overfitting, we use cross-validation (*see* **Note 5**). Cross-validation is a model validation method for estimating how well the results of a machine learning analysis will generalize to an independent dataset [3, 4]. The main application of cross-validation is to estimate the predictive accuracy of the model on unseen data. In a prediction problem, a learning algorithm is trained using known data (training data) and then tested on validation data (testing data), to evaluate predictive accuracy. Cross-validation serves to limit problems like overfitting by giving insight into how the model will generalize to unseen data [3,4]. Overfitting is when a statistical model describes random error or noise instead of the underlying relationship [1, 3, 4]. This occurs when the model is too complex, such as having too many parameters (metabolites) and too few observations (replicates) [16]. An overfit model has poor predictive accuracy, and tends to overreact to minor fluctuations in the training data [4, 16]. When performing machine learning on metabolomics datasets, cross-validation is very important, as overfitting can be a common problem. This is because cross-validation helps reduce the likelihood of overfitting by breaking the data into subsets and testing the accuracy of the algorithm on all possible combinations of training and testing. This strategy helps tune the learning algorithm to be general enough to predict unseen data [3, 17, 18].

Cellular life uses a wide range of metabolic phenotypes to survive while juggling a wide variety of different environmental conditions (*see* **Note 6**). Metabolic phenotypes can be seen to change significantly during growth and exposure to stress [19–21]. These phenotypes can be best discriminated using liquid chromatography mass spectrometry (LC-MS), because of the technologies ability to quantify numerous metabolite targets simultaneously [22–24]. This makes untargeted metabolite profiles very good at identifying biomarkers for disease, stress or environment (*see* **Note 7**). However, normal biological variation in metabolic response can make this difficult [16]. The proper application of machine learning in metabolomics data sets is probably best used in addition to data mining algorithms like principal component analysis or hierarchical clustering. This is primarily because data mining algorithms can be run in an unsupervised manner, which allows the machine to find meaningful relationships between samples with less risk of human bias. This being said, when the problem is too complex for more simple algorithms to predict machine learning algorithms have been successfully implemented using a wide range of experimental sample numbers, e.g., $n = 78, 42, 38, 32$ [25], $n = 34$ [21], and $n = 72$ [17]. Machine learning algorithms represent powerful tools for metabolomics analysis, however, using a personal computer we are limited to simple applications and data formatting as functional use often requires use of substantial computational resources not covered in this chapter.

Here we demonstrate the use multiple algorithms for the analysis of complex biological data.

2 Materials

Installation and data analysis using WEKA was performed on OSX (10.12.4), but should be applicable in general to both Windows and various Linux builds compatible. WEKA and all the algorithm libraries should be compatible on most modern computers. Please refer to WEKA documentation for guidance. Recommended configuration would be Processor (CPU) Intel Core i7 or equivalent, Operating System OSX 10 or higher, Memory: 16 GB RAM, Storage: 500 GB internal hard drive.

3 Methods

3.1 WEKA

The Waikato Environment for Knowledge Analysis (Weka), is a suite of machine learning software for utilizing machine learning written in Java [26] (Fig. 1). It is free software licensed under the GNU General Public License. Weka combines visualization tools with algorithms for predictive modeling. Weka is portable and since it is based on the Java programming language it is compatible with most modern computing platforms. It utilizes a user-friendly graphical user interface for ease of use. We will first validate your Weka installation is operating properly using the iris.arff dataset. The iris flower dataset, or Fisher's Iris data set is a famous dataset from statistics and is heavily used in machine learning. It contains 150 instances (rows) of four attributes and three classes (setosa, versicolor, and virginica).

1. Download appropriate version of weka three from (http://www.cs.waikato.ac.nz/ml/weka/downloading.html).
2. Start Weka, double click weka.jar file or use launcher.
3. Click on the "Explorer" button to launch the weka explorer.
4. Click "Open file..." and open the "data" directory (Weka/weka-3-8-1/data). This directory contains a number of example machine learning datasets (Figs. 2 and 3).
5. Select "iris.arff" to load the iris dataset.
6. Click the "Classify" tab.
7. Click "Choose" and navigate to ZeroR under "Weka/classifiers/rules/ZeroR."
8. Select "ZeroR" algorithm and click start (Fig. 4).

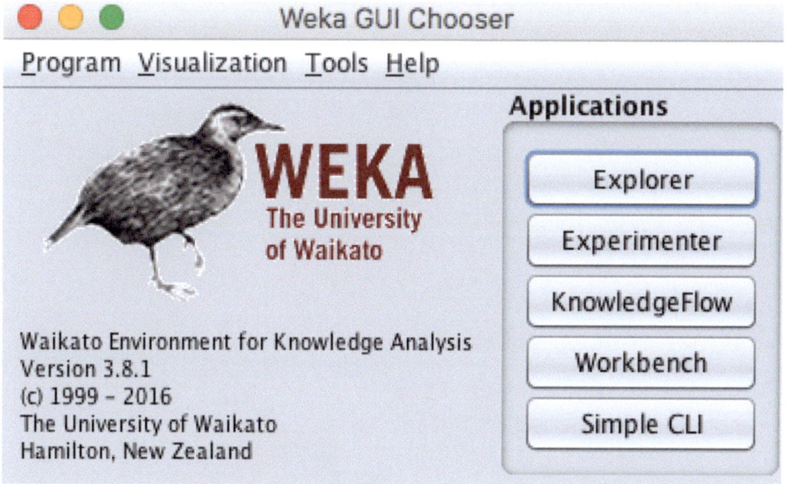

Fig. 1 Weka machine learning software, home screen for using software. For exploring the use of machine learning with metabolomics data we will use the explorer

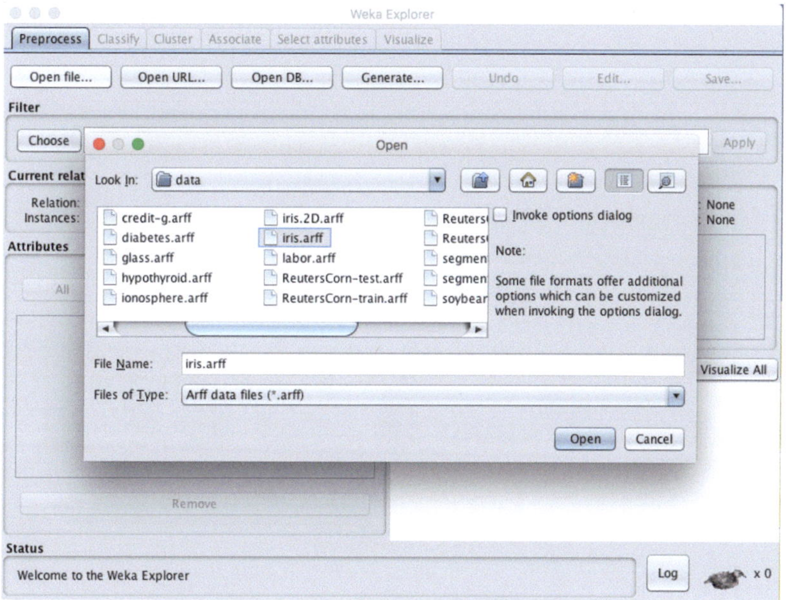

Fig. 2 Loading the example dataset

The ZeroR algorithm predicts which instances belong to the first class in the dataset (setosa). The result as expected contains three classes, each equally represented (50 instances), predicting correctly 33% of the instances belong to setosa. Cross-validation is performed as a tenfold cross-validation. This means that the dataset is split into ten parts, the first nine are used to train the classifier and

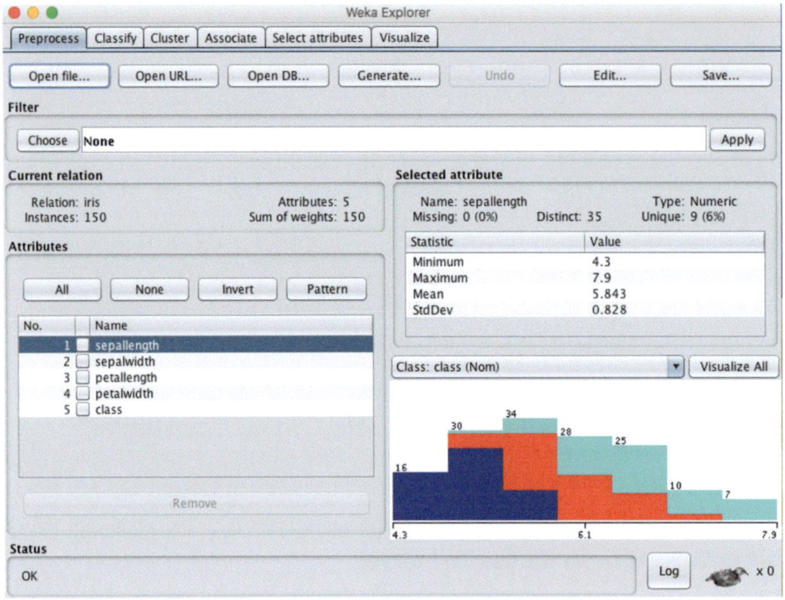

Fig. 3 Dataset overview

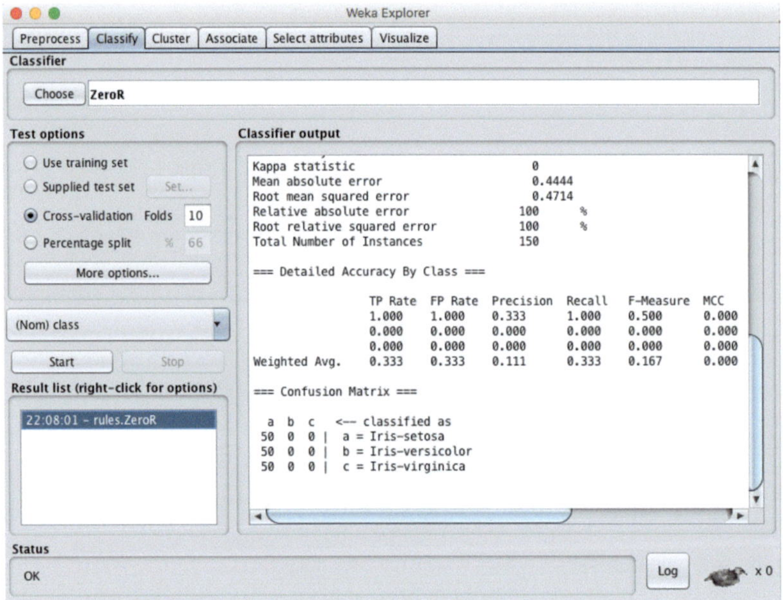

Fig. 4 Output of ZeroR classifier

the 10th is used to test. This process is repeated until each of the ten parts can be tested independently, and then results are pooled.

9. Click the choose button in the "Classifier" section under "Trees" select "J48" algorithm and click start (Fig. 5).

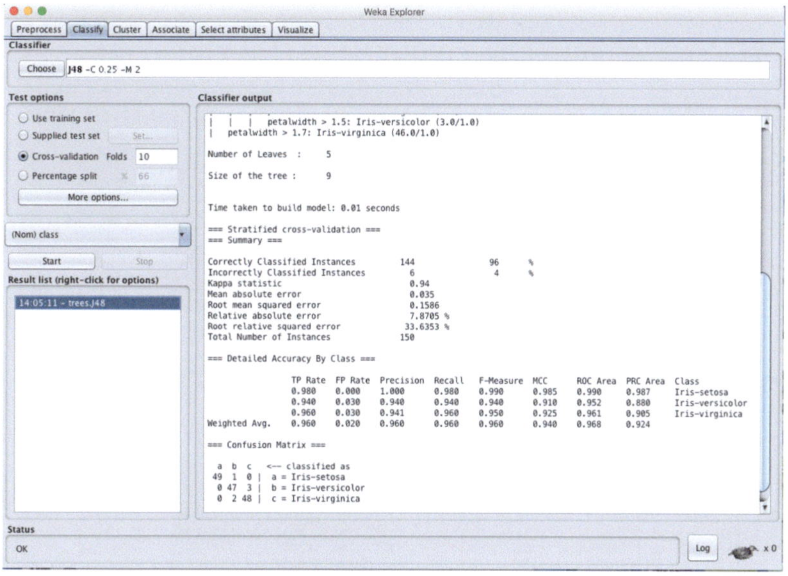

Fig. 5 Output of J48 classifier

J48 is the Weka version of the C4.5 algorithm, a decision tree algorithm. The decision trees used by C4.5 can be used for multi-class classification, whereas the ZeroR algorithm is only a binary classifier. "Classifier output" results from Fig. 3 show J48 is implemented with a tenfold cross-validation just like ZeroR. The result is a summary of those predictions.

3.2 Preparing and Analyzing Metabolomics Data

The (Attribute-Relation File Format) file (i.e., ARFF) is an ASCII text file that describes a list of instances (aka. conditions) sharing a set of attributes (aka. metabolites). The ARFF files as described are used by the machine learning project at the Department of Computer Science in the University of Waikato for use with Weka 3.2 and 3.3, the format is originally attributed to Ian H. Witten and Eibe Frank (**42**). For metabolomics the relative abundance of each mass (m/z) will be considered an attribute, which will be used to compare between different Class (C_i) phenotypes. Our example Classes are: C_{M1_100}, C_{M1_30}, C_{M2_30}, and $C_{control}$, where M1_100 refers to a culture of *S. solfataricus* grown in Media 1, stressed with 100 μM H_2O_2, M1_30 designates Media 1 and 30 μM H_2O_2, and M2_30 designates Media 2 and 30 μM H_2O_2.

In the .csv to .arff format conversion process a header is added to the .arff file, which will be automatically created through the workflow (Fig. 6). The .arff is a text file containing a header and the data in the format of metabolite abundance (a) where each row represents a Class metabolic phenotype (M1_100 control...), which is best described as a mass spectrometry data file containing a list of metabolites and their abundance from xcms (*see* Chapter 16

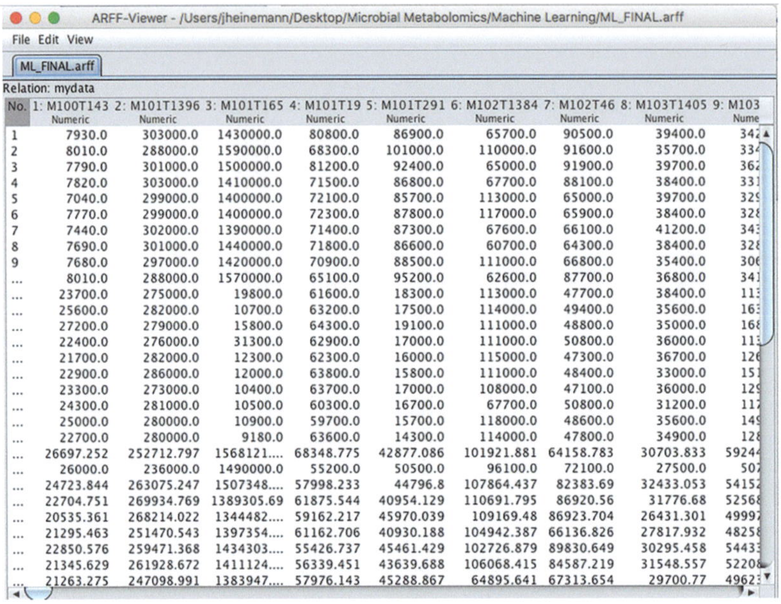

No.	1: M100T143 Numeric	2: M101T1396 Numeric	3: M101T165 Numeric	4: M101T19 Numeric	5: M101T291 Numeric	6: M102T1384 Numeric	7: M102T46 Numeric	8: M103T1405 Numeric	9: M103 Nume
1	7930.0	303000.0	1430000.0	80800.0	86900.0	65700.0	90500.0	39400.0	342
2	8010.0	288000.0	1590000.0	68300.0	101000.0	110000.0	91600.0	35700.0	334
3	7790.0	301000.0	1500000.0	81200.0	92400.0	65000.0	91900.0	39700.0	362
4	7820.0	303000.0	1410000.0	71500.0	86800.0	67700.0	88100.0	38400.0	331
5	7040.0	299000.0	1400000.0	72100.0	85700.0	113000.0	65000.0	39700.0	329
6	7770.0	299000.0	1400000.0	72300.0	87800.0	117000.0	65900.0	38400.0	328
7	7440.0	302000.0	1390000.0	71400.0	87300.0	67600.0	66100.0	41200.0	343
8	7690.0	301000.0	1440000.0	71800.0	86600.0	60700.0	64300.0	38400.0	328
9	7680.0	297000.0	1420000.0	70900.0	88500.0	111000.0	66800.0	35400.0	306
...	8010.0	288000.0	1570000.0	65100.0	95200.0	62600.0	87700.0	36800.0	341
...	23700.0	275000.0	19800.0	61600.0	18300.0	113000.0	47700.0	38400.0	113
...	25600.0	282000.0	10700.0	63200.0	17500.0	114000.0	49400.0	35600.0	163
...	27200.0	279000.0	15800.0	64300.0	19100.0	111000.0	48800.0	35000.0	168
...	22400.0	276000.0	31300.0	62900.0	17000.0	111000.0	50800.0	36000.0	113
...	21700.0	282000.0	12300.0	62300.0	16000.0	115000.0	47300.0	36700.0	126
...	22900.0	286000.0	12000.0	63800.0	15800.0	111000.0	48400.0	33000.0	151
...	23300.0	273000.0	10400.0	63700.0	17000.0	108000.0	47100.0	36000.0	129
...	24300.0	281000.0	10500.0	60300.0	16700.0	67700.0	50800.0	31200.0	117
...	25000.0	280000.0	10900.0	59700.0	15700.0	118000.0	48600.0	35600.0	149
...	22700.0	280000.0	9180.0	63600.0	14300.0	114000.0	47800.0	34900.0	128
...	26697.252	252712.797	1568121....	68348.775	42877.086	101921.881	64158.783	30703.833	5924
...	26000.0	236000.0	1490000.0	55200.0	50500.0	96100.0	72100.0	27500.0	507
...	24723.844	263075.247	1507348....	57998.233	44796.8	107864.437	82383.69	32433.053	5415
...	22704.751	269934.769	1389305.69	61875.544	40954.129	110691.795	86920.56	31776.68	5256
...	20535.361	268214.022	1344482....	59162.217	45970.039	109169.48	86923.704	26431.301	4999
...	21295.463	251470.543	1397354....	61162.706	40930.188	104942.387	66136.826	27817.932	4825
...	22850.576	259471.368	1434303....	55426.737	45461.429	102726.879	89830.649	30295.458	5443
...	21345.629	261928.672	1411124....	56339.451	43639.688	106068.415	84587.219	31548.557	5220
...	21263.275	247098.991	1383947....	57976.143	45288.867	64895.641	67313.654	29700.77	4962

Fig. 6 Data format of an ARFF file, contains class labels at the end of the rows

for formatting mass spectrometry data to .csv format). For conversion of .csv to .arff we must first use R to transpose the data. *See* Chapter 16 for details of R installation.

1. Load R.
2. Change working directory to location of data file.
3. Load our data (*see* Supplemental Files online) with: mydata <- read.table(file = "H2O2_all_final", header = TRUE, sep = ",").
4. Transpose our data: mydata.t < - t(mydata).
5. Save the transposed data: write.table(mydata.t, "~/mydirectory/H2O2_all_final.csv", sep = ","), where "mydirectory" refers to the location you would like to save the file.
6. Label each column header for each class with matching names for each class instance.
7. Open your newly minted CSV in Excel.
8. Delete the first row that contains "V1, V2, V3…".
9. Cut the first column, with the labels and paste to the last column ("ALY" in our example dataset).
10. Delete the now empty column A.
11. Change the labels in column (ALY) so that all replicates have the same designation, i.e., M1_100 for all ten replicates etc. *See* ML_FINAL.arff for reference (*see* Supplemental Files online).
12. Save your newly formatted .csv file as ML_FINAL.csv.

13. Open "Weka GUI Chooser" (home screen).
14. From tools menu choose ARFF viewer.
15. Click "Open file..." and open your .csv file.
16. Next go to File→save as, select Arff data files (*see* **Note 8**).

The ML classifier is supervised, which means it assigns the relative abundance of each metabolite in a list to the Class you define. In our case we have ten replicates of each Class from which to train the algorithm. We should remember the aim of machine learning is rarely to predict the training data, but to predict new cases. This means we would like to generate the right classification for an input instance outside the training set, one which the correct output is not given in the training set. How well a model trained on the training set predicts the right output for new instances is called generalization (*see* **Note 9**).

3.3 Support Vector Machines

Support Vector Machines (SVM) are supervised learning models, we will use the LibSVM support vector machine library for building a learning model. LibSVM utilizes a sequential minimal optimization or SMO-type algorithm to solve a quadratic optimization problem [27]. Although users do not need to understand underlying theory behind SVM, it can be helpful to go through a procedure, which allows an understanding of what can be expected when working with data. A proposed procedure for obtaining familiarity should consist of the following:

- Transform metabolomics data to the format of an SVM package.
- Conduct simple scaling on the data.
- Consider the RBF kernel for complex problems to increase accuracy.
- Use cross-validation to find the bests parameters.
- Use the best parameters to train the whole training set.
- Test.

It is good to go back and review machine-learning concepts to optimize the SVM learning accuracy and prevent misinterpretation of the results.

17. From Weka GUI chooser select the "tools" menu "package manager".
18. In "package search" type "LibSVM".
19. Click on LibSVM (our version is 1.0.10 for this exercise) and click install.
20. Check for "Packages installed successfully".
21. Return to "Weka GUI Chooser" (home screen).
22. Click on the "Explorer" button to launch the weka explorer.

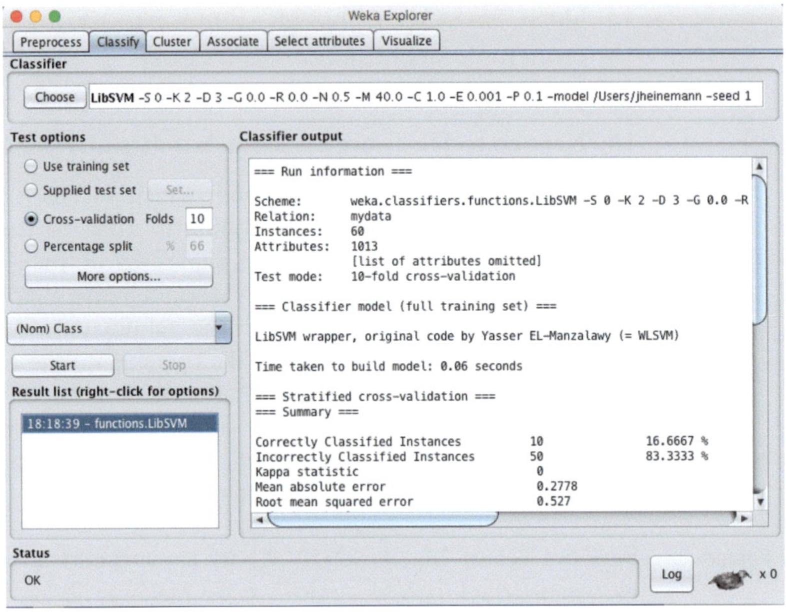

Fig. 7 Output of SVM classifier

23. Click "Open file..." and open the "data" directory.
24. Select "ML_FINAL.arff" to load the Metabolomics dataset.
25. Click the "Classify" tab.
26. Select "Choose" and under/classifiers/functions/select LibSVM.
27. Click "Start" to begin and view the output (Fig. 7).

You will notice that, the number of correctly classified instances is only 16.6667% because the mass spectrometry data needs to be normalized. The relative intensity of the different metabolites spans several orders of magnitude, which can incorrectly bias the learning algorithm.

28. In the "Classifier" section of the "Classify" tab, click on the bold "LibSVM".
29. Change "SVMType" to "nu-SVC (classification)".
30. Change "normalize" to "True".
31. Select "OK".
32. Select "Start" to begin retraining.

The new parameters applied from the LibSVM library should have no problem quickly analyzing your data. The most important number to evaluate is located in the upper right hand screen of the "Classifier output" window. If you used the iris.ARFF data file you will observe that the SVM was 96.6667% accurate with 145 samples classified correctly and only five classified incorrectly out of a total

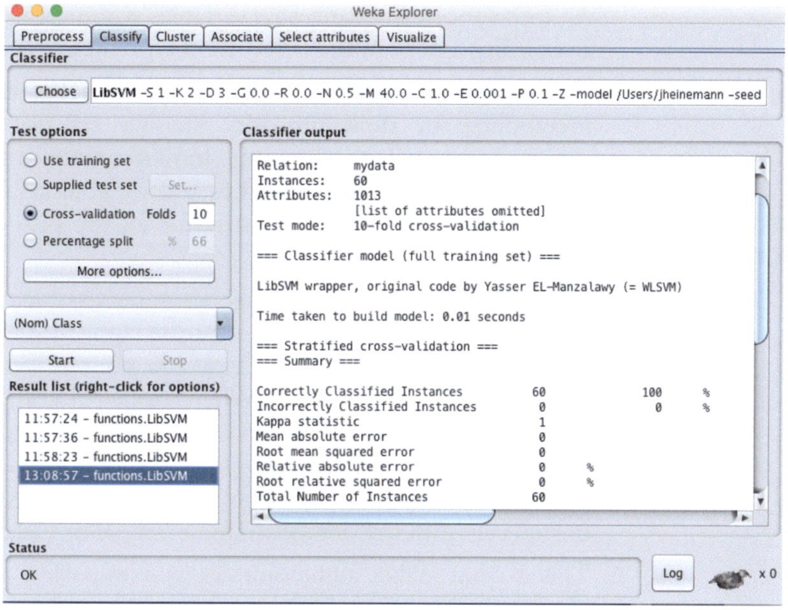

Fig. 8 Output of SVM classifier with normalized data and nu-SVC

of 150 instances. LibSVM has no problem doing multiclass classification with 100% accuracy on the normalized data ML_FINAL.arff as can be visualized from the confusion matrix using nu-SVC (Fig. 8). What you do not see behind the scenes is that the SVM is run recursively over and over again using cross-validation to optimize the decision classification. In the "Classifier" tab where we chose the LibSVM, you can see a long chain of parameters, which were chosen by default. These parameters can be adjusted to improve the accuracy of the algorithm.

4 Notes

1. Dense data refers to each replicate having hundreds to thousands of different variables, which are molecular features or metabolites. This is not common to all data types as often each replicated may only contain a few variables. Clustering mentioned is similar and can include the clustering used in the Cluster Analysis of Untargeted Metabolomics Experiments Chapter 16.
2. Data structures are specialized formats for organizing and storing data for groups. Garbage in = Garbage out. An algorithm's ability to group ultimately depends on the quality of your data, and failure to group is almost certainly a user error. General data structure types include arrays.
3. This refers to the number of instances that should outnumber or at least match the number of variables in all of the datasets.

An imbalance of instances to variables will increase the risk of overfitting and potentially reduce the predictive accuracy.

4. This analysis is similar to PCA which can be used to map metabolites by their similarity like we are using them to group replicates in our example. In metabolomics experiments instances are known as replicates and variables as metabolites. This distinction will remain true for all the purposes of this chapter.
5. Cross validation is only possible for supervised learning as the data categories must be known for the algorithm to run recursively on all possible splits of the data training versus testing. This is different than PCA. Where principal components are arranged in a way where the first component represents the majority of the variance between the samples, the second component the next most variance and so on. PCA is completely unsupervised in the demonstrated example in Chapter 16.
6. This is what makes metabolomics profiles perfect for machine learning applications. That is, the data is so dense that there are many underlying relationships, which are very noisy to external stress and environment. Utilizing the sensitivity of metabolite abundance to stress, disease and environmental variation makes them the perfect medium for creating dense biological data.
7. The ability of LC-MS to detect a wide variety of different biological compounds makes it ideal for identifying new biomarkers and new undiscovered molecules in general. This is due to the ability of mass spectrometry to detect any compound, which holds a charge within a certain mass range. In a sense, LC-MS data is an unsupervised way of scanning the metabolome of living organisms without previous knowledge of what compounds you may find.
8. Note your fields must be separated by a comma and not a semicolon.
9. In the case of untargeted metabolomics data overfitting can be a concern, because having more training data helps only to a certain point. It is important to capture the meaningful data while minimizing the noise.

Acknowledgments

The authors would also like to acknowledge that this work was part of the DOE Joint BioEnergy Institute (http://www.jbei.org) supported by the US Department of Energy, Office of Science, Office of Biological and Environmental Research, through contract DE-AC02-05CH11231 between Lawrence Berkeley National Laboratory and the US Department of Energy.

References

1. Alpaydin E et al (2010) Introduction to machine learning. MIT Press, Cambridge, MA
2. Cortes C, Vapnik V et al (1995) Support-vector networks. Mach Learn 20(3):273–297
3. Kohavi R et al (1995) A study of cross-validation and bootstrap for accuracy estimation and model selection. Proc Fourteenth Int Joint Conf Artif Intell 2(12):1137–1143
4. Hawkins DM et al (2004) The problem of overfitting. J Chem Inf Comput Sci 44(1):1–12
5. Vafaie H, Jong KD et al (1992) Genetic algorithms as a tool for feature selection in machine learning. Proc 1992 I.E. Int Conf on Tools with AI 11:200–203
6. Bartlett MS, Littlewort G, Lainscsek C, Fasel I, Movellan J et al (2004) Machine learning methods for fully automatic recognition of facial expressions and facial actions. Proc 2004 I.E. Int Conf on systems. Man and Cybernetics 10:592–597
7. Russell S, Norvig P et al (2003) Artificial intelligence: a modern approach. Prentice Hall, USA
8. Murtagh F et al (1985) Multidimensional Clustering Algorithms. In: COMPSTAT Lectures 4. Physica-Verlag, Wuerzburg
9. Mardia KV, Kent JT, Bibby JM (1979) Multivariate analysis. Academic Press, London
10. Venables WN, Ripley BD et al (2002) Modern applied statistics with S. Springer-Verlag, Berlin
11. McQuitty LL et al (1966) Similarity analysis by reciprocal pairs for discrete and continuous data. Educ Psychol Meas 26:825–831
12. Gordon AD (1999) Classification, 2nd edn. Chapman and Hall/CRC, London
13. Everitt B (1974) Cluster analysis. Heinemann Educational Books, London
14. Hartigan JA (1975) Clustering algorithms. Wiley, New York
15. Anderberg MR (1973) Cluster analysis for applications. Academic Press, New York
16. Heinemann J, Mazurie A, Tokmina-Lukaszewska M, Beilman GJ, Bothner B et al (2014) Application of support vector machines to metabolomics experiments with limited replicates. Metabolomics 10:1121–1128
17. Guan W, Zhou M, Hampton CY, Benigno BB, Walker LD, Gray A et al (2009) Ovarian cancer detection from metabolomic liquid chromatography/mass spectrometry data by support vector machines. BMC Bioinformatics 10:259
18. Guyon I, Weston J, Barnhill S et al (2002) Gene selection for cancer classification using support vector machines. Mach Learn 46:389–422
19. VeselKov KA, Vingara LK, Masson P, Robinette SL, Want E, Li JV et al (2011) Optimizing preprocessing of ultraperformance liquid chromatography/mass spectrometry urinary metabolic profiles for improved information recovery. Anal Chem 83:5864–5872
20. Lin X, Wang Q, Yin P, Tang L, Tan Y, Li H et al (2011) A method for handling metabonomics data from liquid chromatography/mass spectrometry: combinational use of support vector machine recursive feature elimination, genetic algorithm and random forest for feature selection. Metabolomics 7(4):549–558
21. Bertini I, Calabro A, De Carli V, Luchinat C, Nepi S, Porfirio B et al (2009) The metabonomic signature of celiac disease. J Proteome Res 8:170–177
22. Smith C, O'Maille G, Want EJ, Qin C, Trauger S, Brandon TR et al (2005) METLIN: a metabolite mass spectral database. Ther Drug Monit 27(6):747–751
23. Tautenhahn R, Bo¨ttcher C, Neumann S et al (2008) Highly sensitive feature detection for high resolution LC/MS. BMC Bioinformatics 9:504
24. Yanes O, Tautenhahn R, Patti GJ, Siuzdak G et al (2011) Expanding coverage of the metabolome for global metabolite profiling. Anal Chem 83(6):2152–2161
25. Duan K, Rajapakse JC et al (2005) Multiple SVM-RFE for gene selection in cancer classification with expression data. IEEE Trans Nanobioscience 4:228–234
26. Hall M, National H, Frank E, Holmes G, Pfahringer B, Reutemann P et al (2010) The WEKA data mining software: an update. ACM SIGKDD Explorations Newsletter 11 (1):10–18
27. Asa BH, Horn D, Hava S, Vapnik V et al (2001) Support vector clustering. J Mach Learn Res 2:125–137

Chapter 18

Dynamic ^{13}C Labeling of Fast Turnover Metabolites for Analysis of Metabolic Fluxes and Metabolite Channeling

Mary Abernathy, Ni Wan, Wenqing Shui, and Yinjie J. Tang

Abstract

Dynamic or isotopically nonstationary ^{13}C labeling experiments are a powerful tool not only for precise carbon flux quantification (e.g., metabolic flux analysis of photoautotrophic organisms) but also for the investigation of pathway bottlenecks, a cell's phenotype, and metabolite channeling. In general, isotopically nonstationary metabolic flux analysis requires three main components: (1) transient isotopic labeling experiments; (2) metabolite quenching and isotopomer analysis using LC-MS; (3) metabolic network construction and flux quantification. Labeling dynamics of key metabolites from ^{13}C-pulse experiments allow flux estimation of key central pathways by solving ordinary differential equations to fit time-dependent isotopomer distribution data. Additionally, it is important to provide biomass requirements, carbon uptake rates, specific growth rates, and carbon excretion rates to properly and precisely balance the metabolic network. Labeling dynamics through cascade metabolites may also identify channeling phenomena in which metabolites are passed between enzymes without mixing with the bulk phase. In this chapter, we outline experimental protocols to probe metabolic pathways through dynamic labeling. We describe protocols for labeling experiments, metabolite quenching and extraction, LC-MS analysis, computational flux quantification, and metabolite channeling observations.

Key words LC-MS, Quenching, Metabolism, Isotopic labeling experiments, Flux analysis, Channeling

1 Introduction

^{13}C-metabolic flux analysis (MFA) provides quantitative information about carbon flux through metabolic networks. Traditional ^{13}C-MFA relies on the labeling of proteinogenic amino acids, which have slow turnover rates and require isotopically steady state tracer experiments. On the other hand, transient labeling dynamics of fast-turnover central metabolites can provide snapshots of cell metabolism and offer rich insights into pathway activities. Using experimentally measured labeling dynamics of free metabolites, isotopically nonstationary MFA (INST-MFA) has been developed [1]. Unlike isotopically steady state MFA (cells are cultured with a specific ^{13}C-tracer for a long period),

Edward E.K. Baidoo (ed.), *Microbial Metabolomics: Methods and Protocols*, Methods in Molecular Biology, vol. 1859, https://doi.org/10.1007/978-1-4939-8757-3_18,

INST-MFA requires measuring the change of free metabolite or amino acid labeling after a cell's culture has been switched from nonlabeled medium to labeled medium (e.g., pulse ^{13}C-substrates into nonlabeled cultures). Since metabolite labeling speed is correlated to flux rate and metabolite pool sizes, the knowledge of labeling dynamics of key metabolites can be used to decipher metabolic fluxes and pool sizes through a metabolic network.

INST-MFA has been applied to photoautotrophic organisms that can only utilize CO_2 because steady state CO_2 labeling provides no unique isotopomer data to gain flux information [2]. To perform INST-MFA, transient labeling data are collected immediately following the introduction of a ^{13}C-tracer for multiple time points (Fig. 1; shaded circle). Due to fast metabolite turnover rates, a protocol to rapidly and precisely quench cell metabolism is required. The metabolite labeling patterns are commonly analyzed using LC-MS [3]. The labeling dynamics can be used to quantify metabolic fluxes, assuming that (1) the cells are at metabolic steady state; (2) there is no enzyme preference between ^{12}C and ^{13}C atoms; (3) the culture is homogeneous (no diffusion limitations exist) [4]. For flux estimation, a nonlinear least squares regression is solved where the objective function is to minimize the sum of squared residuals (SSR) between the measured (m) and predicted mass isotopomer distributions ($x(t)$), pool size (C), and metabolic rates (r) [1].

$$\begin{aligned} \text{minSSR} &= \sum \frac{(x(t) - x_{\text{m}}(t))^2}{\sigma_x^2} + \sum \frac{(r - r_{\text{m}})^2}{\sigma_r^2} \\ &\quad + \sum \frac{(C - C_{\text{m}})^2}{\sigma_C^2} \end{aligned} \tag{1}$$

$$\text{s.t.} C \times \frac{dx}{dt} = f_{\text{isotopomer model}}(v, x(t))$$

$$R \times v = r$$

$$S \times v = 0$$

The metabolic network and fluxes (v) are subject to constraints by stoichiometry matrices (S and R) and external metabolic rates (r). Metabolite pool sizes (C) can be measured or fitted using isotopomer data [5]. The elementary metabolite unit (EMU) framework can be used for effective isotopomer calculations [6, 7]. INCA [5] and OpenMebius [8] are excellent dynamic flux analysis software packages, which resolve fluxomes with experimentally determined time-dependent labeling data. The workflow of INST-MFA is shown in Fig. 2.

Isotopic nonstationary labeling provides an enhanced understanding of exchange fluxes and metabolite pool sizes [9, 10] and is

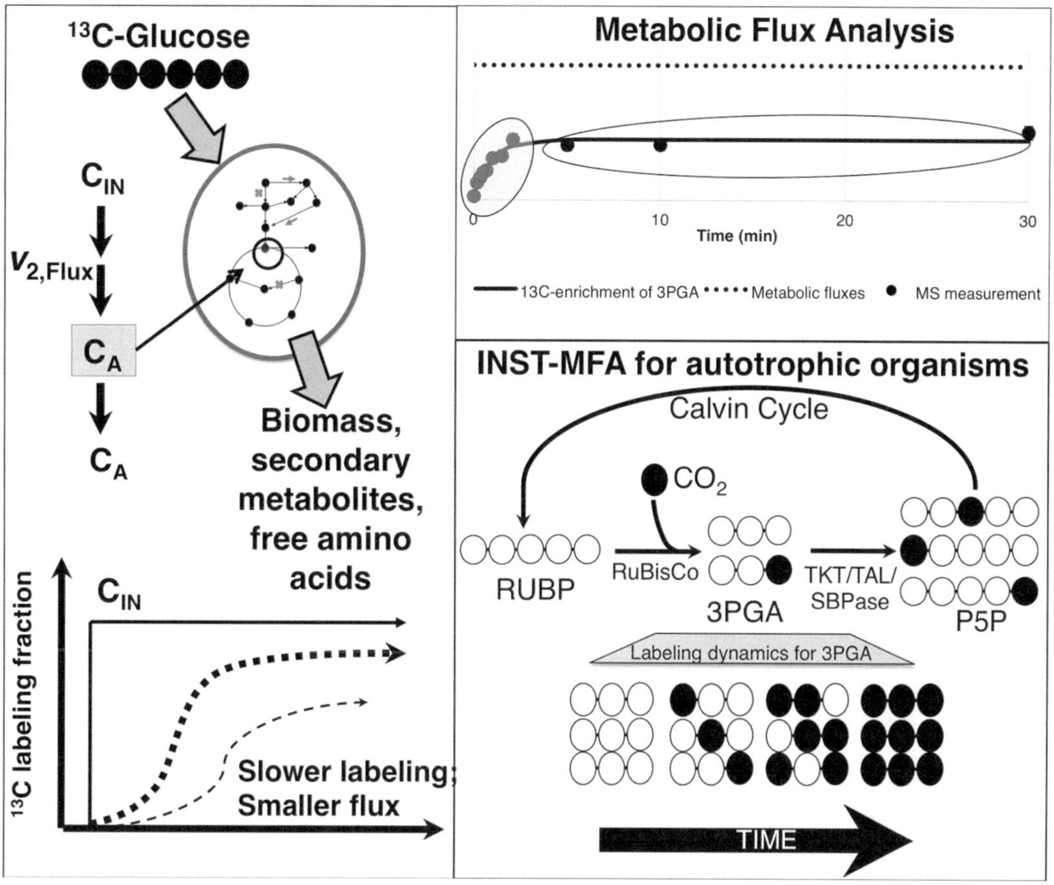

Fig. 1 Isotopically nonstationary metabolic flux analysis (INST-MFA) of steady state metabolism. Left pane: after ^{13}C-substrate pulse (C_{IN}), the flux rate through a key metabolite (e.g., C_A) determines the labeling rate of C_A as seen below. Right pane: in the top panel, the shaded circle represents the measurement time points for INST-MFA, while the non-shaded circle represents measurements taken at isotopic steady state. The dotted line represents metabolic fluxes at steady state. An example of an isotopic labeling experiment is shown in the bottom pane. When labeled CO_2 is introduced into an algal culture, Calvin Cycle metabolites (e.g., 3PGA and P5P) become progressively labeled. The labeling dynamics for 3PGA is illustrated. Once isotopic steady state is reached, the metabolites will all be fully labeled

particularly useful for studying mammalian cells and other higher organisms [11–13]. INST-MFA provides reliable and experimentally determined quantitative measurements of metabolic fluxes [9, 10]. INST-MFA is also useful for elucidating hidden pathway constraints such as metabolic reflux [13], pathway bottlenecks, and metabolite channeling [14]. Metabolite channeling occurs when enzymes form channels to direct fluxes without releasing intermediates into the cytosolic pool. The channeling process creates two metabolite pools: a fast turnover pool (inside the channel) and an inactive pool (outside the channel). In this case, INST-experiments may generate unsequential labeling along cascade metabolites (i.e., downstream metabolites are labeled faster than their precursors).

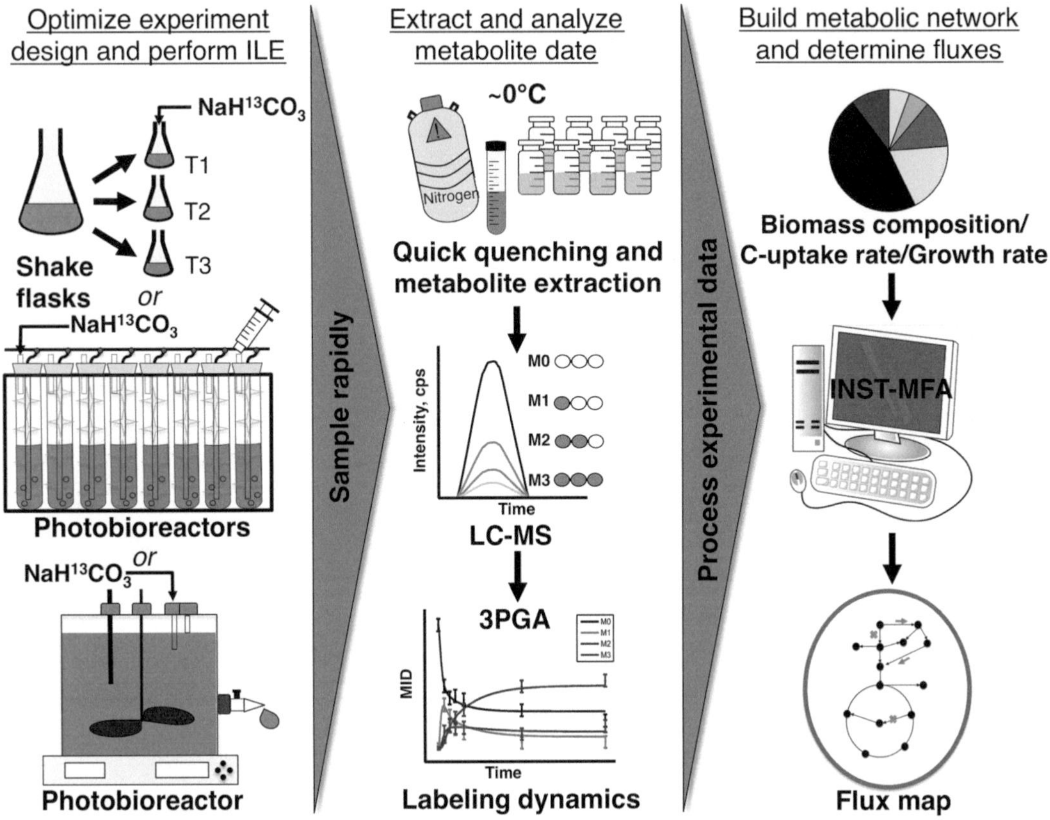

Fig. 2 INST-MFA of cyanobacteria. The workflow consists of three major parts: (1) experiment design and isotopic labeling experiments (ILE); (2) metabolite extraction and dynamic labeling data analysis; (3) metabolic network construction and quantification of fluxes

2 Materials

2.1 Transient Isotope Labeling Experiments

1. Cell culture at metabolic steady state: Minimal media cultures in their exponential growth phase at metabolic steady state. Prepare regular (for seed cultures) and minimal cell media (*see* **Note 1**) specific to the strain being cultured. The volume of culture needed for the experiment is dependent on the strain's cell density, number of experimental time points, and sampling procedure (*see* **Note 2**). Follow specific media instructions for storage conditions.
2. ^{13}C-tracer: Prepare a stock solution of the ^{13}C-tracer (e.g., $NaH^{13}CO_3$ or U-^{13}C-glucose) using Eq. 2 below, where $C_{culture}$ is the desired final culture concentration (*see* **Note 3**), C_{stock} is the concentration of the stock solution being prepared, $V_{culture}$ is the volume of the culture(s), and V_{stock} is the volume of stock solution being pulsed (*see* **Note 4**).

$$C_{\text{culture}} V_{\text{culture}} = C_{\text{stock}} V_{\text{stock}} \quad (2)$$

3. Quenching supplies: Minimal (no carbon) media or phosphate buffered saline and dry ice/liquid N_2 (*see* **Note 5**).
4. Benchtop centrifuge with refrigeration options: Prechill centrifuge prior to the experiment (make sure the centrifuge temperature close to 0 °C). Prechilled centrifuge tubes (50 or 15 mL tubes) are used.

2.2 Analytical Determination of Mass Isotopomer Distributions

1. Extraction solution: The choice of extraction solution depends on the species and the metabolites of interest (*see* **Note 6**). Additionally, the volume of extraction solution can be optimized.
2. Deionized water.
3. Centrifuge filters for LC-MS analysis: Amicon Ultracentrifuge filters—3 kDa (EMD Millipore, USA), Costar Spin-X centrifuge tube filters—0.45 μm cellulose acetate membrane (Corning, USA), or centrifugal filter nylon—0.45 μm (VWR, USA) is recommended depending on instrument technique and metabolites of interest.
4. Vortexer/Lyophilizer/vacuum.
5. Liquid chromatography–mass spectrometer (LC-MS): Two examples of LC settings are given below. All solvents and reagents used should be of HPLC grade. (a) Reverse phase-ion pairing: 10 mM tributylamine and 15 mM acetic acid in water (mobile phase A), 100% HPLC grade methanol (mobile phase B), Phenomenex Synergi Fusion-RP column or an alternative column. (b) Hydrophilic interaction liquid chromatography (HILIC): 10 mM ammonium carbonate in water (mobile phase A) and 10 mM ammonium carbonate in 70% acetonitrile and 30% water (mobile phase B), SeQuant® ZIC®-pHILIC column (EMD Millipore) or an alternative HILIC column. High-resolution mass spectrometers (e.g., TOF or Orbitrap) and/or triple-quadruple instruments for multiple reaction monitoring (MRM) analysis are recommended for reliable metabolite identification. An MS analysis software will be needed along with other LC-MS consumables.
6. Internal standards: Approximately 1 μg/mL analytical metabolite standards should be prepared to aid in peak identification. Additionally, standards should be spiked into biological samples and analyzed to observe matrix effects.

2.3 Metabolic Flux Analysis

1. Computer and software: equipped with the ability to perform isotopically nonstationary metabolic flux analysis and run appropriate software, such as INCA (isotopomer network compartmental analysis) [5].

2. Biomass composition analysis: Dried biomass washed with 0.9% saline solution, analytical instruments and their associated reagents. Common measurements are total protein content, amino acid composition, lipid measurement, and carbohydrate measurement. National Renewable Energy Laboratory developed procedures for standard biomass analysis (http://www.nrel.gov/bioenergy/biomass-compositional-analysis.html) (optional).
3. Growth rate analysis: a growth curve for the species of interest measured at the appropriate absorbance using a plate reader or UV/Vis spectrometer (optional).
4. Carbon uptake and excretion measurements: a steady state culture, an HPLC, a GC, a GC-MS and/or enzymatic kits, appropriate solvents, consumables, and sampling equipment.

3 Methods

3.1 Transient Isotope Labeling Experiments and Fast-Cooling Protocol

1. Prepare cultures. Depending on the strain's cell density and number of time points, prepare proper amount of nonlabeled cultures (*see* **Note 2**). To capture the labeling dynamics where the slope is changing rapidly [15] (i.e., the earlier time points), numerous time-course samples have to be taken after ^{13}C-pulse experiments (*see* **Note 7**).
2. Prepare ^{13}C tracer stock. For example, $NaH^{13}CO_3$ is used to probe photoautotrophic cultures, and U-^{13}C-glucose is often used to study heterotrophic organisms (*see* **Note 8**).
3. Set up for tracer experiment. Before ^{13}C-pulse, obtain dry ice or liquid nitrogen and have the quenching solution and centrifuge tubes prechilled in the appropriate apparatus. Then, deliver the ^{13}C-substrate into the culture and begin timing.
4. Metabolite quenching. At the measured time point, sample the aliquot of culture. If using a quenching solution, the culture should be quickly mixed in with the quenching solution (e.g., an icy cold buffer) then the sample tubes should be submerged in liquid nitrogen (or dry ice) to cool cultures to ~0 °C as fast as possible [14] (*see* **Note 9**). If the quenching temperature is not low enough, central metabolic reactions may be active during sampling processes (Fig. 3). Therefore, it is important to quickly quench cell metabolism at ~0 °C using both liquid nitrogen and cold buffer.
5. Centrifuge cell culture to pellet the quenched biomass at 0 °C. Use appropriate high speed (e.g., 6000 × *g*) and time to pellet cell biomass, while trying to keep centrifuge time short. Cell pellets can be stored at −80 °C before extraction.

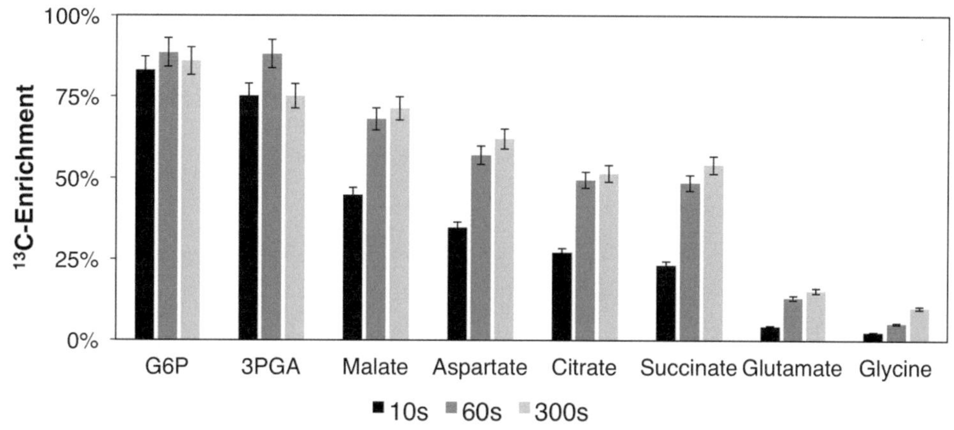

Fig. 3 Labeling dynamics of glycolytic metabolites, TCA cycle metabolites and free amino acids in cold shock *Escherichia coli*. *Escherichia coli* culture was under cold shock (~10 °C) for 60 min. Then fully labeled ^{13}C-glucose was pulsed into the cold shock culture. Biomass was sampled at 10 s, 60 s and 300 s, and centrifuged at ~10 °C. The results indicate glycolytic metabolites were labeled faster than TCA cycle metabolites and free amino acids. Cell catabolism (especially sugar phosphate pathways) remaintains active in cold temperatures. In this case (incomplete quenching), INST-MFA may have poor accuracy

3.2 Analytical Determination of Mass Isotopomer Distributions via LC–MS

1. Extract metabolites from cell biomass. While on ice, add internal standard to cell pellet or biomass (*see* **Note 10**) if trying to quantify metabolites. Mix 1 mL of cold extraction solution (*see* **Note 6**) with biomass then transfer to a new vial. Shake for 4 h to overnight at ~0 °C. After shaking in extraction solution, add 0.5 mL of ddH_2O, vortex, and spin down at 5000 × *g* for 2 min. Pipet out methanol–water layer into a new vial. Repeat extraction with ddH_2O. Filter methanol–water layer using centrifuge filter and transfer filtered sample into a new vial.
2. Concentrate metabolite extraction. Dry down water–methanol samples using lyophilizer/vacuum or N_2 gas. Resuspend the sample in ddH_2O with 100 μL ddH_2O (or ddH_2O-acetonitrile solution, if also looking at water insoluble compounds using the HILIC-MS method).
3. LC-MS Analysis (*see* **Note 11**). Filter all mobile phases before LC-MS analysis. A brief description of reverse phase ion-pairing LC-MS operation is as follows: An injection volume of 10 μL and a recommended column flow rate of 0.2–0.3 mL/min may be used. An LC gradient as follows is suggested: 0% of mobile phase B (0 min), 8% B (8 min), 22% B (18 min), 40% B (28 min), 60% B (32 min), 90% B (34 min), 90% B (37 min), 0% B (39 min), and 0% B (49 min) [2, 15]. The mass spectrometer usually operates with an ESI source under negative ionization mode for analysis of organic acids and sugar phosphates. In targeted metabolite analysis by MRM mode, first optimize the collision energy and define

Table 1
Metabolites of interest for central metabolism flux analysis

Metabolite	Metabolite abbreviation	MRM transition	Number of carbons	Composition (parent ion)
Glycolate	GLY	75/47	1-2	$C_2H_3O_3^-$
Glyceric acid	GA	105/75	1-2-3	$C_3H_5O_4^-$
2-Phosphoglycolic acid	2PG	155/79	1-2	$C_2H_4O_6P^-$
Phosphoenolpyruvate	PEP	167/79	1-2-3	$C_3H_4O_6P^-$
Glyceraldehyde-3-phosphate	GAP	169/97	1-2-3	$C_3H_6O_6P^-$
3-phosphoglyceric acid	3PGA	185/79	1-2-3	$C_3H_6O_7P^-$
Glucose-6-phosphate	G6P	259/97	1-2-3-4-5-6	$C_6H_{12}O_9P^-$
Fructose-6-phosphate	F6P	259/97	1-2-3-4-5-6	$C_6H_{12}O_9P^-$
Glucose-1-phosphate	G1P	259/79	1-2-3-4-5-6	$C_6H_{12}O_9P^-$
Sedoheptulose-7-phosphate	S7P	289/97	1-2-3-4-5-6-7	$C_7H_{14}O_{10}P^-$
Ribulose-1,5-bisphosphate	RUBP	309/97	1-2-3-4-5	$C_5H_{11}O_{11}P_2^-$
Fructose-1,6-bisphosphate	FBP	339/97	1-2-3-4-5-6	$C_6H_{13}O_{12}P_2^-$
Sedoheptulose-1,7-bisphosphate	SBP	369/97	1-2-3-4-5-6-7	$C_7H_{15}O_{13}P_2^-$
Uridine diphosphate glucose	UDP-GLU	565/323	1-2-3-4-5-6	$C_{15}H_{23}N_2O_{17}P_2^-$
Adenosine diphosphate glucose	ADP-GLU	588/346	1-2-3-4-5-6	$C_{16}H_{24}N_5O_{15}P_2^-$

MRM transitions for each metabolite using its chemical standard as seen in Table 1.

A brief description of HILIC-LC-MS protocol developed by the Joint Bioenergy Institute [14] is as follows: A HILIC column, such as a SeQuant® ZIC®-pHILIC (2.1 mm × 150 mm and 5-μm particle size), should be used. For the ZIC®-pHILIC column, the recommended flow rate and injection volume are 0.2 mL/min and 1–2 μL. Metabolites are separated via gradient elution under the following conditions: 100% B (0 min), 60% B (9 min), 60% B (11.8 min), 100% B (12 min), and 100% B (22 min). Additionally, optimize LC and MS conditions with standard solutions.

4. Data analysis. The metabolite labeling pattern is analyzed as mass isotopomer distributions (MID). For example, M + 0 of 3-Phosphoglyceric acid (3PGA) represents the fraction of unlabeled 3PGA, whereas M + 3 represents the fraction of a fully labeled 3PGA (Fig. 1). A metabolite has 2^n isotopomers, where n represents the number of carbons. To determine the fractions of all isotopomers, high resolution MS methods have to be employed for distinguishing authentic labeled isotopomers

from ion interferences. Internal standards and technical replicates are essential for confident identification of each metabolite, integration of all isotopomers (M + 0, M + 1, M + 2, etc.), and an assessment of analytical variations. Metabolite identification is usually based on LC retention time, high-accuracy mass measurement, or specific MRM transitions. Mass spectral libraries such as METLIN or the NIST Mass Spectral Database may be used for compound structural assignment when standards are not available [16, 17]. Next, the MID values can be calculated based on the isotopic envelop detected for each metabolite. The M + 0 distribution will be the peak area of M + 0 divided by the sum of all the isotopomer peak areas for that metabolite. In principle, the sum of all the MID percentages for a metabolite should be equal to one. Biological replicates are needed for deriving the standard deviation for each MID distribution. More accurate MID results can be acquired if using specific algorithms to normalize the natural isotope occurrence [18]. However, this type of normalization is not normally required for LC-MS data processing as the natural abundances of labeled isotopes are much lower than the experimental variation.

3.3 Metabolic Flux Analysis

1. INST-MFA platform. OpenMebius [8], INCA [5], or another appropriate platform can be utilized. INCA integrates transient balances using a custom linear ODE solver that involves computing the matrix exponential of each state-transition matrix. INCA allows for the compartmentalization of metabolites, fitting of metabolite pool sizes, and use of dilution parameters if one observes dilution from unlabeled sources, such as glycogen, biomass reflux, or metabolite channeling. OpenMebius is an open source software, but requires pool size information for precise flux estimation.
2. Metabolic network reconstruction. First, build a network based on the genome of the species of interest and biochemical literature. Useful databases are KEGG, BRENDA, and BioCyc Database Collection. The equations should contain information on reaction stoichiometry, reversibility, and atom transitions. The metabolic network should contain enough information to resolve simulated measurements based on all experimental measurements while also avoiding unnecessary reactions that can lead to overfitting the model. Linear pathways may be grouped to reduce computational costs (*see* **Note 12**). Energy and redox balancing reactions are often omitted from the model.
3. MID data and biomass equations. Flux calculation requires MID data and a biomass equation. The biomass equation represents a proportioned sink (in millimoles per gram of

biomass) of all the metabolites required for biomass growth [1, 19]. The flux values are affected by biomass equations. Although a standard biomass equation is often used as an approximate for other bacterial species, a non-model species may have different biomass compositions (such as amino acid composition, lipids, and carbohydrates) [20]. For example, cyanobacteria tend to have high protein content due to the abundance of its photosystems.

4. Substrate uptake, extracellular metabolite excretion, and growth rate. Additional measurements of inflows and outflows offer constraints for precise flux analysis. Carbon uptake is expressed in units (mmol/gDCW/h). Extracellular metabolite excretions constrain the outflows of metabolic network. Measured growth rate (h^{-1}) can be used to validate flux model predictions. These constraints will improve the accuracy of the model.

5. Estimation of fluxes and pool sizes. INST-MFA model obtains flux solutions by minimizing the difference between measured and simulated mass isotopomer distributions. It is recommended to restart the flux estimation numerous times from different random starting points to obtain a global solution [21]. The goodness-of-fit is based on the sum of squared residuals (SSR) and the degrees of freedom. The confidence intervals of flux values are determined through either parameter continuation or Monte Carlo methods. Parameter continuation is performed using a flux coupling analysis and predictor–corrector method of minimized computational expense. The Monte Carlo method introduces random errors from a normal distribution and flux measurements are repeated with perturbed data sets. In general, parameter continuation is a more efficient method for calculating flux confidence intervals [22]. Moreover, pool size measurement is important for dynamic flux analysis. However, precise pool size analysis is difficult because of metabolite leakage during sample processing and the requirement of internal standards. To overcome this problem, the INCA platform fits the metabolite pool sizes based on isotopomer data. In this case, INST-MFA can also generate a set of predictions of metabolite pool sizes in addition to flux values.

6. Model assessment. One should look at reactions with large confidence intervals, which indicate that the flux cannot be determined precisely. Additional measurements of metabolite pool sizes may improve the flux determinations. A poor squared residual or weighted residual for an individual metabolite could mean there is an error with the measurement or there is a mistake in the metabolic network. Furthermore, it is important to check whether the regression residuals are distributed evenly. Statistical analysis can reveal the importance of a particular measurement for determining certain flux values.

3.4 Consideration of Metabolite Channeling

1. Metabolite channeling affects dynamic ^{13}C-labeling. During channeling, metabolites can be passed through a pathway without mixing with the bulk phase. Channeling may change molecule labeling orientations or bypass the denoted pathway reactions. Certain metabolic nodes of a channeled pathway must be leaky for connecting other pathways that share the same intermediates. At a leaky node, the intermediate can diffuse outside the channel and form an inactive cytosol pool, which will have a slower turnover rate. In this case, dynamic ^{13}C-pulse experiments may be used to identify a "channeling signature": *downstream metabolites are labeled faster than their precursors* [14]. Dynamic ^{13}C labeling can elucidate leaky channels by observing whether there is any unsequential order for labeling cascade metabolites (Fig. 4). To date, there is little knowledge of how microbes channel their native pathways. Metabolic flux analyses (including INST-MFA) are highly sensitive to the effect of channels on molecule orientation and pathway preferences [23]. Any observation of channeling in dynamic labeling experiments should provide caution during construction of metabolic pathways and analysis of labeling data. Since metabolic flux analysis relies on the assumption that all metabolites are homogenously distributed in the cell and no diffusion limitations exist, metabolite channeling may

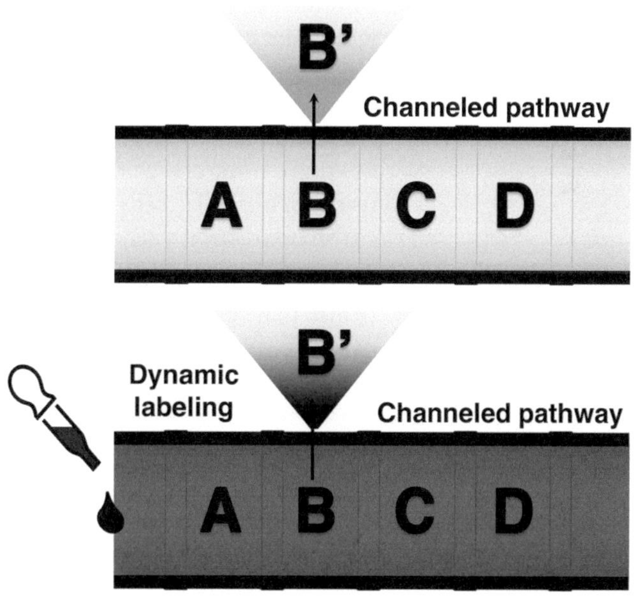

Fig. 4 Metabolite channeling scheme [14]. In a pathway with cascade metabolites A, B, C, and D with B being a channeled metabolite. During ^{13}C-pulse, the presence of B′ outside of channel may dilute the labeling of bulk B. Therefore, the labeling speed of bulk B may be slower than its downstream products C and D if the pool size of B′ is large (high isotopic dilution factor)

confound metabolic flux values or be revealed during INST-MFA.

2. Metabolite channeling in INST-MFA. Channeling establishes microcompartmentation of certain metabolites, which must be considered in MFA. If unsequential labeling of cascade metabolites occurs due to channeling effects (presence of multiple pools of same metabolites), it may cause issues obtaining an accurate and best-fit solution. To improve model fitting quality, dilution parameters are used to account for inactive metabolite pools outside of channels [2]. While the introduction of these dilution factors helps reduce model fitting errors, it also introduces additional degrees of freedom into the model. This may increase undetermined parts of the metabolic network [24]. In INST-MFA, dilution parameters can be introduced into the model [25, 26]. For example, let C_{B} be the active pool of metabolite B (the pool involved in the cascade reaction) and $C_{\mathrm{B}'}$ be the inactive pool of B. $C_{\mathrm{B,\ bulk}}$ is the total sampled pool (mix of C_{B} and $C_{\mathrm{B}'}$). Then three reactions can be added to the model in such form: (1) $C_{\mathrm{B}} \rightarrow C_{\mathrm{B,\ bulk}}$ (2) $C_{\mathrm{B}'} \rightarrow C_{\mathrm{B,\ bulk}}$; (3) $C_{\mathrm{B}} \rightarrow C_{B'}$.

4 Notes

1. Minimal cell media should not include any carbon sources which the cell may uptake in addition to the ^{13}C-tracer. While a rich media may be required to analyze the cell at the desired phenotype, this will create difficulties in analyzing the results later due to multiple nutrition compounds in the culture medium interfering with the intracellular analytes [27]. If using rich media, cells are often washed with a saline solution before extraction [28].
2. Depending on a strain's cell density, different volumes of cell broth may be needed to obtain quality analytical data. For example, typically 5 mL of log phase *E. coli* ($OD_{600} = 1$) is sufficient for LC-MS samples [29], while 25 mL of log phase cyanobacteria ($OD_{730} = 0.4$) may be needed for improved LC-MS peak quality.
3. The final concentration of the substrate added can be estimated from literature uptake values and from the longest time point planned. To achieve metabolic steady-state, the cell should be saturated with labeled substrate during the entire experiment so as to not affect the uptake and growth rates [30].
4. It is best to perform all actions as swiftly as possible: (a) Make sure to have the appropriate tools to inject the entire ^{13}C-stock solution into culture as quickly as possible; (b) High

Table 2
Brief description of quenching methods

Quenching method	Description
Methanol solution	Use a prechilled methanol solution (<−50 °C, either pure methanol or a 40–75% methanol and water solution can be utilized)
Fast-cooling method	Use prechilled minimal media or buffer solution (<−10 °C) while bathing culture in dry ice or liquid N_2
Fast filtration	Set up a vacuum filtration system that can pull liquid through rapidly (<5 s) with a nylon filter (25 mm)
Rapid sampling unit [38, 43]	Use a sampling valve and port injector to rapidly sample from bioreactor; may still need a quenching solvent such as cold methanol [38] or hot ethanol

concentration ^{13}C-substrate stock solution is preferred, but may be limited by substrate solubility.

5. INST experiments require proper quenching methods [31–33]. In this paper, we focus on a fast-cooling method (Table 2), which allows for quenching with high time precision and minimal metabolite leakage. Other methods may also be applicable for rapid quenching procedures.
6. The extraction solution should be optimized to the species and metabolites of interest based on peak height, area, or ion count [3, 34]. Common extraction mixtures are methanol–water, chloroform–methanol, acetonitrile–methanol–water [27, 35]. Additionally, if one is interested in nucleotide triphosphates, it has been shown that the addition of 0.1 M formic acid in acetonitrile improves extraction [36, 37] (after which ammonium hydroxide is combined with the extract to prevent metabolite degradation through neutralization). The amount of extraction solution used is typically between 0.5 and 1.5 mL and is also open to optimization.
7. One can optimize the sampling time points to improve computational quality [30, 38]. In general, it is desired to take more time points during the period of fast labeling changes. One group calculates sampling times through the parameterization of the time points [30] (refer to cited publication for more details).
8. Parallel labeling experiments can provide temporal resolution of certain fluxes and also improve flux precision to a certain point [39]. INCA contains a feature that allows a user to search for optimal parallel tracer combinations that minimize the optimality criterion [5, 40].
9. Use a digital thermometer as the stirring apparatus to help mixing, prevent freezing and inform when the cell culture is quenched (<0 °C).

10. Possible internal standards: uniformly isotopically labeled metabolites, PIPES (parent mass—301) [15], naphthyl phosphate [9], *N*-acetyl-glutamine [34], 2-amino-3-bromo-5-methylbenzoic acid [41].

11. HILIC can achieve separations of common polar metabolites, such as low-abundance metabolites, organic compounds, and amino acids [34], while reverse phase ion-pairing chromatography is particularly good for measuring sugar phosphates [2, 15, 42]. A triple-quadrupole (QqQ) MS instrument can detect a preselected set of targeted metabolites by MRM, which offers supreme sensitivity and accurate quantification, while high-resolution MS provides highly specific analysis of low concentration intracellular metabolites [2, 17]. A common soft ionization technique used in MS analysis is electrospray ionization (ESI), which is particularly useful for chemically unstable metabolites, such as nicotinamide adenine dinucleotide phosphate (NADPH), adenosine triphosphate (ATP), and phosphoenolpyruvate (PEP) [15]. Depending on metabolites of interest and available instruments, multiple combinations of LC and MS systems could be considered.

12. The oxidative pentose phosphate pathway with atom transitions (represented by abcdef) can be grouped as glucose-6-phosphate (G6P) (abcdef) → ribulose-5-phosphate (RU5P) (bcdef) + CO_2 (a) or if 6-phosphogluconic acid (6PG) MID is obtained, the reaction can be broken down into two separate reactions: G6P (abcdef) → 6PG (abcdef), 6PG (abcdef) → RU5P (bcdef) + CO_2 (a).

Acknowledgements

This work was supported by NSF (CBET 1438125).

References

1. Antoniewicz MR (2015) Methods and advances in metabolic flux analysis: a mini-review. J Ind Microbiol Biotechnol 42:317–325. https://doi.org/10.1007/s10295-015-1585-x

2. Young JD, Shastri AA, Stephanopoulos G, Morgan JA (2011) Mapping photoautotrophic metabolism with isotopically nonstationary ^{13}C flux analysis. Metab Eng 13:656–665. https://doi.org/10.1016/j.ymben.2011.08.002

3. Bennette NB, Eng JF, Dismukes GC (2011) An LC–MS-Based Chemical and Analytical Method for Targeted Metabolite Quantification in the Model Cyanobacterium *Synechococcus* sp. PCC 7002. Anal Chem 83:3808–3816. https://doi.org/10.1021/ac200108a

4. Zamboni N, Fendt S-M, Rühl M, Sauer U (2009) ^{13}C-based metabolic flux analysis. Nat Protoc 4:878–892. https://doi.org/10.1038/nprot.2009.58

5. Young JD (2014) INCA: a computational platform for isotopically non-stationary metabolic flux analysis. Bioinformatics 30:1333–1335

6. Antoniewicz MR, Kelleher JK, Stephanopoulos G (2007) Elementary Metabolite Units (EMU): a novel framework for modeling

isotopic distributions. Metab Eng 9:68–86. https://doi.org/10.1016/j.ymben.2006.09.001
7. Young JD, Walther JL, Antoniewicz MR et al (2008) An elementary metabolite unit (EMU) based method of isotopically nonstationary flux analysis. Biotechnol Bioeng 99:686–699. https://doi.org/10.1002/bit.21632
8. Kajihata S, Furusawa C, Matsuda F, Shimizu H (2014) OpenMebius: an open source software for isotopically nonstationary ^{13}C-based metabolic flux analysis. Biomed Res Int 2014:627014. https://doi.org/10.1155/2014/627014
9. Schaub J, Mauch K, Reuss M (2008) Metabolic flux analysis in *Escherichia coli* by integrating isotopic dynamic and isotopic stationary ^{13}C labeling data. Biotechnol Bioeng 99:1170–1185. https://doi.org/10.1002/bit.21675
10. Wahl SA, Nöh K, Wiechert W (2008) ^{13}C labeling experiments at metabolic nonstationary conditions: An exploratory study. BMC Bioinformatics 9:152. https://doi.org/10.1186/1471-2105-9-152
11. Wiechert W, Nöh K (2013) Isotopically non-stationary metabolic flux analysis: complex yet highly informative. Curr Opin Biotechnol 24:979–986. https://doi.org/10.1016/j.copbio.2013.03.024
12. Niklas J, Schräder E, Sandig V et al (2011) Quantitative characterization of metabolism and metabolic shifts during growth of the new human cell line AGE1.HN using time resolved metabolic flux analysis. Bioprocess Biosyst Eng 34:533–545. https://doi.org/10.1007/s00449-010-0502-y
13. Nargund S, Misra A, Zhang X et al (2014) Flux and reflux: metabolite reflux in plant suspension cells and its implications for isotope-assisted metabolic flux analysis. Mol BioSyst 10:1496–1508. https://doi.org/10.1039/c3mb70348g
14. Hollinshead WD, Rodriguez S, Martin HG et al (2016) Examining *Escherichia coli* glycolytic pathways, catabolite repression, and metabolite channeling using Δpfk mutants. Biotechnol Biofuels 9:212. https://doi.org/10.1186/s13068-016-0630-y
15. Ma F, Jazmin LJ, Young JD, Allen DK (2014) Isotopically nonstationary ^{13}C flux analysis of changes in *Arabidopsis thaliana* leaf metabolism due to high light acclimation. Proc Natl Acad Sci 111:16967–16972. https://doi.org/10.1073/pnas.1319485111
16. Jazmin LJ, Young JD (2013) Isotopically nonstationary ^{13}C metabolic flux analysis. Methods Mol Biol Clifton NJ 985:367–390. https://doi.org/10.1007/978-1-62703-299-5_18
17. Zhu Z-J, Schultz AW, Wang J et al (2013) Liquid chromatography quadrupole time-of-flight mass spectrometry characterization of metabolites guided by the METLIN database. Nat Protoc 8:451–460. https://doi.org/10.1038/nprot.2013.004
18. Niedenführ S, ten Pierick A, van Dam PTN et al (2016) Natural isotope correction of MS/MS measurements for metabolomics and ^{13}C fluxomics. Biotechnol Bioeng 113:1137–1147. https://doi.org/10.1002/bit.25859
19. Shastri AA, Morgan JA (2005) Flux balance analysis of photoautotrophic metabolism. Biotechnol Prog 21:1617–1626. https://doi.org/10.1021/bp050246d
20. Pramanik J, Keasling JD (1997) Stoichiometric model of *Escherichia coli* metabolism: incorporation of growth-rate dependent biomass composition and mechanistic energy requirements. Biotechnol Bioeng 56:398–421. https://doi.org/10.1002/(SICI)1097-0290(19971120)56:4<398::AID-BIT6>3.0.CO;2-J
21. Crown SB, Antoniewicz MR (2013) Publishing ^{13}C metabolic flux analysis studies: A review and future perspectives. Metab Eng 20:42–48. https://doi.org/10.1016/j.ymben.2013.08.005
22. Antoniewicz MR, Kelleher JK, Stephanopoulos G (2006) Determination of confidence intervals of metabolic fluxes estimated from stable isotope measurements. Metab Eng 8:324–337. https://doi.org/10.1016/j.ymben.2006.01.004
23. Williams TCR, Sweetlove LJ, Ratcliffe RG (2011) Capturing Metabolite Channeling in Metabolic Flux Phenotypes. Plant Physiol 157:981–984. https://doi.org/10.1104/pp.111.184887
24. van Winden W, Verheijen P, Heijnen S (2001) Possible pitfalls of flux calculations based on ^{13}C-labeling. Metab Eng 3:151–162
25. Kelleher JK, Masterson TM (1992) Model equations for condensation biosynthesis using stable isotopes and radioisotopes. Am J Physiol - Endocrinol Metab 262:E118–E125
26. Antoniewicz MR, Kraynie DF, Laffend LA et al (2007) Metabolic flux analysis in a nonstationary system: Fed-batch fermentation of a high yielding strain of *E. coli* producing 1,3-propanediol. Metab Eng 9:277–292. https://doi.org/10.1016/j.ymben.2007.01.003
27. Madji Hounoum B, Blasco H, Emond P, Mavel S (2016) Liquid chromatography–high-

resolution mass spectrometry-based cell metabolomics: Experimental design, recommendations, and applications. TrAC Trends Anal Chem 75:118–128. https://doi.org/10.1016/j.trac.2015.08.003

28. Murphy TA, Dang CV, Young JD (2013) Isotopically nonstationary ^{13}C flux analysis of Myc-induced metabolic reprogramming in B-cells. Metab Eng 15:206–217. https://doi.org/10.1016/j.ymben.2012.07.008
29. Oldiges M (2004) Stimulation, monitoring, and analysis of pathway dynamics by metabolic profiling in the aromatic amino acid pathway. Biotechnol Prog 20:1623
30. Nöh K, Wahl A, Wiechert W (2006) Computational tools for isotopically instationary ^{13}C labeling experiments under metabolic steady state conditions. Metab Eng 8:554–577. https://doi.org/10.1016/j.ymben.2006.05.006
31. Faijes M, Mars AE, Smid EJ (2007) Comparison of quenching and extraction methodologies for metabolome analysis of *Lactobacillus plantarum*. Microb Cell Factories 6:27. https://doi.org/10.1186/1475-2859-6-27
32. Chen M, Li A, Sun M et al (2014) Optimization of the quenching method for metabolomics analysis of *Lactobacillus bulgaricus*. J Zhejiang Univ Sci B 15:333–342. https://doi.org/10.1631/jzus.B1300149
33. Millard P, Massou S, Wittmann C et al (2014) Sampling of intracellular metabolites for stationary and non-stationary ^{13}C metabolic flux analysis in *Escherichia coli*. Anal Biochem 465:38–49. https://doi.org/10.1016/j.ab.2014.07.026
34. Bajad SU, Lu W, Kimball EH et al (2006) Separation and quantitation of water soluble cellular metabolites by hydrophilic interaction chromatography-tandem mass spectrometry. J Chromatogr A 1125:76–88. https://doi.org/10.1016/j.chroma.2006.05.019
35. Prasad Maharjan R, Ferenci T (2003) Global metabolite analysis: the influence of extraction methodology on metabolome profiles of *Escherichia coli*. Anal Biochem 313:145–154. https://doi.org/10.1016/S0003-2697(02)00536-5
36. Rabinowitz JD, Kimball E (2007) Acidic acetonitrile for cellular metabolome extraction from *Escherichia coli*. Anal Chem 79:6167–6173. https://doi.org/10.1021/ac070470c
37. Bennett BD, Yuan J, Kimball EH, Rabinowitz JD (2008) Absolute quantitation of intracellular metabolite concentrations by an isotope ratio-based approach. Nat Protoc 3:1299–1311. https://doi.org/10.1038/nprot.2008.107
38. Nöh K, Grönke K, Luo B et al (2007) Metabolic flux analysis at ultra short time scale: isotopically non-stationary ^{13}C labeling experiments. J Biotechnol 129:249–267. https://doi.org/10.1016/j.jbiotec.2006.11.015
39. Antoniewicz MR (2015) Parallel labeling experiments for pathway elucidation and 13C metabolic flux analysis. Curr Opin Biotechnol 36:91–97. https://doi.org/10.1016/j.copbio.2015.08.014
40. Möllney M, Wiechert W, Kownatzki D, de Graaf AA (1999) Bidirectional reaction steps in metabolic networks: IV. Optimal design of isotopomer labeling experiments. Biotechnol Bioeng 66:86–103
41. Baran R, Bowen BP, Bouskill NJ et al (2010) Metabolite Identification in *Synechococcus sp. PCC 7002* Using Untargeted Stable Isotope Assisted Metabolite Profiling. Anal Chem 82:9034–9042. https://doi.org/10.1021/ac1020112
42. Arrivault S (2009) Use of reverse-phase liquid chromatography, linked to tandem mass spectrometry, to profile the Calvin cycle and other metabolic intermediates in Arabidopsis rosettes at different carbon dioxide concentrations. Plant J 59:826
43. Link H, Fuhrer T, Gerosa L et al (2015) Real-time metabolome profiling of the metabolic switch between starvation and growth. Nat Methods. https://doi.org/10.1038/nmeth.3584

Chapter 19

Genome-Scale ^{13}C Fluxomics Modeling for Metabolic Engineering of *Saccharomyces cerevisiae*

David Ando and Héctor García Martín

Abstract

Synthetic biology is a rapidly developing field that pursues the application of engineering principles and development approaches to biological engineering. Synthetic biology is poised to change the way biology is practiced, and has important practical applications: for example, building genetically engineered organisms to produce biofuels, medicines, and other chemicals. Traditionally, synthetic biology has focused on manipulating a few genes (e.g., in a single pathway or genetic circuit), but its combination with systems biology holds the promise of creating new cellular architectures and constructing complex biological systems from the ground up. Enabling this merge of synthetic and systems biology will require greater predictive capability for modeling the behavior of cellular systems, and more comprehensive data sets for building and calibrating these models. The so-called "-omics" data sets can now be generated via high throughput techniques in the form of genomic, proteomic, transcriptomic, and metabolomic information on the engineered biological system. Of particular interest with respect to the engineering of microbes capable of producing biofuels and other chemicals economically and at scale are metabolomic datasets, and their insights into intracellular metabolic fluxes. Metabolic fluxes provide a rapid and easy to understand picture of how carbon and energy flow throughout the cell. Here, we present a detailed guide to performing metabolic flux analysis and modeling using the open source JBEI Quantitative Metabolic Modeling (jQMM) library. This library allows the user to transform metabolomics data in the form of isotope labeling data from a ^{13}C labeling experiment into a determination of cellular fluxes that can be used to develop genetic engineering strategies for metabolic engineering.

The jQMM library presents a complete toolbox for performing a range of different tasks of interest in metabolic engineering. Various different types of flux analysis and modeling can be performed such as flux balance analysis, ^{13}C metabolic flux analysis, and two-scale ^{13}C metabolic flux analysis (2S-^{13}C MFA). 2S-^{13}C MFA is a novel method that determines genome-scale fluxes without the need of every single carbon transition in the metabolic network. In addition to several other capabilities, the jQMM library can make model based predictions for how various genetic engineering strategies can be incorporated toward bioengineering goals: it can predict the effects of reaction knockouts on metabolism using both the MoMA and ROOM methodologies. In this chapter, we will illustrate the use of the jQMM library through a step-by-step demonstration of flux determination and knockout prediction in a complex eukaryotic model organism: *Saccharomyces cerevisiae* (*S. cerevisiae*). Included with this chapter is a digital Jupyter Notebook file that provides a computable appendix showing a self-contained example of jQMM usage, which can be changed to fit the user's specific needs. As an open source software project, users can modify and extend the

---

Electronic supplementary material: The online version of this chapter (https://doi.org/10.1007/978-1-4939-8757-3_19) contains supplementary material, which is available to authorized users.

Edward E.K. Baidoo (ed.), *Microbial Metabolomics: Methods and Protocols*, Methods in Molecular Biology, vol. 1859, https://doi.org/10.1007/978-1-4939-8757-3_19,

code base to make improvements at will, allowing them to share their development work and contribute back to the jQMM modeling community.

Key words Flux analysis, Two-scale ^{13}C metabolic flux analysis, -Omics data, Predictive biology, Engineering biology

1 Introduction

The ability to genetically engineer and modify an organism's DNA has radically changed the nature of biology over the last few decades. Synthetic biology is an emerging field which applies genetic engineering techniques toward the systematic design of biological organisms toward specific design and engineering goals [1] using traditional engineering principles such as the design–build–test–learn (DBTL) cycle [2]. Initial efforts in the realm of repurposing biological systems toward the production of biofuels and other chemicals using synthetic and systems biology have succeeded in producing a number of target molecules, although typically at low titers of μg/L to mg/L [3]. However, only the biological production of a few compounds, such as 1,3-propanediol [4], 1–4-butanediol [5], and artemisinin [6], has been produced at high titers and at a commercial scale. This relative scarcity of commercially successful examples showcases the difficulties that the synthetic biology field still faces, among them a lack of predictive tools and models to guide the bioengineering process [7]. Improving the yield, titer, and productivity of engineered bioprocesses to enable commercialization requires detailed manipulation of many processes including microbial physiology, stress response, and metabolism. This chapter focuses exclusively on the particular field of metabolic engineering, and will show how to use metabolomics data to calculate how carbon and energy flow through the cell, which is an effective piece of information to optimize [8] bioproduct formation [3].

The study of metabolic fluxes is of particular interest for engineering microbes that can produce biofuels and other chemicals as fluxes provide a mechanistic understanding for how carbon and energy flow throughout the cell. While often a heuristic and nonsystematic approach is taken toward designing and genetically engineering new organisms [3], metabolic fluxes offer the potential for adding a powerful predictive capability to synthetic biology. Metabolic fluxes can be readily transformed into actionable predictions for improving bioproduct formation via the use of metabolic modeling [8]. Metabolic models include the stoichiometry of an organism's metabolic network and, using a variety of mathematical methods (e.g., COBRA [9]), can predict how fluxes will change through modifications of the metabolic reaction network arising from specific genetic engineering strategies.

The DBTL cycle is one of the common practices of engineering transplanted into synthetic biology. This cycle starts with the design (D) of a biological system in the form of a best guess, given initial data and biological understanding, on how to achieve a desired particular outcome (e.g., fuel production at a given yield). This biological design is then built (B) in the next phase of the cycle from DNA parts and an appropriate microbial chassis using the tools of synthetic biology. The next phase involves testing (T) whether the biological system performs as desired in the original design using a variety of assays (e.g., production measurement or/and omics profiling). Given the complex nature of biological systems, it is extremely unlikely that the first design would behave as desired, so further attempts will most likely be needed to meet the desired specification. It would be desirable to use a validated predictive model based on data generated in previous rounds to systematically converge toward engineering goals more quickly. This task is referred to as the learn (L) phase of the DBLT cycle and is, arguably, the hardest and least supported step in current metabolic engineering practice [2]. However, given recent increasingly available high-throughput workflows for data generation, there is an opportunity to use these copious amounts of data to validate and create models which can provide *actionable* items for metabolic engineers: that is, suggestions that can be acted upon with the available synthetic biology tools (e.g., to change a particular gene's ribozyme binding site (RBS) or knock out a particular gene in order to reroute metabolic fluxes).

In this chapter we will show how to use metabolomic data obtained from ^{13}C labeling experiments to generate actionable items to increase ethanol production in *S. cerevisiae*. We will demonstrate how to use the Joint BioEnergy Institute (JBEI) Quantitative Metabolic Modeling library (jQMM) [10] to calculate cellular fluxes and make predictions. The jQMM library is currently capable of measuring and predicting internal metabolic fluxes using a variety of techniques: ^{13}C metabolic flux analysis (^{13}C MFA) [11], flux balance analysis (FBA) [9], and two-scale ^{13}C metabolic flux analysis (2S-^{13}C MFA) [12]. First, we will provide a brief background on metabolomics and more specifically ^{13}C labeling experiments, which consist of cellular cultures in which the feed source (e.g., glucose) is labeled with carbon atoms that have an extra neutron (i.e., carbon isotopes) at selected and specific positions. We will then succinctly describe how to measure the ensuing carbon isotope labeling in the metabolites of the studied. Next, we will describe in detail how to use the labeling data from different metabolites in the cell to determine the cell's internal metabolic fluxes through a technique called two-scale ^{13}C metabolic flux analysis (2S-^{13}C MFA). 2S-^{13}C MFA is a novel technique which introduces the capability to use experimental ^{13}C labeling data to constrain comprehensive *genome-scale* models (rather than with small models of central metabolism as done with traditional ^{13}C MFA) by taking into account the system-wide balances of metabolites [8]. This is possible by assuming that flux flows from core to peripheral metabolism and does not flow back. This assumption stems from the bow tie structure of cellular metabolism [2], and is further justified by the fact that traditional ^{13}C MFA, using only core metabolism models, can convincingly explain labeling patterns for amino acids and intracellular metabolites for model organisms [13–16]. We then perform an External Labeling Variability Analysis (ELVA), a self-consistency test of this modeling approach which gauges the effect of the ignored noncore reactions [12]. ELVA finds the maximum impact that the unknown inflow metabolite labeling can have on the measured labeling pattern, and provides evidence of biophysical consistency of the calculated metabolic fluxes. Finally, we will show how to use the COBRA (Constraint Based Reconstruction and Analysis) [17] methods of MoMA (Minimization of Metabolic Adjustment) [18] and ROOM (Regulatory On/Off Minimization) [19] to predict, based on the measured flux profiles, which gene knock outs will increase acetate production in *S. cerevisiae*.

2 Materials

The general workflow for determining and visualizing metabolic fluxes of cellular systems is shown in Fig. 1. First, a microbial culture is grown with ^{13}C labeled glucose feed. Next, mass spectroscopy is used for the analysis of the distribution of ^{13}C in metabolites extracted from cell culture to obtain the Mass Distributions Vectors (MDVs) [20] for each measured metabolite, which give the relative frequency of isotopologs. Each isotopolog of a metabolite has a different number of carbons with an extra neutron, and each isotopolog can exhibit the heavy carbon(s) substituting any of the carbon(s) located within the metabolite. For example, a metabolite with five carbon atoms could have five different isotopologs. The different labeling patterns for the measured metabolites ultimately provides enough information to determine the internal flux profile [12] via knowledge of how reactions in the metabolic network rearrange carbon atoms within metabolites (so called carbon transitions). MDVs then need to be then formatted for input into the jQMM, so that metabolic fluxes can be calculated at the genome scale using the 2S-^{13}C methodology. For genome scale models, there are thousands of reactions, so we recommend that fluxes be visualized and more easily understood via a plotting environment called “Arrowland” which is located at the following website: http://public-arrowland.jbei.org. Fluxes can also be more simply plotted within the jQMM.

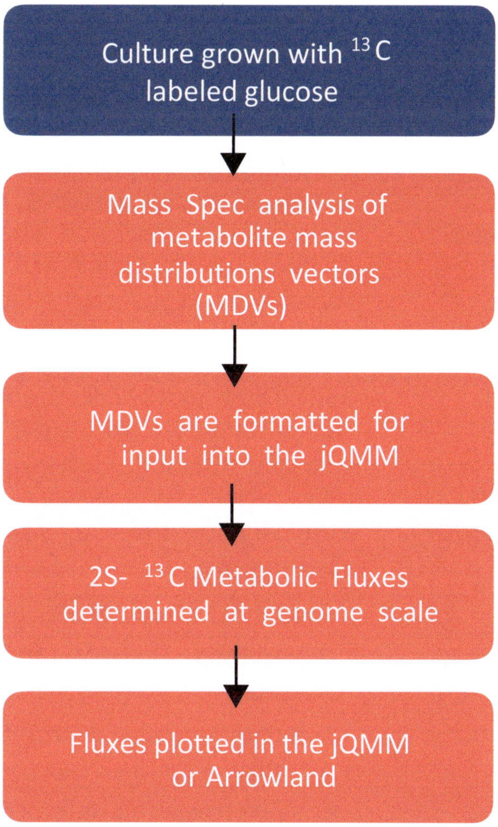

Fig. 1 Overview of the workflow for ^{13}C two-scale metabolic flux analysis. The test phase of the workflow is in blue while the learn phase steps are in light red

2.1 Computational Requirements

To calculate fluxes using the jQMM, a desktop computer or server system that is capable of running heavy computational loads for extended periods of time is necessary. The more available cores/CPUs in the system the better, as calculations within the jQMM library often take many hours to days (for a desktop computer), although the library is parallelized and can leverage additional cores/CPUs to reduce computation time. Windows, Mac, and Linux operating systems are all compatible with running python code, which the jQMM is written in. Error-correcting code RAM (ECC RAM) reduces incorrect memory read and write operations to system RAM and is recommended to reduce the probability of faulty calculations over lengthy computations due to memory access errors. Xeon processors from Intel, for example, are specially designed for long continuous computations at high CPU load, unlike the consumer series of desktop processors sold by Intel such as the i7 and i5 series (circa 2016), and therefore we also recommend the use of Xeon like processors.

2S-^{13}C MFA computations require the following:

1. At least 32 GB of RAM and 500 GB of free disk space.
2. Ownership of both the GAMS and CONOPT solver licenses.
3. Python version 2.7 installed and configured.

2.2 Software Installation and Configuration

Use of the jQMM requires installation of accessory libraries (such as numpy, as described below), which can be done in two different ways: the traditional way, which includes a self-installation on the hardware and operating system at hand, or via the use of a preconfigured and preinstalled Docker container (http://www.docker.com). We recommend using our preconfigured Docker container for use of the jQMM as the flux modeling environment will be immediately usable once a GAMS and CONOPT license are purchased. However, for expert users or those who wish to use a custom installation, a self-installation is readily achieved as follows.

2.2.1 Self-Installation

First install the following jQMM dependencies to the Python 2.7 environment:

1. libSBML: available at http://sbml.org/Software/libSBML.
2. matplotlib: available at http://matplotlib.org/users/installing.html.
3. numpy: available at http://www.scipy.org/scipylib/download.html.
4. jupyter: available at http://jupyter.org/.

For self-installation *see* **Note 1**.

2.2.2 Preconfigured and Preinstalled jQMM Library

Docker (https://docs.docker.com/) is a technology that is based on Linux containers that allows for the building, running, testing, and deploying applications such as the jQMM library into a complete file system that contains everything it needs to run: code, runtime, system tools, system libraries, and supporting data files. This guarantees that it will always run correctly and in the same way, regardless of the system environment it is running in. The jQMM docker container format is standardized and can be run on virtually any cloud computing service such as AWS (Amazon Web Services), Google Cloud Platform, Microsoft Azure, etc. Additionally, the jQMM Docker container can be conveniently run on a personal computer running either Microsoft Windows or the Mac operating system, although we discourage this practice for anything other than training purposes given how slowly the jQMM will run on personal computers. For further information on web-based platforms please *see* **Note 2**.

The prebuilt jQMM docker container, which is available for download at https://github.com/JBEI/jqmm, has all of the software needed to run the jQMM library preinstalled and preconfigured to work out of the box. The GAMS and CONOPT solvers (http://www.gams.com/, http://www.conopt.com/), are required to do flux analysis, but their usage within the Docker container will require purchase of GAMS and CONOPT licenses separately. Once these licenses are included with the GAMS and CONOPT solver installations the jQMM docker container will be fully functional.

2.3 ^{13}C Labeling Experiments and Mass Spectrometry Data Analysis

The initial step for any ^{13}C based metabolic flux experiment is performing a ^{13}C labeling experiment with the organism of interest. We will only give a brief description here, the reader can refer to previously published protocols for more information [21, 22]. Cultures should use minimal media to avoid nonlabeled carbon sources, and can be grown using different types of labeling for the feed (*see* **Note 3**). A common choice is to use 20% U-^{13}C glucose (universal labeling of carbon atoms) and 80% 1-^{13}C glucose (first carbon atom only is labeled) (for example *see* **Note 4**). Briefly, we recommend that culture samples are prepared for measurement on an LC-MS system by a rapid quenching step to halt cellular metabolism immediately after removing an aliquot of cells from culture, which is then followed by extraction of intracellular metabolites and LCMS sample preparation [21, 22].

The quantification of relative cellular metabolite isotopolog concentration, consisting of MDVs for metabolites in cellular metabolism (*see* Fig. 2 for an example), is done by determining the relative concentration of each metabolite's isotopologs via mass spectrometry. Once chromatograms from a mass spectrometry instrument have been analyzed and integrated such that the relative frequency of each metabolite's mass distribution has been determined, these data need to be prepared and formatted for input into the jQMM library so that metabolic fluxes can be determined.

2.4 Input Data Preparation

2.4.1 ^{13}C Labeling Data

Once the measured MDVs have been determined and normalized such that their sum is equal to one for each metabolite, they have to be entered into a text document with the following format so that the data can be read by the jQMM:

```
Amino acid Mass distribution
 m0 m1 m2 m3 m4 m5 m6 m7 m8
Gly M-0 0.7426 0.0592 0.1982 - - - - -
Ala M-0 0.3903 0.3884 0.0221 0.1992 - - - -
Thr M-0 0.3260 0.3416 0.1670 0.1513 0.0000 - - -
Asp M-0 0.3192 0.3564 0.2003 0.1035 0.0206 - - -
```

The first two lines should remain fixed and are ignored by the jQMM library, while the following lines need to include the MDV information for every metabolite for which isotopolog data exists in the following format:

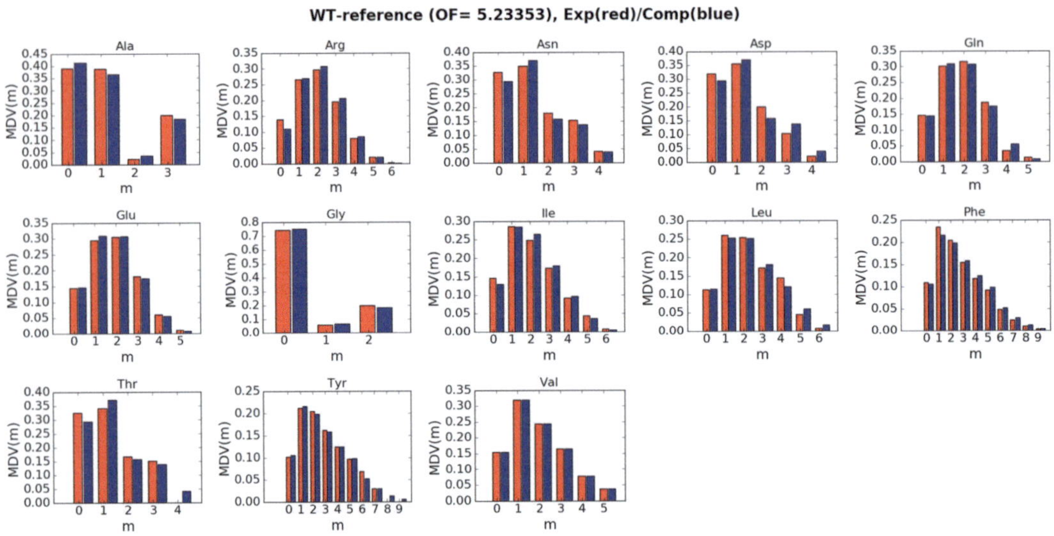

Fig. 2 Plot of wild-type *S. cerevisiae* MDVs from Ghosh et al. [8]. Experimentally measured MDVs are plotted in red bars for each metabolite, while the MDVs implied by the predicted fluxes are shown in the blue bars. Computational and experimental data match closely implying that the model is quantitatively correct

1. First the metabolite name using the metabolite abbreviation used in the BIGG database (http://bigg.ucsd.edu) is specified and followed by a tab character or space(s).
2. Next tab or space separated relative frequencies of each metabolite isotopolog are specified. The sum of all isotopolog frequencies needs to add to 1 for each individual metabolite.
3. Isotopologs which do not exist for a particular metabolite, should be represented by a "-" (dash), and must not be entered as a zero value, which would indicate a different meaning in the jQMM. Zero values indicate that an isotopolog can exist but is not present in the sample.

2.4.2 Choose, Edit, and Define the Metabolic Model

To define a metabolic model for the jQMM to use in the modeling of the particular organism being studied, we recommend using the BIGG database (http://bigg.ucsd.edu) and downloading a SBML version of the metabolic model which is appropriate to the problem being studied. Smaller models run much faster in the jQMM, while more comprehensive genome scale models are necessary for problems involving peripheral metabolism or which utilize the 2S-^{13}C MFA analysis methodology. Retooling of the jQMM library code may be required for the library to understand metabolite names which do not follow the naming convention used in the iMM904 metabolic model [23] format, although sample code is already included in the jQMM for using metabolite names from the iJO1366 [24] and iAF1260 [25] models, which is located in the "sbmlio.py" python code file located in the jQMM "code/core" directory. Network reactions which come from a heterologous pathway that has been engineered into a microbe should be manually copied into the downloaded SBML model.

2.4.3 Defining Exchange Reaction Fluxes

Both measured fluxes through exchange reactions and the biomass flux that correspond to the time that a sample was taken for ^{13}C isotopolog analysis are detailed in a FLUX.txt file. A sample exchange flux file should be formatted as follows for input into the jQMM library:

```
biomass_SC5_notrace: 0.24 [==] 0.24
GLCt1: 8.5 [==] 8.5
EX_glc(e): -8.5 [==] -8.5
EX_ac(e): 0.45 [==] 0.45
```

The Biomass flux is in units of 1/h and is normalized to equal the growth rate while all other exchange fluxes are in units of mMol/gdw/h (millimoles/grams of dry weight/hour). The glucose uptake rate can be measured for example by a HPLC determination of glucose concentration at two different times around the ^{13}C analysis sample time. If the glucose concentration is measured as g_1 and g_2 (in millimoles/volume) at times t_1 and t_2, the glucose flux can be approximated as $(g_2 - g_1)/gdw_1/(t_2 - t_1)$, where gdw_1 is the grams of dry weight of cells per unit volume at time t_1. Similarly, HPLC measurements can be used to determine fluxes of acetate, lactate, and other organic acids excreted by the cell.

2.4.4 Defining Carbon Transitions in the Metabolic Model

Atom transitions can be used to represent the fate of each carbon in a reaction [11]. In the example reaction:

```
A (abc) -> B (ab) + C (c)
```

Uppercase letters in a carbon transition reaction represent the metabolites present in a reaction, while the lowercase letters in parentheses represent the carbon atom transitions, *see* **Note 5**. Irreversible reactions are denoted by a '->' or '=>' arrow and reversible reactions are denoted by a '<->' or '<=>' arrow. In the jQMM, multiple reactions are separated by carriage returns or by placing a semicolon at the end of each reaction equation.

For example, in the following reaction from the iMM904 metabolic model for *S. cerevisiae*:

```
AKGDam akg_m --> sdhlam_m + co2_m abcde : bcde + a
```

akg (2-Oxoglutarate) gets split into sdhlam (Succinyldihydrolipoamide) and co2 (carbon dioxide) in reaction AKGDam, with the last four carbons going to sdhlam and the first carbon going to co2.

For 2S-^{13}C based metabolic flux analysis in the jQMM, one needs to create a REACTIONS.txt file which contains the carbon transition information for core reactions in the metabolic network and which has the following format:

```
# Metabolic Exchange fluxes
&MLIST etoh 0
&MLIST glucon 0
&MLIST lac-L 0

# Amino acid metabolite fluxes
&MLIST ala-L[c] 0
&MLIST asp-L[c] 0
&MLIST asn-L[c] 0
&MLIST gly[c] 0
&MLIST val-L[c] 0
&MLIST arg-L[c] 0

# Carbon source
&SOURCE glcD[e]

# Input Reactions
EX_glc(e) glc-D[e] <==> glcDEx abcdef : abcdef
GLCt1 glc-D[e] --> glc-D abcdef : abcdef
HEX1 glc-D --> g6p abcdef : abcdef

# Carbon Transitions
PGI g6p <==> f6p abcdef : abcdef
PFK f6p --> fdp abcdef : abcdef
FBA fdp <==> g3p + dhap abcdef : cba + def
```

See **Note 6** for metabolite abbreviations and **Note 7** for additional REACTIONS.txt file specifications.

2.4.5 Defining Feed Labeling

The type of labeled glucose used in the experiment, together with its concentration relative to the amount of unlabeled glucose, is detailed in a FEED.txt file. A sample feed specification file contains the following line:

```
1.5% Glucose: 80% 1-C 20% U 0% UN
```

See **Note 8** for file specification details.

3 Methods

An example Jupyter notebook which demonstrates metabolic flux analysis on *S. cerevisiae* is included in the Supplementary Information and can be run directly within the jQMM (*see* Electronic

Supplementary Material, *see* **Note 9**). This notebook is also reproduced in this section as an example of how to use the FluxModels module in the jQMM to do 2S-^{13}C MFA and then predict the outcomes on ethanol production of different reaction knockouts. The Jupyter notebook provides a convenient way to reproduce results and is easily modified to fit the user's specific needs, and it is divided into seven different steps for turning experimental data into actionable predictions for increasing a targeted biochemical via genetic engineering:

1. Setting up python environment.
2. Gathering input data.
3. Creating the Reaction Network.
4. Creating the two-scale metabolic model.
5. Calculating internal metabolic fluxes through 2S-^{13}C MFA.
6. Performing an ELVA consistency check.
7. Visualizing flux profiles.
8. Predicting which genes to knockout using MoMA and ROOM.

```
-- Jupyter Notebook Start --
```

^{13}C-2S Metabolic Flux Analysis for *S.cerevisiae* with ELVA methods: This Jupyter notebook presents a computable step-by-step description of how to use metabolite data from ^{13}C labeling experiments to determine fluxes for *S. cerevisiae* with quality control of flux results via ELVA methods. We also use MOMA and ROOM methodologies to produce actionable insights to improve ethanol production.

3.1 Setup

The first step involves specifying the correct path for the library to be loaded:

```
In [1]:
%matplotlib inline
import sys, os
path = "/scratch/user"
pythonPath = path + "/quantmodel/code/core"
if pythonPath not in sys.path:
 sys.path.append(path + '/quantmodel/code/core')
os.environ["QUANTMODELPATH"] = path +'/quantmodel'
```

We then need to import the needed classes for the notebook:

```
In [2]:
from IPython.display import SVG
import FluxModels as FM
```

```
import enhancedLists, ReactionNetworks, predictions, copy,
core, os
```

and then move to a defined working directory where output and intermediate files will be kept:

```
In [3]:
cd ~/tests
```

Test that GAMS and CONOPT solvers are installed and working:

```
In [5]:
status = os.system('gams')
if status == 0:
 print "GAMS call working."
else:
 print "GAMS is not installed."
GAMS call working.
```

3.2 Gathering Input Data

As part of the test (T) phase of the DBTL cycle, we gather all the relevant experimental data from the ^{13}C labeling experiments (*see* "Materials" Subheading 2). These data involve:

- A base *genome-scale model* that will act as the reference for all other data types.
- *Exchange fluxes* containing the measured fluxes of metabolites being exchange by cells with the environment.
- *Transition information* on the fate of each carbon in the core reaction network.
- *Metabolite labeling* information in the form of Mass Distribution Vectors (MDVs).
- *Metabolite labeling error* information.
- *Feed labeling* information on the type of labeled glucose the cell culture was fed.

Discussion of these data types can be seen in Subheading 2.

For this demonstration, we will use the data from Ghosh et al. [8] for the wild type strain grown using a glucose feed:

```
In [6]:
datadir = os.environ['QUANTMODELPATH']+'/data/tests/yeast2S/
wyr1/'
strain ='wyr1Glc'
#A base genome-scale model:
BASEfilename = datadir + 'SciMM904wyr1GlcTS.xml'
#Exchange fluxes
```

```
FLUXESfilename = datadir + 'FLUX'+strain+'.txt'
#Transition information
TRANSITIONSfilename = datadir + 'REACTIONS'+strain+'-core-too-
small.txt'
#Metabolite labeling
MSfilename = datadir + 'LCMS'+strain+'.txt'
#Metabolite labeling error
MSSTDfilename = datadir + 'LCMSerr'+strain+'.txt'
#Feed labeling
FEEDfilename = datadir + 'FEED'+strain+'.txt'
```

3.3 Creating the Reaction Network

Once we have gathered all the needed input files, we can condense all this information into a single sbml file. We will do this using a reaction network from the ReactionNetworks module in the jQMM library. A reaction network contains all information related to the metabolic reaction network used for the simulation:

```
In [7]:
# Load initial SBML file
reacNet = ReactionNetworks.TSReactionNetwork(BASEfilename)
# Add Measured fluxes
reacNet.loadFluxBounds(FLUXESfilename)
# Add carbon transitions
reacNet.addTransitions(TRANSITIONSfilename,translate2SBML=-
True)
# Add measured labeling information
reacNet.addLabeling(MSfilename,'LCMS',MSSTDfilename,
minSTD=0.001)
# Add feed labeling information
reacNet.addFeed(FEEDfilename)
# Limit fluxes to 500
reacNet.capFluxBounds(500)
# Create sbml file to store the two-scale model.
# All input files are combined in a tuple of the type:
(fileName, string of contents)
SBMLfile = ('SciMM904'+strain+'TS.xml',reacNet.write('to-
String'))
```

3.4 Creating the Two-Scale Metabolic Model

The next step is to use the SBML file we just created to create a two-scale model [12] that we will use to calculate fluxes through 2S–^{13}C MFA:

```
In [9]:
TSmodel = FM.TwoSC13Model(('SciMM904'+strain+'TS.xml',reac-
Net.write('toString')))
```

TSmodel now contains all the information needed to calculate fluxes along with the methods to do this calculation and other analysis [12].

3.5 Calculating Internal Metabolic Fluxes through 2S-^{13}C MFA

We can now use the *findFluxesRanges* method in *TSmodel* to find the fluxes that best fit the experimentally obtained metabolite labeling data (MDVs) and find the ranges of fluxes compatible with this labeling data and the corresponding experimental error:

```
In [10]:
%%time
fluxNames =
TSmodel.reactionNetwork.C13ReacNet.reactionList.getReaction-
NameList(level=1)
TSresult =
TSmodel.findFluxesRanges(Nrep=30,fluxNames=fluxNames,proc-
String='proc')Wall time: 1.75 hrs
```

Nrep represents the number of replicates used for the calculation. Since the problem to be solved is a nonconvex problem there is no guarantee that a single run will find the best global fit. Hence, we run 30 independent processes and keep the one that best fits the data. *fluxNames* indicates the fluxes for which full flux confidence intervals will be calculated. *procString* indicates that the data (for this case) needs no derivatization correction. For large models, especially with eukaryotic organisms, these flux calculations can be lengthy (on the order of several hours).

We can check how accurate the model is by comparing the measured labeling distribution (MDVs, red) with the one predicted through the computational model (blue) by using the `plotExpvsCompLabelFragment` method:

```
In [11]:
TSresult.plotExpvsCompLabelFragment(titleFig='WT-reference')
```

See Fig. 2.

or by using `plotExpvsCompLabelXvsY`, if we prefer to see these fits as an X vs. Y plot:

```
In [12]:
TSresult.plotExpvsCompLabelXvsY(titleFig='WT-
reference')
```

See Fig. 3.

The closeness of the fit data and the experimental data validate the use of this model.

Results are stored in a reaction network inside *TSresult* and can be explored through the reactionList methods. For example, we can print the fluxes that best fit the data:

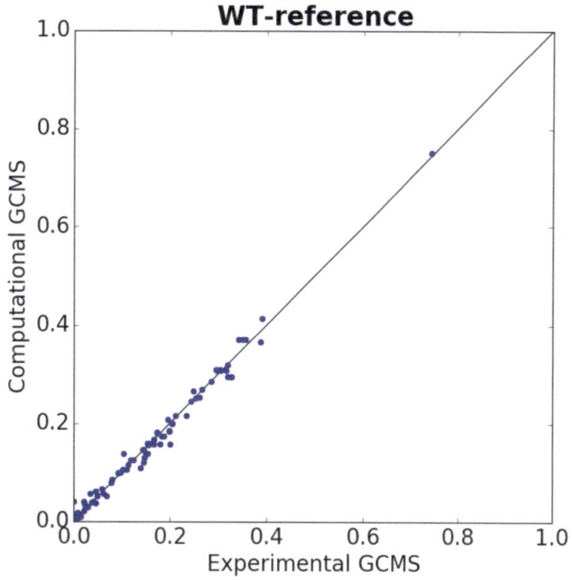

Fig. 3 Plot of experimentally measured MDVs versus the computationally predicted MDVs in an x-y plot (different way to plot data in Fig. 2), which demonstrates that experimental and predicted MDVs are comparable

```
In [13]:
TSresult.reactionNetwork.reactionList.printFluxes(brief="-
True",names="exchange")
EX_co2_e_: 14.2007023028
EX_glc_e_: -8.5
EX_etoh_e_: 8.5
EX_h2o_e_: 7.11362355138 EX_o2_e_: -4.90795595138
EX_h_e_: 2.72395044862
EX_glyc_e_: 2.45 EX_nh4_e_: -1.3431504
EX_glx_e_: 0.999214848625
EX_ac_e_: 0.45 biomass_SC5_notrace: 0.24
EX_pi_e_: -0.0474479999997
EX_ttdca_e_: 0.03
EX_ddca_e_: 0.02
EX_so4_e_: -0.018552
EX_hdca_e_: 0.0143
EX_ocdca_e_: 0.00193
EX_for_e_: 0.0012408
```

3.6 Performing an ELVA Consistency Check

The 2S-^{13}C methods assume flux flows from core metabolism to reactions in the periphery, and does not flow back. This assumption of one way flow out of central metabolism stems from the bow-tie structure of cellular metabolism [2], and is further justified by the fact that traditional ^{13}C MFA, using only core metabolism models, can convincingly explain labeling patterns for amino acids and intracellular metabolites for model organisms [13–16].

ELVA [12] is a test that checks that the error incurred by ignoring noncore reactions is negligible. If the ELVA indicates that the noncore contributions are large, then the core set of reactions which include carbon transition information is expanded. This is repeated until a self-consistent result is found that produces ELVA flux ranges which are small and compatible with the experimental data.

In general, the ELVA optimization problem finds the maximum impact that the unknown inflow metabolite labeling can have on the measured labeling pattern. ELVA determines if the fluxes obtained from the fit are consistent with the two-scale approximation [12]. This extra step sets the method apart from standard ^{13}C metabolic flux analysis (MFA), which does not check the effect of ignored reactions. ELVA considers only core metabolism with the impact of noncore metabolism being represented through inflow metabolites, dummy metabolites with unfixed labeling since their labeling is, by definition, unknown. Essentially, using the previously obtained genome-scale flux solution as a constraint, ELVA allows the labeling for the inflow metabolites to vary and, for each metabolite with measured labeling, uses each element of the mass distribution vector (MDV) as the objective function to be maximized or minimized. By this method, we obtain a confidence interval that represents the maximum possible difference in labeling that could be attributed to noncore reactions for the current solution. The reactions that contribute an unacceptable amount of uncertainty are then added to the core set and the procedure can be repeated as necessary, until a core set of reactions is found which fully justifies the two-scale approximation. ELVA results can be calculated as follows:

```
In [14]:
resultELVA = TSmodel.ELVA(TSresult)
```

ELVA results can be plotted in an x-y graph showing the experimentally determined isotope labeling which defines a confidence interval that represents the maximum possible difference in labeling that could be attributed to noncore reactions for the current solution. If there are reactions that contribute an unacceptable amount of uncertainty they can be added to the core set and the procedure can be repeated as necessary, until a core set of reactions is found which fully justifies the two-scale approximation. In this example, 13 metabolites have large fluctuations as predicted by computational labeling (y-axis error bars).

```
In [15]:
resultELVA.plotExpvsCompLabelxvsy(titleFig="WT",outputFileNa-
```

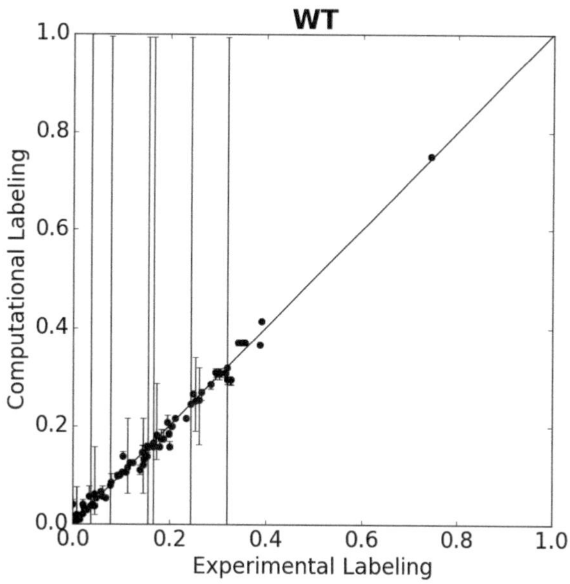

Fig. 4 ELVA plot that shows an x-y graph of the experimentally determined isotope labeling versus the computationally predicted labeling. The vertical error bars define the computational error that represent the maximum possible difference in labeling that could be attributed to noncore reactions for the current solution. In this example, several reactions display large computational errors, indicating that noncore reactions do significantly contribute to core metabolite labeling [12]. More reactions with carbon transition information need to be added to the core reactions list to remedy this problem and satisfy the two-scale assumptions

```
me="ELVAComparisonWT.txt",save="ELVA-W.eps")
```

See Fig. 4.

The error bars in the y axis (computational error) are NOT of the same order of magnitude as the experimental error, and therefore we need to rework the set of core reactions until we can satisfy the two-scale assumptions of no flux back into the core network [12]. We therefore add additional reactions to the REACTIONS file, and repeat the two-scale modeling process. We can see from Fig. 4 that there are 13 metabolites for which the fluctuation in computational labeling was large, which are described in more detail in an ELVAfluxReport that is generated by the jQMM.

The file "*ELVAfluxReport.txt*" is automatically saved to our current working directory /~tests and indicates which metabolites had large fluctuations. The format of the ELVAfluxREPORT file is as follows: first a metabolite is listed, which is then followed by the reactions that flow into it from noncore metabolism, then the reactions which flow into that metabolite from core metabolism.

The following ELVA report has been truncated for space, please see the full Jupyter notebook for complete details.

```
In [16]:
cat ELVAfluxReport.txt
RI3mob_c(0.0104):
        ----> Non Core:
        3MOBtm [ 0.0881793]
        ----> Core:
        IPPS [-0.0246753]
        VALTA [-0.063504]
RIakg_c(0.0019):
        ----> Non Core:
        HSTPT [ 0.015912]
        ----> Core:
        AATA [ 0.068688]
        ALATA_L [ 0.11842353]
        ASPTA [-0.69541853]
        GLUDyi [-1.1814192]
        ICDHy [ 1.230851]
        ILETA [ 0.046248]
        LEUTA [ 0.071136]
        PHETA1 [ 0.032136]
        PSERT [ 0.1367712]
        SACCD2 [ 0.068688]
        TYRTAi [ 0.02448]
        VALTA [ 0.063504]
RIhdca_c(0.0002):
        ----> Non Core:
        FACOAL160 [ 0.01623]
        HDCAt [-0.0143]
        ----> Core:
        FAS180 [-0.00193]
RIpyr_c(0.0008):
        ----> Non Core:
        ANS [ 0.006816]
        ----> Core:
        AGTi [ 0.00831153]
        ALATA_L [-0.11842353]
        PC [-2.9406824]
        PYK [ 12.82902984]
        PYRDC [-9.54111443]
        PYRt2m [-0.24393701]
RO3mob_m(0.0104):
        ----> Non Core:
        3MOBtm [-0.0881793]
        ----> Core:
        DHAD1m [ 0.13464]
        IPPSm [-0.0464607]
```

See **Note 6** for metabolite abbreviations. As a result of the ELVA report file which indicates that metabolites akg_c, co2_c, co2_m, fum_c, g3p_c, pyr_c, accoa_m, asp_L_c, f6p_c, g6p_c, glu_L_c, glx_c, and glyc3p_c produce large computational errors, we have added the reactions for Valine and Isoleucine synthesis and reactions 3C4MOPtm, 3MOBtm, OMCDCm to the REACTIONS file containing carbon transitions for core reactions. After expanding the core set of reactions we now need to rerun the two-scale ^{13}C MFA with the corrected core set of reactions.

```
In [17]:
datadir = os.environ['QUANTMODELPATH']+'/data/tests/yeast2S/
wyr1/'
strain ='wyr1Glc'
#A base genome-scale model:
BASEfilename = datadir + 'SciMM904wyr1GlcTS.xml'
#Exchange fluxes
FLUXESfilename = datadir + 'FLUX'+strain+'.txt'
#Corrected REACTIONS file with an expanded core set of reac-
tions
#Transition information
TRANSITIONSfilename = datadir + 'REACTIONS'+strain+'-fixed.
txt'
#Metabolite labeling
MSfilename = datadir + 'LCMS'+strain+'.txt'
#Metabolite labeling error
MSSTDfilename = datadir + 'LCMSerr'+strain+'.txt'
#Feed labeling
FEEDfilename = datadir + 'FEED'+strain+'.txt'
```

3.6.1 Re-Creating the New Reaction Network for Fixed Reactions File

Now that we have gathered all the needed input files for recalculating fluxes with the expanded set of core reactions, we again condense all this information into a single SBML file. We will do this using a reaction network from the ReactionNetworks module in the jQMM library. A reaction network contains all information related to the metabolic reaction network used for the simulation:

```
In [21]:
# Load initial SBML file
reacNet = ReactionNetworks.TSReactionNetwork(BASEfilename)
# Add Measured fluxes
reacNet.loadFluxBounds(FLUXESfilename)
# Add carbon transitions
reacNet.addTransitions(TRANSITIONSfilename,translate2SBML=-
True)
# Add measured labeling information
reacNet.addLabeling(MSfilename,'LCMS',MSSTDfilename,
```

```
minSTD=0.001)
# Add feed labeling information
reacNet.addFeed(FEEDfilename)
# Limit fluxes to 500
reacNet.capFluxBounds(500)
# Create sbml file to store the two-scale model.
# All input files are combined in a tuple of the type:
(fileName, string of contents)
SBMLfile = ('SciMM904'+strain+'TS.xml',reacNet.write('to-
String'))
```

3.6.2 Re-Creating the Two-Scale Metabolic Model

We again use the SBML file we just created to create a two-scale model [12] that we will use to calculate fluxes through 2S-^{13}C MFA:

```
In [22]:
TSmodel = FM.TwoSC13Model(('SciMM904'+strain+'TS.xml',reac-
Net.write('toString')))
```

TSmodel now contains all the information needed to calculate fluxes along with the methods to do this calculation and other analysis [12].

3.6.3 Recalculating Internal Metabolic Fluxes Through 2S-^{13}C MFA

We can now use the *findFluxesRanges* method in *TSmodel* to find the fluxes that best fit the experimentally obtained metabolite labeling data (MDVs) and find the ranges of fluxes compatible with this labeling data and the corresponding experimental error:

```
In [23]:
%%time
fluxNames =
TSmodel.reactionNetwork.C13ReacNet.reactionList.getReaction-
NameList(level=1)
TSresult =
TSmodel.findFluxesRanges(Nrep=30,fluxNames=fluxNames,proc-
String='proc')
```

Again, *Nrep* represents the number of replicates used for the calculation. Since the problem to be solved is a nonconvex problem there is no guarantee that a single run will find the best global fit. Hence, we run 30 independent processes and keep the one that best fits the data.

We can now explore the new results through the reactionList methods. For example, we can print some of the new fluxes:

```
In [24]:
TSresult.reactionNetwork.reactionList.printFluxes(brief="-
```

```
True",names="exchange")
EX_co2_e_: 13.7769121853
EX_glc_e_: -8.5
EX_etoh_e_: 8.5
EX_h2o_e_: 6.90234889266
EX_o2_e_: -4.69606089266
EX_h_e_: 2.93522510734
EX_glyc_e_: 2.45
EX_nh4_e_: -1.3431504
EX_glx_e_: 1.21173030734
EX_ac_e_: 0.45
biomass_SC5_notrace: 0.24
EX_pi_e_: -0.0474479999999
EX_ttdca_e_: 0.03
EX_ddca_e_: 0.02
EX_so4_e_: -0.018552
EX_hdca_e_: 0.0143
EX_ocdca_e_: 0.00193
```

Again, we test to make sure that the assumptions used in 2S-^{13}C MFA hold using ELVA.

```
In [25]:
resultELVA = TSmodel.ELVA(TSresult)
```

We now plot ELVA results as an x-y graph.

```
In [26]:
resultELVA.plotExpvsCompLabelxvsy(titleFig="WT",outputFileNa-
me="ELVAComparisonWT.txt",save="ELVA-W.eps")
```

See Fig. 5.

The error bars in the y-axis (computational error) are now of the same order of magnitude as the experimental error (x-axis), which implies that our carbon reactions for the core of metabolism now satisfy the two-scale model assumptions [12]. We now have biophysical consistency in the calculated fluxes and can move on to analyzing and visualizing fluxes. If a metabolite still had large fluctuations at this point in the analysis we would need to add more reactions to the core set and repeat until fluctuations in the ELVA plot were small.

3.7 Visualizing Flux Profiles

Once the metabolic fluxes have been calculated they can be understood visually via their plotting on a flux map. In the jQMM library fluxes can be plotted via the commands:

```
In [27]:
TSresult.drawFluxes('wt.svg',svgInFileName='SciMM904TKs.svg',
```

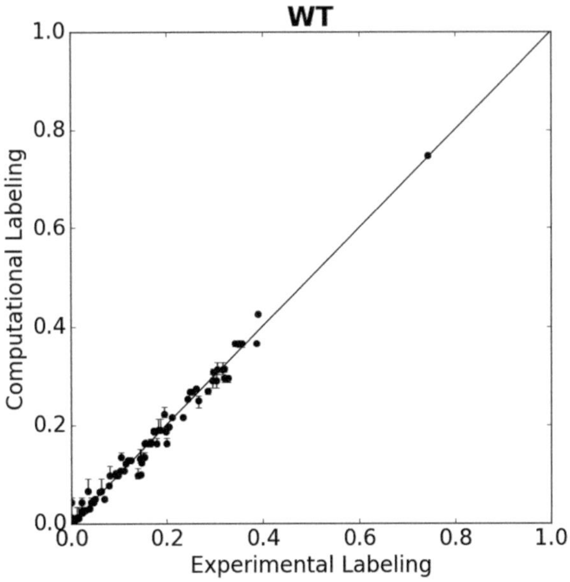

Fig. 5 ELVA plot which shows an x-y graph of the experimentally determined isotope labeling versus the computationally predicted labeling. Since more reactions have been added to the core, all reactions now have only small computational errors in predicted computational labeling, confirming that noncore reactions do not significantly contribute to core metabolite labeling

```
norm='EX_glc_e_')
svgin: /scratch/david.ando/quantmodel/code/core/SciMM904TKs.
svg
```

where 'SciMM904TKs.svg' is the base flux map contained in the jQMM library [10]. The *drawFluxes*() method will indicate the flux magnitude on the base flux map in two ways: visually by changing the flux arrow width according to the flux magnitude through a reaction, and also numerically by showing the net flux value (with confidence intervals) next to the reaction:

The command 'SVG' displays the flux map in the Jupyter notebook which is contained in the svg file which was saved locally.

```
In [28]:
SVG(filename='wt.svg')
Out[28]:
```

See Figs. 6a and b.

Fluxes can also be displayed in a more sophisticated manner using the web browser-based flux plotting library Arrowland (http://public-arrowland.jbei.org).

3.8 Predicting Which Genes to Knockout Using MoMA and ROOM

So far, we have used targeted metabolomic data from ^{13}C labeling experiments to infer the underlying internal metabolic fluxes in the cell. We will now use these inferred fluxes along with two COBRA methods to predict which genes to knock out in order to increase the production of ethanol. These methods are MoMA (Minimization of Metabolic Adjustment [18]) and ROOM (Regulatory On/Off Minimization [19]). MoMA provides an approximate solution for a sub-optimal growth flux state after a knockout has been made to an organism, which is nearest in flux distribution to the unperturbed state. On the other hand, ROOM aims to

a **iMM904**

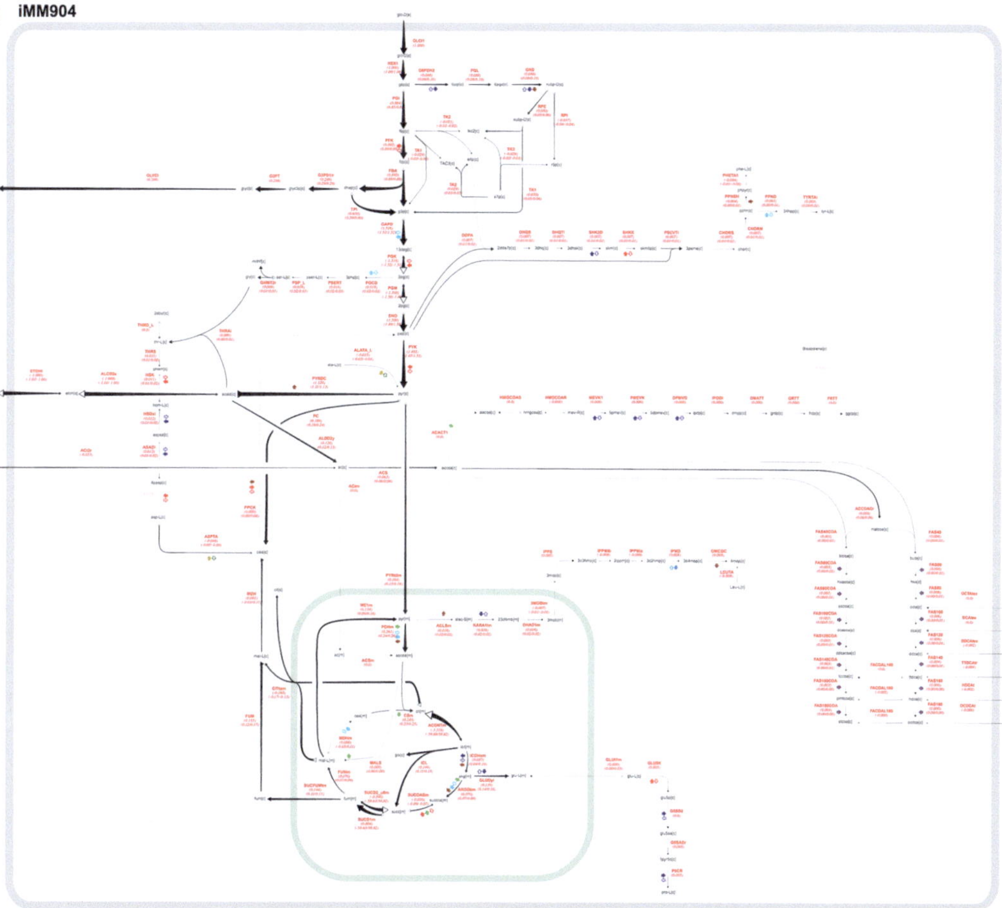

Fig. 6 a) Entire flux map of SciMM904TKs.svg, which displays some of the key 2S-^{13}C metabolic fluxes that are calculated from the iMM904 model for S. cerevisiae. Reactions in the green box are those which are located in the mitochondria. **b**) A zoomed-in plot of these fluxes focused on the upper glycolysis as calculated by the jQMM library using the SciMM904TKs.svg base flux map for a wild type strain of *S. cerevisiae* from Ghosh et al. [8]. This map can be interactively studied using Arrowland (https:/publicarrowland.jbei.org/, wry1Glc). Below each reaction is a single number in parenthesis which indicates the flux value which best fits the experimental data, while the two numbers below the best fit flux represent minimum and maximum fluxes compatible with the MDVs provided by the user

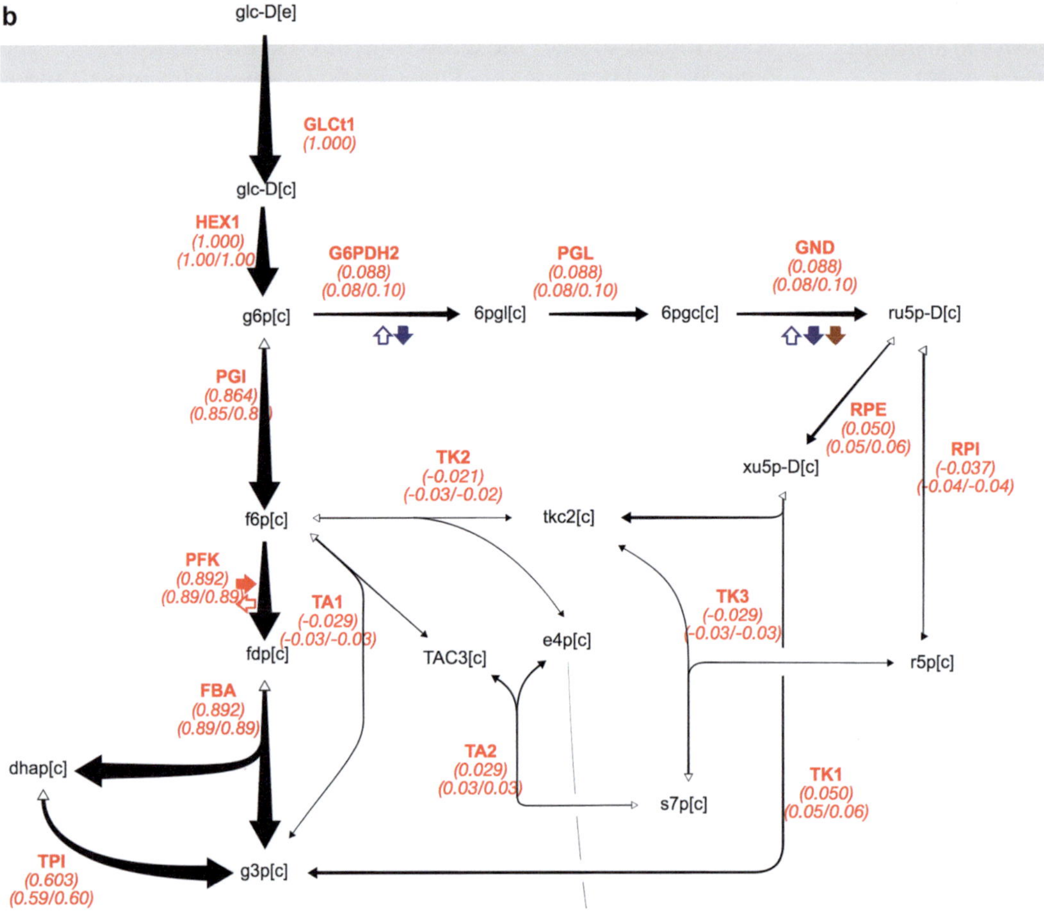

Fig. 6 (continued)

minimize the number of significant flux changes with respect to the wild type to predict resultant fluxes from a knockout of a reaction.

First, we need to specify flexible flux bounds for the final solution in order to avoid biasing the knockout predictions:

```
In [29]:
reactionNetwork = copy.deepcopy(TSmodel.reactionNetwork)
reactionNetwork.changeFluxBounds('GLCt1',core.fluxBounds
(0,10,False)[1])
reactionNetwork.changeFluxBounds('EX_glc_e_',core.fluxBounds
(-10,-0,True,True)[1])
reactionNetwork.changeFluxBounds('biomass_SC5_notrace',core.
fluxBounds( 0,0.25,False)[1])
reactionNetwork.changeFluxBounds('EX_ac_e_',core.fluxBounds
(0,0.5,True,True)[1])
reactionNetwork.changeFluxBounds('EX_etoh_e_',core.fluxBounds
(0,10,True,True)[1])
```

Then we can calculate the base flux profiles for MoMA and ROOM:

```
In [30]:
TSresult = TSmodel.findFluxesStds(Nrep=30,Nrand=10)
```

We then specify a list of reactions to knock out and determine resultant fluxes.

```
In [31]:
KOs = ['PYRt2m','RPI']
```

For reference, we determine the amount of ethanol production in the base WT strain:

```
In [32]:
fluxDict = TSresult.reactionNetwork.reactionList.getReaction-
Dictionary() print       'predicted ethanol flux = ',fluxDict
['EX_etoh_e_'].flux.net.best
predicted ethanol flux = 8.5
```

This implies that before any reaction knockouts that the base strain is releasing 8.5 mmol/gdw/h of ethanol into the environment.

Perform MOMA and ROOM predictions over the set of specified knockouts:

```
In [33]:
for KO in KOs:
 print KO,'knockout:'
 TS13CMOMA = predictions.predict(TSresult, KO, 'MOMA', reac-
tionNetwork.getSBMLString())
 TS13CROOM = predictions.predict(TSresult, KO, 'ROOM', reac-
tionNetwork.getSBMLString())
 fluxDict =
TS13CMOMA.reactionNetwork.reactionList.getReactionDictionary()
  print  ' MoMA predicted ethanol flux = ',fluxDict['EX_e-
toh_e_'].flux.net.best
 fluxDict =
TS13CROOM.reactionNetwork.reactionList.getReactionDictionary()
  print  ' ROOM predicted ethanol flux = ',fluxDict['EX_e-
toh_e_'].flux.net.best
 print '--------------------'
 print ''PYRt2m knockout:
 MoMA predicted ethanol flux = 8.48554566176
 ROOM predicted ethanol flux = 8.756
--------------------RPI knockout:
 MoMA predicted ethanol flux = 8.15179595791
 ROOM predicted ethanol flux = 8.244
--------------------
```

As can be observed, knocking out the PYRt2m reaction is predicted to increase ethanol production by 3.0% according to the ROOM methodology and decrease ethanol production by less than 1% when using the MoMA methodology. As can be seen with an RPI knockout, both MoMA and ROOM predict a decline in ethanol production.

```
-- Jupyter Notebook End --
```

4 Notes

1. Self-installation of the jQMM can be achieved by first downloading the jQMM library from https://github.com/JBEI/jqmm and unpacking the downloaded file. For reproducibility we recommend that the jQMM be used through an interactive python (iPython) notebook server called Jupyter Notebook (http://jupyter.org/) which is a web application that allows for the interactive execution, visualization, and documentation of python code. This integrated computing format greatly enhances repeatability and reproducibility of results within the computational modeling community. Next, usage of the jQMM library can begin by logging into the local Jupyter server using a web browser and navigating to the jQMM folder, and then running some of the example Jupyter notebooks contained within the jQMM. If desired, the Jupyter server can be run directly from the command line within a linux terminal via the command "jupyter notebook". Both GAMS and CONOPT licenses are needed for the jQMM to be fully functional and must be purchased separately.
2. If choosing to run the jQMM docker container on a web-based platform one will avoid the upfront costs of having to purchase an expensive high performance server system, but over the lifetime of a purchased server it may be more inexpensive than a web-based solution (depending on intended usage intensity). Pricing for cloud computing services is typically based on usage, which allows for the ability to automatically adjust the usage of computational services to as much or as little as needed, at any time. When choosing a type of instance to use on a cloud based system we recommend instances which focus on computational speed and not RAM size or disk drive access speed. On the AWS, this includes the instances of type M4 and C4, with the C4 instance currently featuring the highest performing processors and the lowest price/compute performance ratio offered by AWS (circa 2016).

3. Proper quality control of MDV data is crucial to proper determination of fluxes. Experiments should be designed to include internal controls, and should include several biological and technical replicates.
4. The use of labeled glucose of the form U-^{13}C glucose (universal labeling of carbon atoms) together with 1-^{13}C glucose (first carbon atom only is labeled) in an 20:80 ratio is convenient because the 1-^{13}C glucose provides resolution on the activity of the pentose phosphate cycle, while the U-^{13}C glucose provides insight into the activity of reactions which mix carbon backbones of substrates. The pentose phosphate pathway loses the first labeled carbon as CO_2, while other reactions within central metabolism that mix carbon backbones can be resolved by having a combination of U-^{13}C and 1-^{13}C glucose present.
5. It is not necessary to follow this naming convention of using uppercase for metabolites and lowercase for atom transitions. The parentheses delimit the start and end of each atom transition. Any alphabetic, numeric, or underscore character comprising the regular expression [a-zA-Z0-9] can be included in the metabolite names and atom transitions.
6. Metabolite abbreviations used:
   (a) akg, 2-Oxoglutarate,
   (b) co2, carbon dioxide,
   (c) fum, fumarate,
   (d) g3p, glyceraldehyde 3-phosphate,
   (e) pyr, pyruvate,
   (f) accoa, acetyl-CoA,
   (g) asp, aspartate,
   (h) f6p, D-fructose 6-phosphate,
   (i) g6p, D-glucose 6-phosphate,
   (j) glu, glutamate,
   (k) glx, glyoxylate,
   (l) glyc3p, glycerol 3-phosphate,
   (m) glc-D, glucose,
   (n) fdp, D-fructose 1,6-bisphosphate,
   (o) g3p, glyceraldehyde 3-phosphate,
   (p) 3mob, 3-methyl-2-oxobutanoate,
   (q) hdca, hexadecanoate.
7. Reactions that must be specified in the REACTIONS.txt file, in terms of carbon atom transition information, are those that utilize metabolites for which ^{13}C isotopolog data are input into the jQMM library and for reactions which are considered to be

at the core of the metabolic network. Finally, this file also specifies a metabolite which serves as the ^{13}C carbon source (typically glucose, i.e., glcD[e], via the &SOURCE command) and input reactions which bring this ^{13}C labeled metabolite into the cell.

8. The number (1.5% in our example FLUX.txt file) at the beginning of a feed definition specifies the total glucose percentage of the initial cell culture. The percentage of 1-C glucose, which has its first carbon atom labeled, is specified next, together with the percentage of U glucose, which is uniformly labeled among all the glucose carbon atoms, and finally the percentage of normal glucose which is completely unlabeled (UN).
9. Free open-source software tools, such as the jQMM, provide for universal accessibility and unlimited modification and customization. Overall, we wish that the community can also support the jQMM's further development by submitting bug fixes to the github repo (https://github.com/JBEI/jqmm) and including any functional extensions that different research groups have achieved. Community driven development will help in the adoption of fluxomics tools like the jQMM which are free, flexible, and cater to the needs of the metabolic engineering community.

Acknowledgments

This work was part of the DOE Joint BioEnergy Institute (http://www.jbei.org) supported by the US Department of Energy, Office of Science, Office of Biological and Environmental Research, through contract DE-AC02-05CH11231 between Lawrence Berkeley National Laboratory and the US Department of Energy. The US Government retains a nonexclusive, paid-up, irrevocable, worldwide license to publish or reproduce the published form of this chapter, or allow others to do so, for US Government purposes; and the publisher, by accepting the chapter for publication, acknowledges the same.

References

1. Kitney R, Freemont P (2012) Synthetic biology—the state of play. FEBS Lett 586:2029–2036. https://doi.org/10.1016/j.febslet.2012.06.002
2. Nielsen J, Keasling JD (2016) Engineering cellular metabolism. Cell 164:1185–1197. https://doi.org/10.1016/j.cell.2016.02.004
3. Chubukov V, Mukhopadhyay A, Petzold C, Keasling J (2016) Synthetic and systems biology for microbial production of commodity chemicals: from target selection to scale-up. npj Syst Biol Appl 16009:1–11. https://doi.org/10.1038/npjsba.2016.9
4. Nakamura CE, Whited GM (2003) Metabolic engineering for the microbial production of 1,3-propanediol. Curr Opin Biotechnol 14:454–459. https://doi.org/10.1016/j.copbio.2003.08.005

5. Yim H, Haselbeck R, Niu W et al (2011) Metabolic engineering of Escherichia coli for direct production of 1,4-butanediol. Nat Chem Biol 7:445–452. https://doi.org/10.1038/nchembio.580
6. Paddon CJ, Westfall PJ, Pitera DJ et al (2013) High-level semi-synthetic production of the potent antimalarial artemisinin. Nature 496:528–532. https://doi.org/10.1038/nature12051
7. Van Dien S (2013) From the first drop to the first truckload: commercialization of microbial processes for renewable chemicals. Curr Opin Biotechnol 24:1061–1068. https://doi.org/10.1016/j.copbio.2013.03.002
8. Ghosh A, Ando D, Gin J et al (2016) 13C metabolic flux analysis for systematic metabolic engineering of S. cerevisiae for overproduction of fatty acids. Front Bioeng Biotechnol 4:76. https://doi.org/10.3389/fbioe.2016.00076
9. Lewis NE, Nagarajan H, Palsson BO (2012) Constraining the metabolic genotype-phenotype relationship using a phylogeny of in silico methods. Nat Rev Microbiol 10:291–305. https://doi.org/10.1038/nrmicro2737
10. Birkel G, Ghosh A, Vinay K et al The JBEI quantitative metabolic modeling library (jQMM): a python library for modeling microbial metabolism. Microb Cell Factories In review
11. Wiechert W (2001) 13C metabolic flux analysis. Metab Eng 3:195–206. https://doi.org/10.1006/mben.2001.0187
12. Garcia Martin H, Kumar VS, Weaver D et al (2015) A method to constrain genome-scale models with 13C labeling data. PLoS Comput Biol. https://doi.org/10.1371/journal.pcbi.1004363
13. Antoniewicz MR, Kraynie DF, Laffend LA et al (2007) Metabolic flux analysis in a nonstationary system: fed-batch fermentation of a high yielding strain of E. coli producing 1,3-propanediol. Metab Eng 9:277–292. https://doi.org/10.1016/j.ymben.2007.01.003
14. Schaub J, Mauch K, Reuss M (2008) Metabolic flux analysis in Escherichia coli by integrating isotopic dynamic and isotopic stationary 13C labeling data. Biotechnol Bioeng 99:1170–1185. https://doi.org/10.1002/bit.21675
15. Moxley JF, Jewett MC, Antoniewicz MR et al (2009) Linking high-resolution metabolic flux phenotypes and transcriptional regulation in yeast modulated by the global regulator Gcn4p. Proc Natl Acad Sci 106:6477–6482. https://doi.org/10.1073/pnas.0811091106
16. Kajihata S, Matsuda F, Yoshimi M et al (2014) 13C-based metabolic flux analysis of Saccharomyces cerevisiae with a reduced Crabtree effect. J Biosci Bioeng 120:140–144. https://doi.org/10.1016/j.jbiosc.2014.12.014
17. Schellenberger J, Que R, Fleming RMT et al (2011) Quantitative prediction of cellular metabolism with constraint-based models: the COBRA toolbox v2.0. Nat Protoc 6:1290–1307. https://doi.org/10.1038/nprot.2011.308
18. Segrè D, Vitkup D, Church GM (2002) Analysis of optimality in natural and perturbed metabolic networks. Proc Natl Acad Sci U S A 99:15112–15117. https://doi.org/10.1073/pnas.232349399
19. Shlomi T, Berkman O, Ruppin E (2005) Regulatory on/off minimization of metabolic flux changes after genetic perturbations. Proc Natl Acad Sci U S A 102:7695–7700. https://doi.org/10.1073/pnas.0406346102
20. Suthers PF, Burgard AP, Dasika MS et al (2007) Metabolic flux elucidation for large-scale models using 13C labeled isotopes. Metab Eng 9:387–405. https://doi.org/10.1016/j.ymben.2007.05.005
21. Toya Y, Ishii N, Hirasawa T et al (2007) Direct measurement of isotopomer of intracellular metabolites using capillary electrophoresis time-of-flight mass spectrometry for efficient metabolic flux analysis. J Chromatogr A 1159:134–141. https://doi.org/10.1016/j.chroma.2007.04.011
22. Zamboni N, Fendt S-M, Rühl M, Sauer U (2009) 13C-based metabolic flux analysis. Nat Protoc 4:878–892. https://doi.org/10.1038/nprot.2009.58
23. Reed JL, Vo TD, Schilling CH, Palsson BO (2003) An expanded genome-scale model of Escherichia coli K-12 (iJR904 GSM/GPR). Genome Biol 4:R54. https://doi.org/10.1186/gb-2003-4-9-r54
24. Orth JD, Conrad TM, Na J et al (2011) A comprehensive genome-scale reconstruction of Escherichia coli metabolism--2011. Mol Syst Biol 7:535. https://doi.org/10.1038/msb.2011.65
25. Feist AM, Henry CS, Reed JL et al (2007) A genome-scale metabolic reconstruction for Escherichia coli K-12 MG1655 that accounts for 1260 ORFs and thermodynamic information. Mol Syst Biol 3:121. https://doi.org/10.1038/msb4100155

Index

Edward E.K. Baidoo (ed.), *Microbial Metabolomics: Methods and Protocols*, Methods in Molecular Biology, vol. 1859,
https://doi.org/10.1007/978-1-4939-8757-3, © Springer Science+Business Media, LLC, part of Springer Nature 2019

N

P

Q

R

S

T

MIX
Papier aus verantwortungsvollen Quellen
Paper from responsible sources
FSC® C105338

If you have any concerns about our products,
you can contact us on
ProductSafety@springernature.com

In case Publisher is established outside the EU,
the EU authorized representative is:
Springer Nature Customer Service Center GmbH
Europaplatz 3, 69115 Heidelberg, Germany

Printed by Libri Plureos GmbH
in Hamburg, Germany